I0056973

Underground Operators Conference 2025

7–9 April 2025
Adelaide, Australia

The Australasian Institute of Mining and Metallurgy
Publication Series No 1/2025

≋ AusIMM

Published by:
The Australasian Institute of Mining and Metallurgy
Ground Floor, 204 Lygon Street, Carlton Victoria 3053, Australia

© The Australasian Institute of Mining and Metallurgy 2025

No part of this publication may be reproduced, stored in a retrieval system or transmitted in any form by any means without permission in writing from the publisher.

All papers published in this volume were peer reviewed before publication.

The AusIMM is not responsible as a body for the facts and opinions advanced in any of its publications.

ISBN 978-1-922395-45-0

Advisory Committee

Alastair Grubb
MAusIMM(CP)
Conference Advisory Committee Chair

Alex Campbell
FAusIMM(CP)

Chris Carr
FAusIMM(CP)

Katrina Crook
FAusIMM(CP)

Anne-Marie Ebbels
MAusIMM(CP)

Dave Kilkenny
FAusIMM(CP)

Erin Langworthy

Joe Luxford
FAusIMM(CP)

Adrian Penney
FAusIMM(CP)

Adrian Pratt
FAusIMM(CP)

Iain Ross
MAusIMM

Farell Simanjuntak
MAusIMM

John Stanton
MAusIMM

Claudia Vejrazka
MAusIMM

AusIMM

Julie Allen
Head of Events

Fiona Geoghegan
AAusIMM
Senior Manager, Events

Raha Karimi
Program Coordinator, Events

Reviewers

We would like to thank the following people for their contribution towards enhancing the quality of the papers included in this volume:

Sophia Adamopoulos

Fiona Bodycoat

Terry Burns

Alex Campbell

Geoff Capes

Rui Carboni

Chris Carr

Gregor Carr

Peter Christen

Katrina Crook

Andrew Derrington

Chris Desoe

Stephen Duffield

Geoff Dunstan

Anne-Marie Ebbels

Chris Elliott

Desmond Fitzgerald

Andrew Glastonbury

Ed Gleeson

Donald Grant

Alastair Grubb

Andrew Harris

Robert Jele

Anthony Keers

Dave Kilkenny

Greg Laing

Euan Leslie

Charles Lilley

Joe Luxford

Stephen Manning

David McMillan

Adrian Penney

John Player

Simon Pope

Adrian Pratt

Walter Rojas

Iain Ross

Erin Savage

Clint Scott

Delia Sidea

Farell Simanjuntak

John Stanton

Robyn Stonell

Iain Thin

Claudia Vejrazka

Nadine Wetzel

Dean Will

Foreword

On behalf of the Advisory Committee, we are delighted to welcome you to the Underground Operators Conference 2025.

The conference brings together more than 1500 mining professionals and leading experts to connect, learn, share and collaborate on issues facing underground operations. Through this network we will showcase and celebrate success, find new solutions to shared problems and discover how advances in knowledge and technology can improve our operations in the future.

The Advisory Committee has put together an impressive range of presentations across a wide variety of topics, with technical sessions, panel discussions and keynote speakers spread over three days, showcasing leading innovations and research. The conference provides delegates numerous opportunities to share their expertise, meet new contacts and rejuvenate past connections.

Through the support of our Principal Partners: Business Events Adelaide, Adelaide Convention Centre, and the Government of South Australia, the 2025 Conference will take place against the backdrop of the beautiful city of Adelaide.

We would like to thank the Advisory Committee, the AusIMM Events team, authors and presenters, paper reviewers, attending delegates and our many wonderful sponsors and exhibitors listed in the following pages.

We welcome you to Underground Operators Conference 2025 and trust that you will find it an enjoyable and rewarding event.

Yours faithfully,

Alastair Grubb *MAusIMM(CP)*

Underground Operators Conference 2025 Advisory Committee Chair

Sponsors

Major Conference Sponsor

BHP

Principal Partners

ADELAIDE CONVENTION CENTRE

Business Events Adelaide

SOUTH AUSTRALIA

Government of South Australia
Department of the Premier and Cabinet

Mining Services Partner

BYRNECUT

Sustainability Partner

CATERPILLAR®

Blasting Technology Partner

DYNO NOBEL

Innovation Partner

ORICA

Platinum Sponsors

Epiroc

SANDVIK

Sponsors

Gold Sponsors

Barminco

BECK ENGINEERING

Enaex Australia

MP MINING PLUS

Silver Sponsors

AMC consultants
mine smarter

KOMATSU

SANDVIK

Conference Dinner Sponsor	Networking Hour Sponsor
BYRNECUT	SANDVIK

Exhibition Lounge Sponsor	Meeting Hub Sponsor
MASPRO	HILLGROVE RESOURCES

Welcome Reception Sponsor	Name Badge and Lanyard Sponsor
micromine	ORICA

Sponsors

Lunch Sponsor

REDPATH

Delegate Bags Sponsor

MACMAHON

Conference Proceedings Sponsor

AECOM

Notepad and Pens Sponsor

ROCKTOOLS LHS AUSTRALIA

Hydration Sponsor

MX3

Conference App Sponsor

JENNMAR

Charging Station Sponsors

ABB

DYNO NOBEL

Coffee Cart Sponsors

BECK ENGINEERING

DYNO NOBEL

Epiroc

Robit
FURTHER. FASTER.

AECOM

Trusted underground mining solutions — Australian innovation delivered globally.

For over 30 years, we have delivered safe, reliable, and energy-efficient infrastructure for underground mining. Our Australian-based team provides global clients with industry-leading solutions in materials handling, crushing, shaft haulage, and integrated mine infrastructure services— tailored to your needs.

Delivering a better world

Carrapateena Mine, South Australia

Connect with us

aecom.com

Contents

Collaboration between suppliers, operations, and alternative industries

Feasibility studies and mine design

Technology and innovation

Keynote

Out from underground – planning to produce: operational ESG for mine development and operations

B Harvey[1] and G Deans[2]

1. Director, Resolution88, Melbourne Vic 3000. Email: bruce.harvey@resolution88.com.au
2. Director, Modifying Factors, Adelaide SA, 5000. Email: geoff.deans@modfyingfactors.com

INTRODUCTION

This paper supports a keynote presentation at the AusIMM Underground Operators Conference in Adelaide, April 2025. The paper itself is intended as a short guidance note on ESG matters in the extractive sector and ESG-related accountability for mining professionals, with specific reference to underground minerals project development and operations. The keynote material is covered in the Extended Abstract.

EXTENDED ABSTRACT

ESG, shorthand for Environment, Social and Governance, is pervasive in business and financial commentary. Unfortunately, confusion about its meaning, purpose and application dilutes its utility to manage project and operational risk. The most glaring misunderstanding for mine operators arises because of its use in the financial world to describe securities portfolios that are weighted to sectors and businesses with deemed virtuous purposes and products. This is 'portfolio ESG' and mining is rarely invited. Misunderstanding also arises from media and company use as a proxy term for singular popularist issues such as 'net zero'. Such attempts at virtue signalling or greenwashing around issues of pre-determined importance can lead to businesses seriously misjudging the risks that are most important to them in their operating and business context.

ESG impacts and project risk – fat and long tails: This paper does not address 'portfolio ESG', rather it sets out to address 'operational ESG' – the practical management of material ESG-related risk factors at projects and operations that can impair, constrain or enable real performance. In short, how to make ESG management integral to life-of-mine planning and plans.

External variables with fixed costs: How mines (including underground) are designed and operated has far greater impact on ESG-related risks than financial labelling, sustainability reporting and any marketing campaign. The work of mining engineers effectively operationalises many of the material ESG factors that have life-of-mine impacts and consequences that are embedded in project designs, consents and licenses. In the domain of megaproject management, such impacts are described as having 'fat and long tails' (Flyvbjerg, 2003). Just as (say) excavations for underground crushers or conveyors are consequential and irreversible, so are many ESG-related matters.

Managing the essential versus the good: Operational ESG is about overall or 'total' asset performance – ensuring that what matters 'above ground' and to the 'outside' world, (essentially affected stakeholders and the environment) is matched against what is operationally feasible. In total, there are some 40–50 ESG elements commonly referenced in performance standards and due diligence instruments. This includes the proposed updates to the Australasian Joint Ore Reserves Committee (JORC) minerals reporting code, and specifically its requirements regarding ESG-related Modifying Factors. Irrespective of which checklist is used, confusion and self-deception can prevail when project proponents and operators attempt to manage them all with equal attention in a regulatory tick box approach or at a macro level, when only a limited number of discrete impacts apply locally.

Focusing on risk: Effective management requires focusing on the right risk at the right time. Underground mining engineers, with the aid of detailed geotechnical studies, focus on critical risks, such as unwanted ground collapse. Geologists and mining engineers understand that each mineral deposit is unique and that mine development requires orebody-specific optimisation. This focusing principle is equally true for ESG-related matters, and the many potential issues cannot be focused on all at once. In many cases the factors are common, but they differ in their exact likelihood and impact consequence. What is needed is thorough project-level materiality assessment of each ESG element, relevant to each phase of project development and operationalised in front-line work packages – in short, a 'first principles' rather than a regulation- or standards-driven approach. Many ESG elements will not be relevant or will not carry high risk in context.

Whack-a-mole ESG management: Systematic risk thinking inclusive of ESG-related factors requires an understanding of what factors are consequential and how reversible they are (Grant, 2023). Even with ESG-related risks reduced to as few as (say) five for priority attention, they can interact in unique and unpredictable ways to create multi-factor risks that can lead to a game of 'whack-a-mole' to manage. This can lead to compounding risk, increasing mitigation costs and rework. Critically, many ESG matters that are material to project development are outside the control of proponents, often involving complex dynamic interactions with external stakeholders and their behaviours, hence things can literally pop up out of nowhere.

Cause and effect – 'kicking shins' – outbound impacts cause inbound risks: All mineral projects and operations have outbound impacts. Metaphorically, these impacts are like kicking people in the shin. The shin impact will normally illicit a reaction, but exactly how a person will react is uncertain, indeed often unknowable. There are range of possibilities, including that they will walk away, physically retaliate or take legal action. Any of these responses (inbound risks) are possible but not knowable. Such uncertainty constitutes the inbound risk that projects and mines incur from their outbound impacts (Figure 1). ESG-related risk exposures are uniquely context and time-dependent. To inform decision-making, the challenge is to understand the range of likely impact consequences. If outbound impacts are consequential and irreversible, then very careful planning indeed is required.

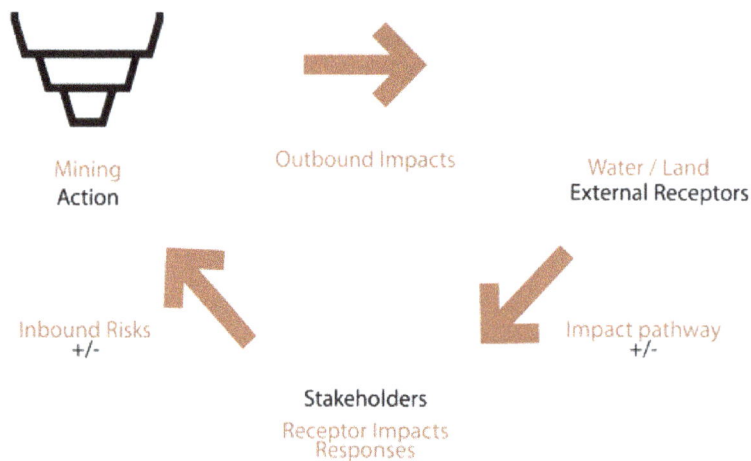

FIG 1 – Outbound impact and inbound risk.

The various forms of interrelated outbound impact and associated inbound risk are classic attributes of complex systems, where small actions or changes can result in large impacts. Causal relationships in such systems are frequently nonlinear, so standard analytical approaches such as sampling, extrapolation and trend analysis do not work. Theory can seldom be deployed to predict outcomes and reference classes that exactly match a project's situation rarely exist. In such circumstances, it is tempting to surrender to fate and attempted narrative control to try to achieve a satisfactory outcome.

Trust in science: Despite the generally non-deterministic nature of ESG issues, the well-versed approach of scientific management commonly deployed in the minerals sector provides a good foundation for heuristic problem identification and solving of ESG-related challenges. This includes good governance architecture, knowledge base studies, outbound impact, inbound risk and materiality assessments, proper planning, vectored work packaging and assurance activities – all very familiar at well-run minerals projects and operations. With such a foundation in place, solutions to ESG-related issues can usually be proactively found and deployed by teams of diligent, experienced and intelligent professionals working collaboratively.

Net project requirements: Mines are not just mines – amongst other things they require ancillary infrastructure, such as power, water, transport and communications, much of it needing to be consented, approved, developed and/or provided by other parties. All such ancillary developments have labour source, suppliers and contractor factors in common, with each projecting their own impacts on host landscapes and stakeholders. Where these so-called ancillary matters were once secondary, they are now potential show-stoppers – no land access, no water, no acceptable waste disposal, no transport corridor – then no mine.

Two sides of the same coin – technical and ESG factors: At its core, ESG is about risk management. As with any risk management process, elimination and substitution are the most value accretive and effective controls. In short, the earlier proper assessments are undertaken, the earlier potential ESG-related threats can be 'designed out' and opportunities 'engineered in'. Early and genuine design adaption and subsequent good management to eliminate material ESG-related risk is key, very different to the Decide-Announce-Defend (DAD) approach that results in delays, rework and project impairment. For all mining professionals, managing and working within constraints is a core competency, recognising that a critical issue is if constraints are ill-defined, they are very difficult to manage.

Examples of ESG-related multi-factor issues for underground projects and operations include land tenure, water and waste. These are common ESG issues that challenge underground mine design, especially mining methods such as caving (sublevel or block) that create a subsidence zone.

- **Land** – gaining tenure depends on having acceptable and approved above ground land use, usually a precondition for development. What is located on the surface above a deposit is a constraining factor. The presence of a heritage site, national park or significant civic infrastructure can limit what is possible.

- **Water** – too much water, not enough water, quality of water, storage of water, third party users (including environment) and discharge of water are all matters carrying concurrent technical, social, environmental, financial and regulatory risks.

- **Waste** – mining and processing minerals produces waste, usually in large volumes. This waste must be stored safely, involving challenges that are interconnected with water and land. Tailings storage facilities usually require materials that must be 'won' from further afield than the orebody. In underground operations, materials balances and qualities are significant matters of ESG performance.

First Principles versus 'The Vibe': Much of the industry and media commentary and cherry picking of salient (trending) 'global' issues and the associated narrative management that is presented as ESG is distracting at best and counterproductive at worst. ESG-related factors such as water, waste, regulation and stakeholder acceptance have always been risk factors in mining. The ESG label, whilst expanding the range of potential elements to consider, has not changed the fundamentals.

The superficial ESG 'vibe' that emerged in much of the western world over the past decade is shifting back to renewed respect for reality, resilience and performance-based management. ESG is nothing special, not exotic and has no utility when deployed as strategic banner waving that does not address risk, resilience and insurability.

Play the ball, not the umpire – reducing regulatory impact and rework: Laws and regulation define minimum acceptable legal standards that reflect societal expectations. Standards change in response to impacts deemed unacceptable (such as the Global Industry Standard on Tailings Management developed in response to catastrophic waste storage failures). Standards in turn affect how projects are assessed and developed, affecting project design criteria, the risk of rework and regulatory oversight.

However, just as when playing football (any code will do), players don't watch the umpire, they focus on the ball (what they can control). Success requires players to understand the rules, play to a strategy that optimises their performance to match or exceed the minimum standards, reducing the risk of a loss while gaining competitive advantage to win. In real life, standards are supposed to protect downside risk and should be calibrated to meet minimum requirements. This provides little room for error or redundancy, Conversely, driven by political responses to social and environmental sentiment outside the full control of any company, there is high risk in a dynamic world that regulatory changes will mount up over time into over regulation that stifles innovation and genuine performance.

The appropriate response from the mineral sector is to demonstrate that it can achieve better ESG-related outcomes through competent self-management based on self-interest. This requires diligent, well-resourced ESG-competent people working in operations-directed roles. Accordingly, underground operators need a good sense of how their work and its knock-on effects interact with the world above ground, and how they need to contribute to derisking and optimising development pathways and operational performance. In summary, if you manage work packages that directly, indirectly or cumulatively impact things such as land use, waste or water matters, whether you know it or not, you are involved in operational ESG. By supporting a first principles approach aligned to enlightened self-interest this can help mining companies demonstrate in good faith that (over-) regulation is the least effective way of ensuring desirable ESG outcomes.

WHAT IS ESG AND WHERE DID IT COME FROM?

The term ESG was first used to designate Environment, Social and Governance in a UN publication called 'Who Cares Wins' in 2004. Over the past decade it has become prominent in business commentary, indicating the corporate and financial world's acceptance of it as a preferred term for the notion of 'sustainability', which itself was first referred to as Sustainable Development in the United Nations (UN) 1987 publication 'Our Common Future' (Brundtland, 1987). The publication laid the groundwork for the 1992 Rio de Janeiro Earth Summit, the first of a burgeoning field of international talkfests on sustainability.

Essentially and broadly, operational ESG concerns the things referred to in orthodox economics as 'externalities' – things for which costs can be disproportionately borne by parties other than the primary party. For two decades business was tepid in its participation in the global 'sustainability zeitgeist', perhaps because the concept lacked the rigour of business management and quantifiable metrics to inform a business case. This changed in the past decade as the finance and corporate world has tacitly signalled acceptance in the form of a more business-driven approach under the moniker of ESG. This involves an increasingly standardised management approach with measurable (mostly lagging) performance indicators, categorised in E, S and G dimensions, to be tracked and reported alongside an economic dimension.

ESG as a generalised concept

Depending how you slice it, there are currently some 40–50 commonly recognised (E)ESG elements with associated metrics that are relevant to the minerals sector (Figure 2).

Economic dimension
- Local employment
- Local enterprise linkages
- Technology transfer
- Energy profile
- Infrastructure partnering
- Local investment
- Equitable distribution of wealth
- Taxes and other government payments
- Research and development (R&D) support
- Circular economy

Environmental dimension
- Air emissions
- Noise and vibration
- Wastes, discharges and contaminants
- Surface and groundwater use and impacts
- Land, soil and agriculture impacts
- Rare, endangered and protected fauna
- Native vegetation
- Protected areas and habitats
- Climate adaption and weather events
- Pollution controls

Governance dimension
- Transparency and reporting
- Record keeping and data security
- Commercial and partnership arrangements
- Supply chain standards
- Consumer protection
- Anti-corruption, -bribery and -competition
- Legal compliance
- Corporate governance
- Risk management systems
- ESG management systems
- Environmental/Social Impact Assessment
- Ethical considerations
- Asset and cyber-security

Social dimension
- Cultural competency
- Labour rights, DEI, skills development, fair wages
- Non-discrimination and no harassment
- Workplace and public safety, health and security
- Human rights due diligence (incl. supply chain)
- Stakeholder opposition and engagement
- Local compensation and benefits
- Indigenous and ethnic minority rights
- Gender considerations
- Resettlement
- In-migration
- Cultural heritage
- Complaints, conflict and grievance

FIG 2 – (E)ESG elements relevant to the minerals sector.

The listed elements are not plucked from the air; they are adapted from a list of the most frequently referenced (salient) ESG issues identified in a meta-study of 150 investment due diligence instruments (Sauvant and Mann, 2019). Unfortunately, as with all generalised concepts, this leads to abstraction as analysts and commentators presume all listed elements are important everywhere and all at once, needing to be addressed with equal attention. Inevitably, this leads to a 'tick box' approach that detracts from any real analysis, mitigation and performance improvement in local context. In short, a generalised approach that misses the most material local risks.

Generalised abstraction also leads to confusion about applicability and public conception. For instance, within the world of Exchange Traded Funds (ETFs) and related financial activity, ESG is used to describe securities portfolio management that is weighted to sectors and businesses with deemed virtuous purposes, values and products. Minerals businesses, clumped in the same bucket as tobacco, armaments, old growth forestry and gambling, are rarely included in ESG ETF portfolios. While some in the minerals industry are responding by seeking virtuous labelling, such as 'critical minerals' for the 'energy transition', most people see this as special pleading and not relevant in securing customers for what are undifferentiated mineral commodities.

Another finance sector application is company ESG scoring by ratings agencies based on comparisons of declared performance against broad spectrum indicators. Also, ESG is frequently used as a way of referring to single headline issues with a journalistic emphasis on pop culture' catch-all's such as 'net zero', 'diversity, equity, and inclusion' (DEI) and 'human rights'. None of this has anything to do with properly managing ESG-related matters at mineral projects and operations. Attempts at attention grabbing around single issues of pre-determined importance can lead to businesses seriously misjudging the risks that are most important to them in local context.

CONTEXT IS EVERYTHING

The context of a project informs the materiality of impacts. Understanding impact in context is critical to understanding risk.

For instance, when considering water as a technical and ESG risk factor, there is a big difference between water impacts and risks in an arid context (such as much of South Australia) versus a tropical context (such as Indonesia). Too little water versus too much water is a critical technical and ESG consideration for mine studies, development and operations, and equally for existing water users, such as communities, industries and the environment.

Minerals projects and operations impact their context, and their context brings reciprocating and evolving risk. While the complexity of managing risk in context is a common challenge for all industry sectors, it is more crucial in minerals development as orebodies are locked in place. Unlike a solar installation or a factory, there is no option to relocate a minerals project to a less risky location or jurisdiction.

A study of complex orebodies by Valenta et al (2019) highlights some multi-factor risks that are potential technical and ESG constraining factors for minerals project development. While many of these factors are common across all projects, their precise impacts are unique to their local context in an interrelated and interdependent way (Figure 3).

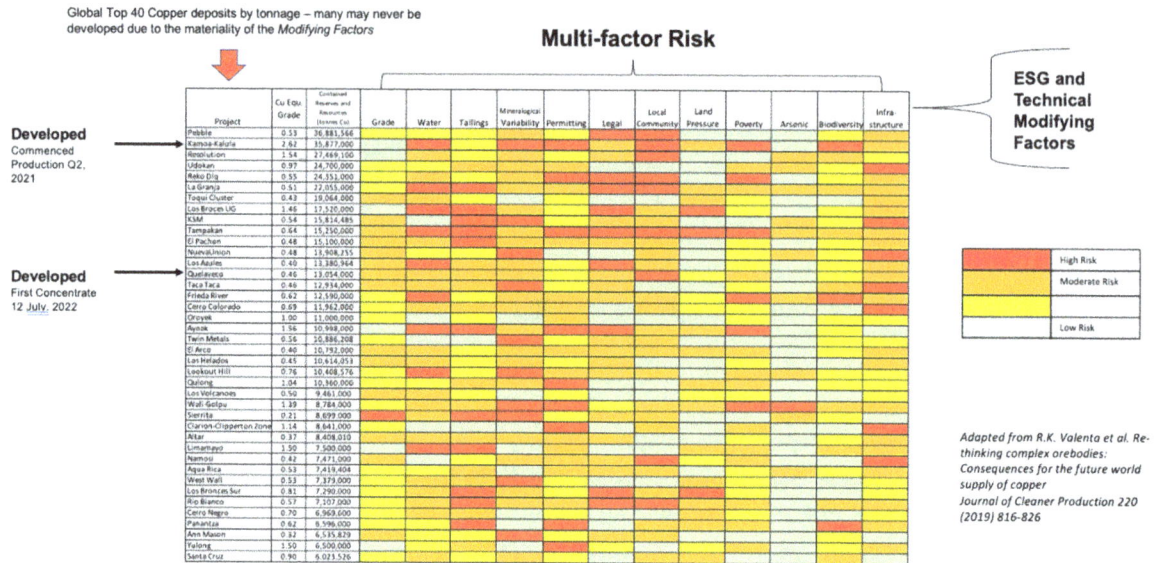

FIG 3 – Multi-factor risk for 40 largest undeveloped copper deposits (adapted from Valenta *et al*, 2019).

Underground mining constraints

All mining developments have common modifying factors that act as constraints to project design and viability. Underground engineers by nature regularly deal with complicated systems, structures and devices to meet functional requirements, while also considering limitations imposed by regulation, safety, cost, and more. ESG factors are constraints and enablers in the 'more' category. The work of Valenta *et al* (2019) demonstrates that these can be the material determinants of project viability. Underground mine designers and operators must understand these to develop an 'optimal possible' design solution for mining an orebody. Hence, appropriately managing things that can lead to ESG impacts is core business for underground engineering. As these are seldom unconstrainted in themselves, understanding thresholds and trade-off's as part of the planning process is crucial to ensure there is a Reasonable Prospect for Economic Extraction (RPEE; JORC, 2024a) and to reduce or eliminate rework.

Some real examples from actual underground projects and mines are:

- An underground mine being unable to gain traditional owner consent under the Australian Native Title Act until heritage values involving the underground presence of a mythical being were resolved.

- A very large underground copper mine having to extend tailings containment structures beyond its granted tenure to ensure multi-decadal operations. There are in fact many examples of this.

- Underground workings breaching water-bearing zones leading to propagation and surface subsidence compromising a public road.

- A large underground mine decades into production and seeking to expand, opting to reduce aquifer drawdown in favour of desalination and a long pipeline to future proof its water needs.

- A proposed large-scale block cave mine held up indefinitely by not being able to guarantee its subsidence would not breach the surface into a public camping area.

- A mining application rejected by a mining minister not willing to risk the backlash of local wine makers.

Why managing ESG matters well is important

A range of reasons are commonly cited for why good management of operational ESG secures overall business benefit. These include reputational gain, easing regulatory approvals, attracting and retaining talent who can choose where to work, securing access to debt and equity finance, demonstrating insurer risk requirements, and 'social licence' more broadly. What is not commonly appreciated is the intrinsic business benefit gained by an overall reduction in business variability. In short, managing ESG matters well as a form of risk management leads to a reduction of operational and business uncertainty.

Managing uncertainty – reconciling inside and outside views

The greatest source of uncertainty for any business lies in its external environment, the things that are seemingly beyond its control. Minerals development occurs in a dynamic economic, social, legal and environmental context Just as understanding things such as geotechnical properties and underground aquifers are material to mine design, so are regulatory, environmental and social factors. Accordingly, managing ESG matters well is a way of achieving more control and hence less variance in business interactions with the outside world. Overall, reconciling internal needs (inside view) within the project context (outside view) leads to a more stable and robust business (Figure 4).

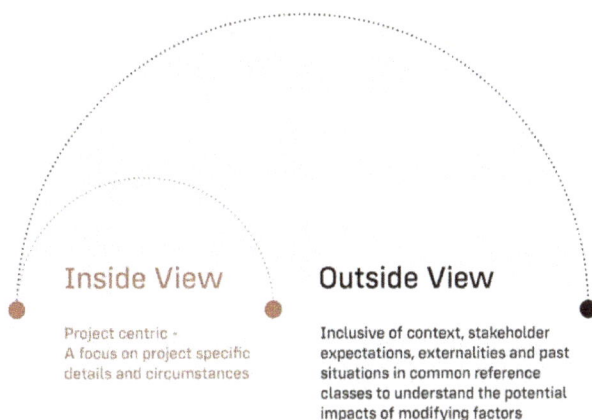

Inside View

Project centric -
A focus on project specific
details and circumstances

Outside View

Inclusive of context, stakeholder
expectations, externalities and past
situations in common reference
classes to understand the potential
impacts of modifying factors

FIG 4 – Expanded scope of knowledge and analysis.

ESG considerations as Modifying Factors

The term 'Modifying Factor', a defined term in the Australasian Joint Ore Reserves Committee (JORC) mineral reporting code, is common to many industry sectors and is deemed to be anything that can affect business and product performance or viability (Figure 5). In the minerals sector for well over a decade this has included '*mining, processing, metallurgical, infrastructure, economic, marketing, legal, environmental, social, and governmental*' considerations (JORC, 2012).

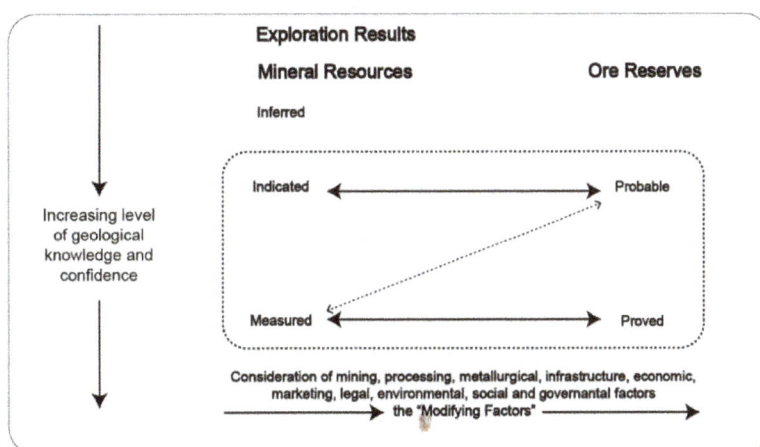

FIG 5 – General relationship between Exploration Results, Mineral Resources and Ore Reserves (JORC, 2012).

It is important to note that the JORC Code is not a prescriptive regulatory requirement; rather it is a public disclosure requirement. In this context, it is also important to appreciate that 'the law' is a lagging indicator of societal expectation, hence miners and project developers draw false comfort from managing risk through heavy-handed regulation and by following the 'letter of the law' (NB: A good example of this is the great financial and reputational damage that Rio Tinto suffered from destroying the Juukan Aboriginal heritage site in the Pilbara in 2020, despite being within legal boundaries in the way it went about it.) A more responsible path is to undertake comprehensive risk assessment of ESG-related matters on a first principles basis with the same degree of rigour that technical and financial risks are assessed. This is what lies behind the escalated importance of publicly reporting Modifying Factors and the risk context of resource projects in the proposed updated JORC Code (JORC Code, 2024a, 2024b). For instance, matters such as land access approvals and closure considerations that were once regarded as secondary are now primary concerns. The ability to identify, manage and appropriately mitigate a full intersectional suite of Modifying Factors, including ESG-related matters, is now a reporting requirement and an important competitive capability (Harvey and Deans, 2024).

While technical matters are generally well understood at most resource and infrastructure projects, ESG and other Modifying Factors usually remain less quantified or absent from early-stage analysis and assessment. At a time when more and much larger infrastructure projects are being proposed and (occasionally) built, many have strikingly poorly quantified financial, environment and public support risk profiles (Flyvbjerg and Gardner, 2023). This trend continues to intensify as investor and political risk appetites are trending down, while project complexity and societal expectations are trending up (Figure 6).

Increasing Complexity

Decreasing Risk Appetites

Increasing Stakeholder Expectations

FIG 6 – Countervailing trends (Deans and Harvey, 2024).

The proposed updated JORC Code (JORC Code, 2024a) reflects how ESG-related Modifying Factors, such as cultural heritage clearance, land access (eg native title) approvals, local content considerations, the Global Industry Standard on Tailings Management (GISTM), decarbonisation targets, climate resilience disclosures, sovereign risk and political instability, are now more material to project viability than ever before (Deans and Harvey, 2024). In fact, Modifying Factors properly tackled present an integrated analytical framework that can help inform good management of impact and risk frequency, consequence and materiality preconditions for project approvals (Figure 7). The framework is presented in the proposed JORC Code – Guidance Notes: Exposure Draft (JORC, 2024b), and a consistent framework is set out in Chapter 2, *Permitting, Environment and Social*, of AusIMM's Monograph 35 – *Study Processes Handbook* (AusIMM, 2024).

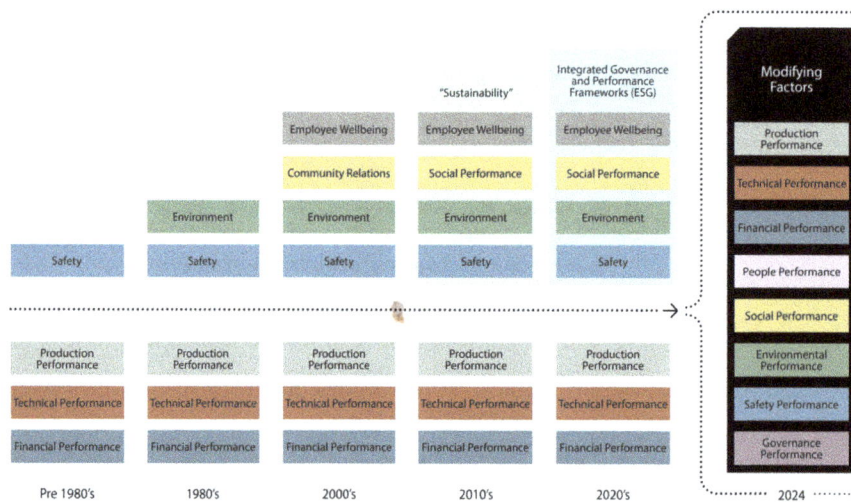

FIG 7 – Emergence of Modifying Factors as material to project approval (Deans and Harvey, 2024).

'Technically feasible' versus 'optimally possible'

In mineral project development there is an inherent tension between what is 'technically feasible' and what is 'optimally possible'. The traditional view of project proponents is that there should be a primary focus on technical (and financial) viability, and that work on 'other' considerations can take place in subsequent phases of development. This techno-rational perspective oversimplifies what is at stake. The reality is that many of the ESG matters that and outside the control of project proponents and material to project development are identifiable during exploration phases. To optimise project value, study teams as part of life-of-asset planning must consider and attempt to trade-off numerous interdependent inputs such as land, power, water, mine design and projected operating costs across the project development life cycle (currently on average almost 18 years – Manalo, 2024). Reducing (ideally eliminating) inputs, outbound impacts and regulatory requirements (including closure bonds) through mine design can be the best way to eliminate in-bound risks. The mining method, for instance, greatly effects the nature of inputs, outputs, impacts and risks. In relatively densely populated, scenically attractive or protected environment areas, underground mining has far greater social and politically acceptability than open cut mining.

Data-driven analyses of the many ESG-factors in play needs to happen to aggregate an overall assessment of what is 'technically feasible'. A major material risk is when ESG-related Modifying Factors that were not visible and/or not considered in early studies will eventually come into play and many solutions that might be deemed 'feasible' are often not 'permissible'. Through all this, cut-off grades and mineable boundaries change, and project value is steadily eroded (Figure 8).

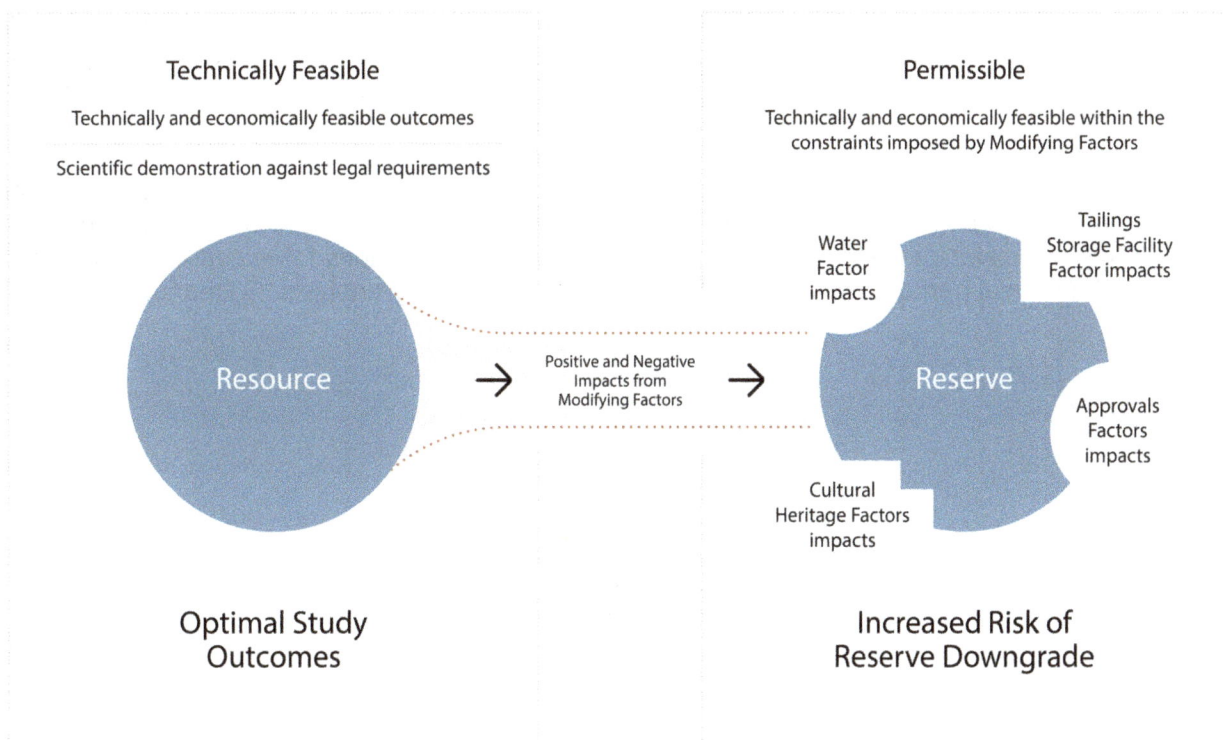

Technically Feasible

Technically and economically feasible outcomes

Scientific demonstration against legal requirements

Permissible

Technically and economically feasible within the constraints imposed by Modifying Factors

Resource → Positive and Negative Impacts from Modifying Factors → Reserve

Water Factor impacts

Tailings Storage Facility Factor impacts

Approvals Factors impacts

Cultural Heritage Factors impacts

Optimal Study Outcomes

Increased Risk of Reserve Downgrade

FIG 8 – Erosion of Resource to Reserve during project evolution (Deans and Harvey, 2024).

Multi-factor and meta-risk

Minerals project development is capital intensive with multi-decade timelines being normal. Risks handled in isolation from project evolution and from each other represent a cartoonish view of reality. Evolving multi-factor risk is present and emerging all along the way, with the substantially compounding costs arising from these risks often not identified, let alone mitigated, at early stages. In theory, ESG-related risks drawn from a checklist (Figure 2) help, however they do not behave in isolation of each other. Project study approaches frequently adopt a view that individual risks can be isolated out from other risks, that ancillary infrastructure can be figured out retrospectively and that with enough 're-design-on-the-run', ESG-related risks can be ironed out in due course.

Experience indicates otherwise. It shows that over long time frames, and particularly where innovation and/or non-standard technology is being considered and there are multiple interested parties, it is not possible to plan in detail for everything upfront. As a project progresses, external context and stakeholder expectations inevitably change – without adaptive capacity in design and systems, these changes lead to a damaging fight to control escalating redesign work and costs. The changes to cost and/or schedule, and/or the necessary insurance and penalty costs that ensue, mean projects are more and more expensive to deliver.

Meta-risks (adapted from Gray, 2002) are the implicit qualitative risks, beyond the scope of explicit technical and financial risks, born out of complex interactions between the behaviours of individuals, societies and organisations. Minerals operations are not just mines. Mines require power, water, transport, communications and waste storage infrastructure, all of which have labour source, suppliers and contractor factors in common, whilst projecting their own unique impacts on host landscapes and stakeholders. Beyond orebody knowledge factors, such as commodity mix and prices, location and depth, maximising orebody and project value is driven by cause-and-effect relationships between 'input factors' (eg cost of capital and materials, labour, energy, licenses and access agreements) and 'output factors' (eg technical performance, production rates, marketing, sales, royalties, taxes and fees). The permutations are dynamic, seemingly endless and unique to each project.

Secondary matters are now a primary concern

In decades past, these ancillary matters were secondary to the primary technical considerations of ore grade, mine design and beneficiation recovery. This is no longer the case, what were formerly ancillary matters are now potential show-stoppers – no land access, no water, no acceptable waste disposal, no transport corridor access – then no mine. Perhaps, ironically, the renewables and civic infrastructure sectors are currently experiencing this as much or more than the minerals sector. Furthermore, the competing and intensifying priorities between different sectors add to congestion or 'chokepoints', particularly where there are common stakeholders and external constraints. For instance, to develop some classes of linear infrastructure there are up to 29 Modifying Factors in common with the minerals sector that are preconditions or considerations for acceptable economic, social, environmental performance and permitting (Figure 9; primary analysis: G Deans).

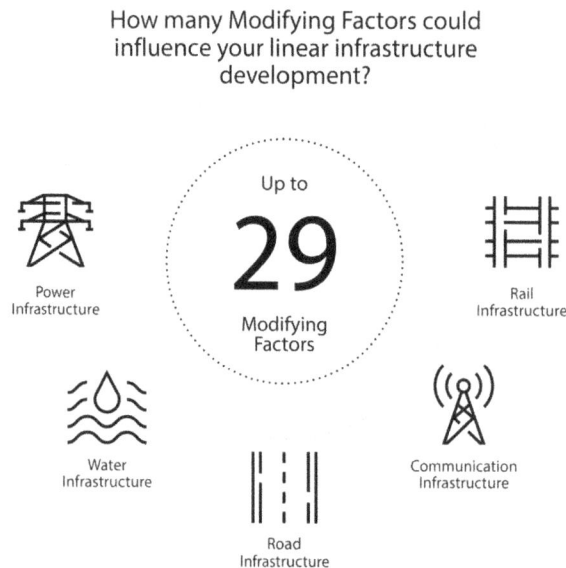

FIG 9 – Land-based Modifying Factors for linear infrastructure.

The various forms of interrelated outbound impact and inbound risk described above are classic attributes of complex systems. A small change or input can lead to a major response. For instance, a relatively small outbound impact can be emotionally amplified by affected stakeholders. Attempts to rationally downplay the likely physical impact will frequently further increase the sense of emotional outrage they feel, leading to major escalation of inbound risk to a project and its proponent. Causal relationships in complex systems are frequently nonlinear, so standard analytical approaches, such as sampling, extrapolation and trend analysis, do not work. In such situations, theory cannot be deployed to predict outcomes, and reference cases that exactly match context-specific complexity simply do not exist. In the face of all this ESG-related complexity what should project developers do?

HOW TO MANAGE ESG-RELATED COMPLEXITY?

The most effective risk control is elimination or substitution. Traditional approaches to dealing with ESG-related interactions, such as Decide Announce Defend (DAD), no longer work in an increasingly scrutinised and politically activated world. Fortunately, despite the generally non-deterministic nature of ESG-related issues, the well-versed approach of scientific management commonly deployed in the minerals sector provides a good foundation for heuristic problem solving on ESG-related matters.

With reference to the *source-pathway-receptor* model (described in Peirce, Vesilind and Weiner, 1990), well defined critical controls at *source* and *pathway* are key to avoiding costly and often compensatory consequences at the *receptor* end of things. For instance, water is a common multi-user critical risk (Figure 10).

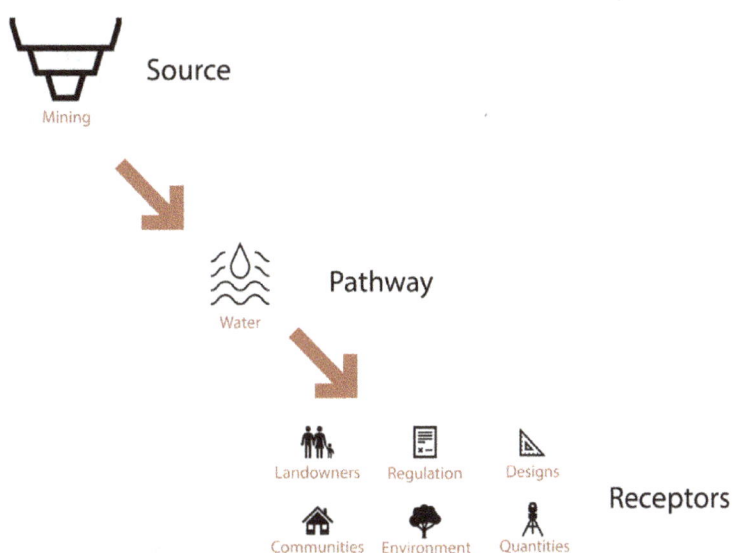

FIG 10 – Source, pathway, receptor model for water impact management (AusIMM, 2021).

Most ESG-related issues are best managed by preventing negative impacts and enhancing positive opportunities at the operational point where they originate, not by managing issues after they have occurred.

When it comes to working on matters above ground and outside the mine fence, a model for is presented in Figure 11 (in fact, as a model, it can relate to almost any of area of work at a minerals project). Step one is to build a baseline and maintain a knowledge base; there are many analytical and assessment methods for this. Step two is to foster relevant relationships and partnerships, and to engage competent people who know how to undertake the work that will be required. Step three is to design the work activity, and resource and execute it using competent management approaches that are already well known in the minerals industry.

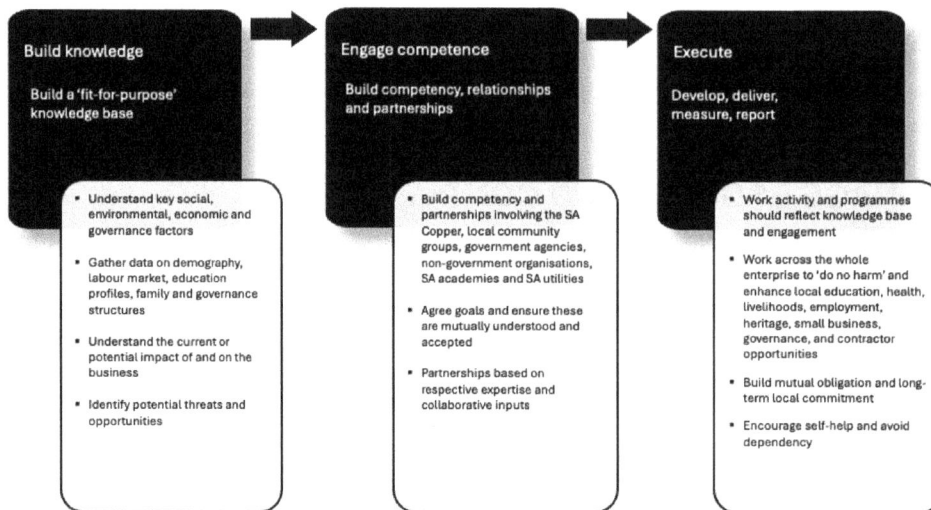

FIG 11 – A model for approaching ESG-related work.

Unfortunately, very often when it comes to ESG-related work, teams and projects commonly dive straight into step three without building a robust baseline, maintaining a current knowledge base and engaging appropriately competent people to test core input assumptions. As with any analysis or assessment, if material information is omitted from the model, results are ineffectual and counterproductive.

The work models described above should be familiar to experienced minerals professionals; it is how many technical work streams in the minerals sector are tackled, expanded upon in the following section.

Corporate governance

For a complex business to endure it must have systemic control; that is, it must be able to ensure all its workforce, including contractors, understand precisely what is expected of them in the way they carry out their work and interact with other people. This requires a cascaded set of instruction and guidance. This usually involves high-level statements of purpose and values, tightly articulated standards and assurance systems, and a voluminous body of instructive material with names like guidance notes, specifications, standard operating procedures and the like. Deeper discussion is beyond the scope of this paper; suffice to say nothing that follows can occur without good governance architecture in place.

Achieving operational focus – the essential few versus the many good

Good management requires focus and the ESG matters referred to in Figure 2 are not all important, or even relevant, in every context. Furthermore, the right work should be done at the right time in the project life cycle as different impacts and their associated risks can come to life during different phases of the project life cycle. Hence, the first step in Operational ESG is to reduce the list of ESG issues down to what matters most in context against the project development life cycle, to be reflected in a project delivery framework.

Getting the wrong questions correct

Context is critical – too often project proponents and operators are asking the wrong questions and responding to the wrong stimuli. Local conditions and issues are frequently overshadowed by salient (most mentioned) issues in global social media, leading to operational distraction and neglect of the most material issues locally. The logical vector of analysis leading to focused attention on ESG issues that matter most in an operational context is:

1. Outbound impact assessment.

2. Inbound reciprocating risk assessment.

3. Materiality assessment.

4. Gap analysis.

In practice, these analyses are iterative and there can be flexibility in whether impact or risk assessment comes first (Figure 1).

ESG planning – doing the right work at the right time, in context

The preparation of an Operational ESG plan is the culmination of the analyses described above. Planning using validated data, expertise, options analysis, and multi-disciplinary debate during assessment is where all the value is generated. The plan merely records the results and allows for allocation of *source* and *pathway* control accountabilities into functional department plans and an overall operations plan. Integrated planning and plans are key to tight Operational ESG. The identification of performance tracking metrics against ESG-related actions, resourcing and accountabilities is also integral to success. Taking a lead from health, safety and well-being, metrics for each of the ESG performance areas are rapidly becoming standardised, as described in the Global Reporting Initiative (GRI) Mining Sector (Reporting) Standard discussion draft (GRI, 2024) and increasingly published in ESG data books made available by leading companies in their integrated reporting.

Needless to say, specific and cumulative costs for redesign and adaption rapidly escalate as projects progress. The later a risk is discovered and mitigated in a project schedule, the more the re-work and expense that accrues (Figure 12).

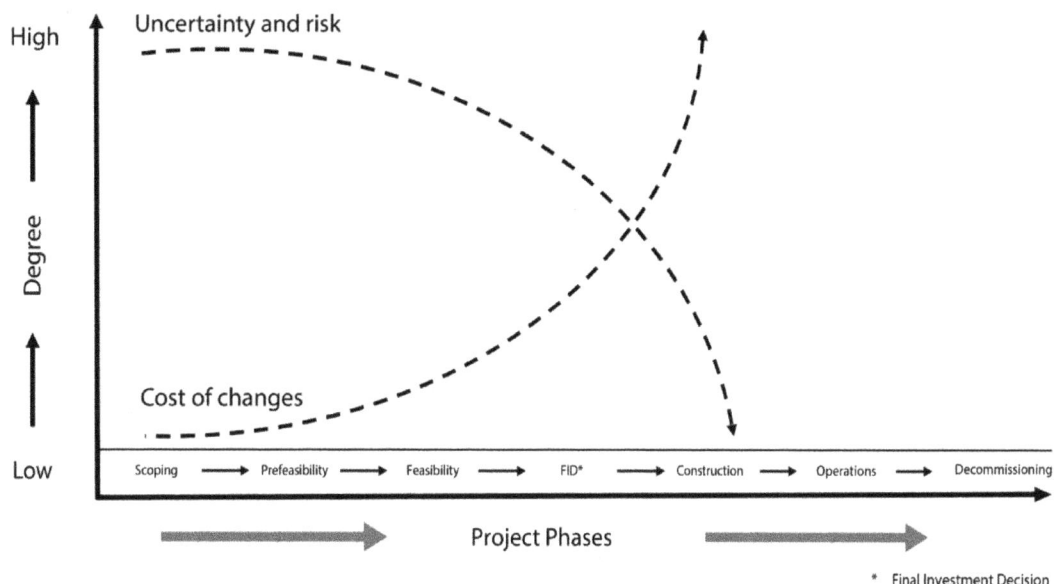

FIG 12 – The costs of change increases over the project life cycle (adapted from De Abreu, 2020).

Outbound impact assessment

Direct and indirect impacts, defined as the uncertainty and effects experienced by people and the biophysical environment due to proposed activities, should be rigorously baselined and assessed as a first step in managing ESG-related issues. This might involve regulatory Environmental and Social Impact Assessment (ESIA), typically a condition of permitting, noting that regulated scope and consequent word volume can impede penetrating insight. Driven by enlightened self-interest and insight, robust ESG impact assessment starts with a good environmental and social knowledge base(line) before progressing to analysis. The ESG matters listed in Figure 2 can serve as high level prompts for impact assessment, however tailored granularity on critical matters is recommended. Stakeholder analysis and engagement is a vital component. The involvement of experienced environmental and social performance professionals is essential in this and the assessments that follow.

Inbound risk assessment

Rebounding ESG-related risk, defined as the uncertainty for a business arising from its interactions with the social and biophysical environment, can be undertaken only after a business understands how it is impacting others and how they may respond. This should include regulatory, compliance, legal and contractual response. Standard risk assessment methods should be applied with proper facilitation and multi-disciplinary participation. The exercise should be directed to identifying opportunity as much as threat and inform the design of mitigating actions that will reduce variability of outcome. A good operator wants to know as precisely as possible what the outcome from undertaking a fully costed action or set of actions will be, acknowledging that prediction always involves uncertainty variability.

Materiality assessments – if not, why not? If so, why so?

Underground mining professionals are scientists. Assessing risk and materiality with a scientific approach, as described above, is critical to systematic decision-making. Once outbound impacts and inbound risks are identified and triangulated with stakeholders in context, materiality assessment aiming to distinguish the most important ESG issues on which to focus on at each stage of the project life cycle can be undertaken. Typically, materiality assessment is illustrated by a matrix such as the schematic example in Figure 13. The relative importance of issues for stakeholders are informed by outbound impact assessment and stakeholder engagement. The importance of issues from a business perspective are informed by the inbound ESG risk assessment and the strategic direction set out in the higher-level elements (purposes, values, standards) of a business's governance model.

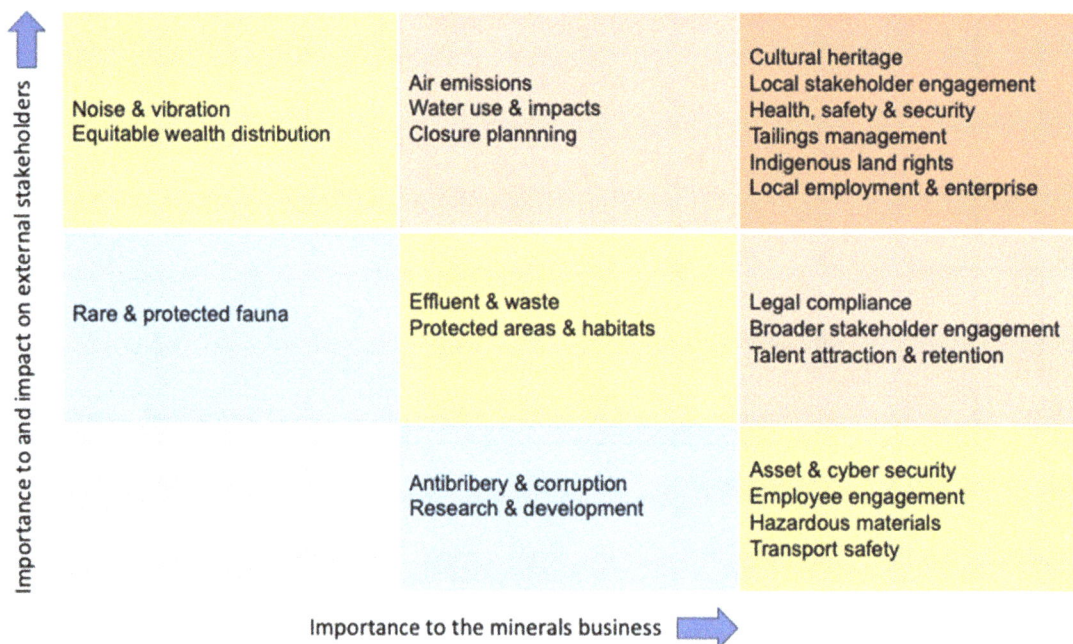

FIG 13 – Example of a Materiality Assessment Diagram.

Note that material issues should align with ESG-related Modifying Factors in the context of project financing, insurance, valuation and JORC and VALMIN Code disclosure.

Gap analysis

Following a ranked understanding of which ESG-related issues are most material in context and require priority attention, a comparison of the desired state of control versus current state can be undertaken using orthodox gap analysis. This should progress to a list of responsive actions, broken down to 'what', 'where', 'who', 'how' and 'when' to form the basis of an operational ESG plan.

ESG assurance

A 'Three Lines' assurance approach is common in the minerals sector and ESG-related matters should get the same level of scrutiny and verification as the rest of a business. For most value, a combination of audit (backward looking) and review (forward looking) elements should be present in assurance activities, properly facilitated within a collegiate atmosphere involving people experienced and competent in respective E, S and G matters. Validation of measurable performance data is vital to preserve integrity in the ESG disclosure that is now expected, summed up in the phrase – *Don't trust us, measure and verify us.*

ESG PROFESSIONAL DEVELOPMENT – AUSIMM CONTRIBUTIONS

The professional competencies required for rigorous management of ESG-related matters are not well catered for in mining schools and are not consistently taught in mainstream tertiary institutions. To help rectify this, the AusIMM in the past six years has generated and contributed to substantial ESG-related professional development materials, monographs and codes. These include:

- The Institute's Social Responsibility Framework and Statement (AusIMM, 2020).

- Social and Environmental Performance Area of Practice professional competencies (AusIMM, 2022).

- Chartered Professional recognition for Social Performance professionals.

- Professional Certificates ESG and Social Responsibility online course (AusIMM, 2021).

- Chapter 2 – Permitting, Environment and Social of the Study Processes Handbook – Monograph 35 (AusIMM, 2024).

- The Draft JORC Code, August 2024 (JORC, 2024a).

- The JORC Code – Guidance Notes: Exposure Draft, August 2024 (JORC, 2024b).

Importantly, the content and language of these materials are consistent. This is critical for coherent ESG-related work and the development of professionally competent practitioners. General understanding and management of ESG-related matters are equally important.

CONCLUSION

Whether they know it or not, mining engineers are as accountable as any mining professional for managing ESG factors and related risks. Adopting an approach to operational ESG as described in this paper is how they and minerals businesses wishing to gain broad-based support, business certainty, permitting, financing, insurance and an engaged workforce will achieve success. ESG management does not require special approaches – it can be based on existing common analytical and systemic management approaches. Although this will not generate ready-made solutions to complex ESG-related issues, it will provide an essential basis for iterative ESG-related harm reduction and value addition based on tested thematic experience under the direction of ESG-competent professionals.

Conversely, much of the current commentary and cherry picking of salient (trending) 'global' issues and narrative management that is presented as ESG is distracting at best and counterproductive at worst. This distraction results in management failure at *source* and *pathway*, requiring expensive re-work, compensation and business suspension to conciliate *receptors*.

Fortunately, the superficial 'vibe' that has prevailed in much of the western world for the past decade, often presented as ESG, is shifting to a re-emerging respect for reality and performance management. Equally, heavy-handed regulation of ESG-related matters needs to give way to competent self-management. The approach described in this paper, one requiring professional judgement and assurance, aligns well with this shift. ESG, however referred to, is not exotic and has no utility when subjected to rigid regulation or deployed as strategic banner waving – it requires diligent, well-resourced competent people working in operations-directed roles. This includes underground operators, particularly those wishing to pursue expanded career options, with a good sense of how their work and its impacts affect the world above ground.

REFERENCES

AusIMM, 2020. AusIMM Social Responsibility Framework and Statement, Governance, AusIMM. Available from: <https://www.ausimm.com/about-us/governance/social-responsibility-framework-and-statement/#:~:text=AusIMM%20expects%20all%20its%20members,societies%20affected%20by%20their%20work>

AusIMM, 2021. ESG and Social Responsibility [online course], Professional Certificates, AusIMM. Available from: <https://www.ausimm.com/courses/professional-certificates/esg-and-social-responsibility/>

AusIMM, 2022. Social Performance and Environmental Professional Development Toolkit, AusIMM. Available from: <https://www.ausimm.com/globalassets/career-hub/esg-mapping-tool/environment-and-social-performance-mapping-tool.pdf>

AusIMM, 2024. *Study Processes Handbook, Monograph 35*, Publications, AusIMM. Available from: <https://www.ausimm.com/publications/monograph/monograph-35---study-processes-handbook/>

Brundtland, G H, 1987. Our Common Future: Report of the World Commission on Environment and Development, Geneva, UN-Document A/42/427.

De Abreu, A, 2020. The Use of Schedule Risk Analysis in Construction Projects its and Application in Delay Analysis, Master's thesis, Robert Gordon University, 49 p. https://doi.org/10.13140/RG.2.2.14859.52002

Deans, G and Harvey, B, 2024, 15 March. Can you dig it? Modifying Factors and multi-factor risk, *AusIMM Bulletin*. Available from: <https://www.ausimm.com/bulletin/bulletin-articles/can-you-dig-it-modifying-factors-and-multi-factor-risk/>

Flyvbjerg, B and Gardner, D, 2023. *How big things get done: The surprising factors that determine the fate of every project, from home renovations to space exploration and everything in between* (Penguin Random House).

Flyvbjerg, B, 2003. *Megaprojects and risk: An anatomy of ambition* (Cambridge University Press).

Global Reporting Initiative (GRI), 2024. Sector Standard for Mining, GRI 14: Mining Sector 2024, Standards development, Global Reporting Initiative. Available from: <https://www.globalreporting.org/standards/standards-development/sector-standard-for-mining/>

Grant, A, 2023. *Hidden potential: The science of achieving greater things* (Random House Large Print).

Gray, J, 2002. What matters is meta-risks, *The Treasurer Magazine,* June 2002 (The Association of Corporate Treasurers, London, England).

Harvey, B and Deans, G, 2024, 26 September. JORC Code 2024 – Modifying Factors, including ESG and Risk, *AusIMM Bulletin.* Available from: <https://www.ausimm.com/bulletin/bulletin-articles/jorc-code-2024--modifying-factors-including-esg-and-risk/>

JORC, 2012. The JORC Code, 2012 edition, Joint Ore Reserves Committee (JORC) of The Australasian Institute of Mining and Metallurgy, Australian Institute of Geoscientists and Minerals Council of Australia. Available from: <https://www.jorc.org/docs/JORC_code_2012.pdf>

JORC, 2024a. Draft JORC Code - August 2024, JORC. Available from: <https://www.jorc.org/code-update/>

JORC, 2024b. The JORC Code – Guidance Notes: Exposure Draft, August 2024, JORC. Available from: <https://jorc.org/docs/Draft_JORC_Guidance_Note_01Aug2024_readonly.pdf>

Manalo, M, 2024. Average lead time almost 18 years for mines started in 2020–23 in S&P Global, Research, S&P Global. Available from: <https://www.spglobal.com/market-intelligence/en/news-insights/research/average-lead-time-almost-18-years-for-mines-started-in-2020-23>

Peirce, J J, Vesilind, P A and Weiner, R, 1990. *Environmental Pollution and Control* (Elsevier Inc).

Sauvant, K P and Mann, H, 2019, 17 December. Making FDI More Sustainable: Towards an Indicative List of FDI Sustainability Characteristics, *The Journal of World Investment and Trade*, pp 916–952. https://doi.org/10.1163/22119000-12340162

Valenta, R K, Kemp, D, Owen, J R, Corder, G D and Lèbre, É, 2019. Re-thinking complex orebodies: Consequences for the future world supply of copper, *Journal of Cleaner Production*, 220:816–826.

Case studies and operating practice

Overslept sublevel cave production rings – a post-weather event analysis

M Bouwmeester[1]

1. Mining Engineer, Ernest Henry Operations, Cloncurry Qld 4824.
   Email: matt.bouwmeester@evolutionmining.com

ABSTRACT

Evolution Mining's Ernest Henry Operations (EHO) is a large-scale underground copper and gold mining operation 38 km North-East of Cloncurry, Queensland, Australia. EHO employs a sublevel cave (SLC) mining method where pre-charging production rings is critical to the safety and scheduling performance of the mining operation. On 8 March 2023, a weather event resulted in an ingress of water to the underground workings, impacting production for nearly two months. As a result, the firing of 74 pre-charged production rings was delayed during this period, creating risks and uncertainties around the condition of the explosives and blast performance of the overslept rings.

A literature review was conducted indicating a potential world-first event for firing overslept rings after such a period. EHO's site procedure for sleeping production rings is a maximum of 28 days before firing, whilst rings that sleep longer become classified as overslept rings. Consultations with the explosives contractor ensued and specifications for the products were reviewed and evaluated. After conducting an in-depth risk assessment, vibration blast monitors were installed in ore drives, and an overslept rings management guide was developed in anticipation of firing the first post-weather event production ring.

On 24 April 2023, the first overslept production ring was fired at 77 days slept, with blast monitor vibration data captured and analysed upon detonation. Over the course of the next 37 days, all 74 overslept rings, including a combination of wireless blasting technology as well as conventional millisecond (MS) non-electric detonators were blasted, with the longest overslept ring fired at 104 days slept. This paper provides a detailed post-weather event analysis, outlining the approach, controls and results the business undertook to ensure the firing and extraction of 74 overslept production rings was conducted as safely as possible.

INTRODUCTION

Pre-charging production rings in a sublevel cave (SLC) is critical to the performance of the mining operation as it allows flexible scheduling of drill stocks, charge stocks and available drawpoints for the loaders to bog from. Upon completion of each charged ring, the explosives enter a period known as 'sleep time'. After 28 days, the ring is deemed overslept and must be safely fired as soon as possible to avoid any potential explosives degradation which may compromise the outcome of the firing.

On 8 March 2023, a weather-event resulted in water ingress to the mine, delaying production for nearly two months. As access became re-established, questions around a potential restart to production began to arise. Many unknown scenarios were presented, not least of all the length of time the production rings had been overslept. Additionally, explosives retention, blast performance, mud rush risks, and various other hazards and controls were identified and outlined in an in-depth risk assessment. For the purposes of this paper, not all rings have been analysed, instead a focus on some of the main issues observed upon production commencement are covered, and results and learnings discussed.

BACKGROUND

Evolution Mining's Ernest Henry Operations (EHO) is located approximately 38 km north-east of Cloncurry in north-west Queensland (Figure 1). The orebody is a porphyry copper and gold deposit dipping at 45°. Currently a large-scale underground SLC operation, production targets of 6.8 Mt per annum are achieved via a 1000 m deep hoisting shaft that transports ore to the surface. The open pit operated from 1997–2014, whilst the first cut of the decline was taken in February 2008. Underground production began in December 2011 when the first level, 1650RL (RL is sea level plus

2000 m), was transitioned from the open pit. Levels are stepped out at 25 m intervals. At the time of the weather event in March 2023, production levels were 1275, 1250 and 1225, whilst development was occurring at 1200, 1175 and 1150 levels, along with two new declines.

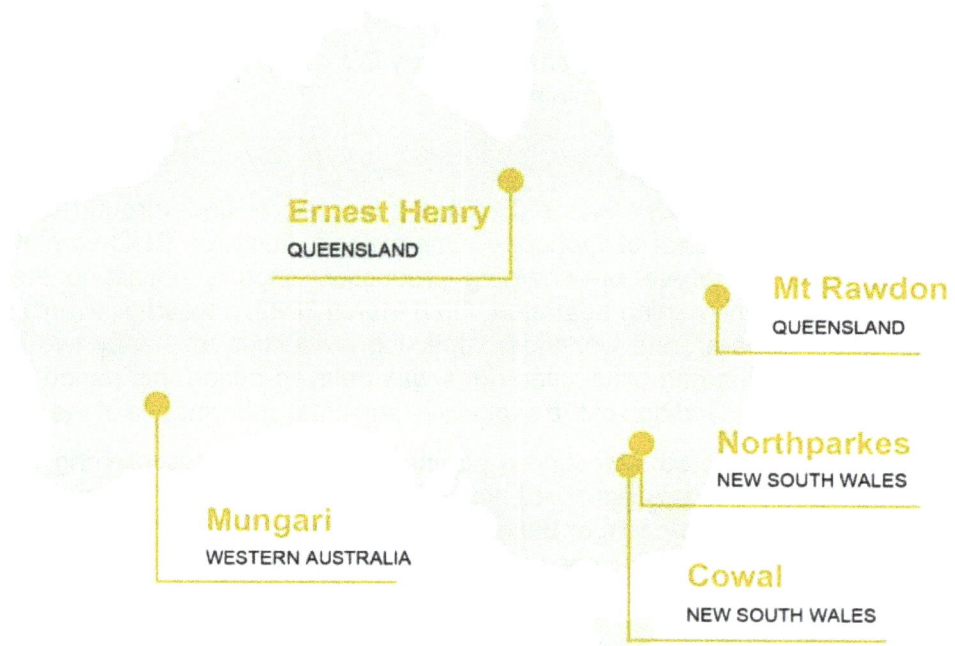

FIG 1 – Location of Ernest Henry Operations (Evolution Mining, 2024).

Each level of EHO's SLC has a series of ore drives that are separated by pillars on the hanging wall, whilst infrastructure is placed on the footwall drive. As seen in Figure 2, each level is divided into western and eastern zones, each consisting of a fresh air rise (FAR), return air rise (RAR) and electrical recesses. Western and eastern orepasses (WOP and EOP) transfer ore via a series of finger passes down to the crusher, where both a western and eastern bogger tip ore into a central crusher. Each level's ore drives are mined in a southerly direction and extracted back towards the footwall drive, causing subsidence to the south wall on the surface.

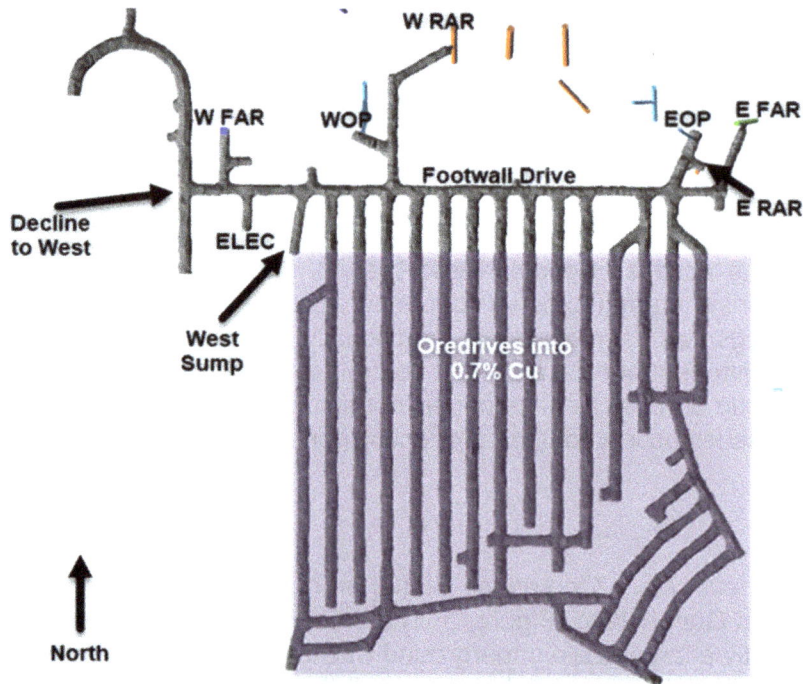

FIG 2 – Typical layout of a SLC level.

Centrelines of the ore drives are 15 m apart with dimensions of 6.0 m wide by 4.8 m high. Levels may vary; however, most contain around 15 main ore drives, as well as some transfer drives, subject to the orebody wireframe. Production rings are typically designed as eight × 102 mm diameter upholes in a fan shaped pattern with a ring spacing burden of 2.6 m and a ten-degree dump forward (Figure 3). Ore drives can consist of over 100 rings, with each standard ring consuming approximately 160 drill metres, 1200 kgs of emulsion and providing around 2000 t of broken ore stocks upon firing. Considering the time it takes to consume each ring and the number of rings available, drills can be positioned in one ore drive for several shifts, making it easier to schedule drill and blast stocks. Rings are pre-charged to a minimum of three rings before firing the next ring at the cave front. At the time of the weather event, there were 27 drawpoints available for bogging across three production levels.

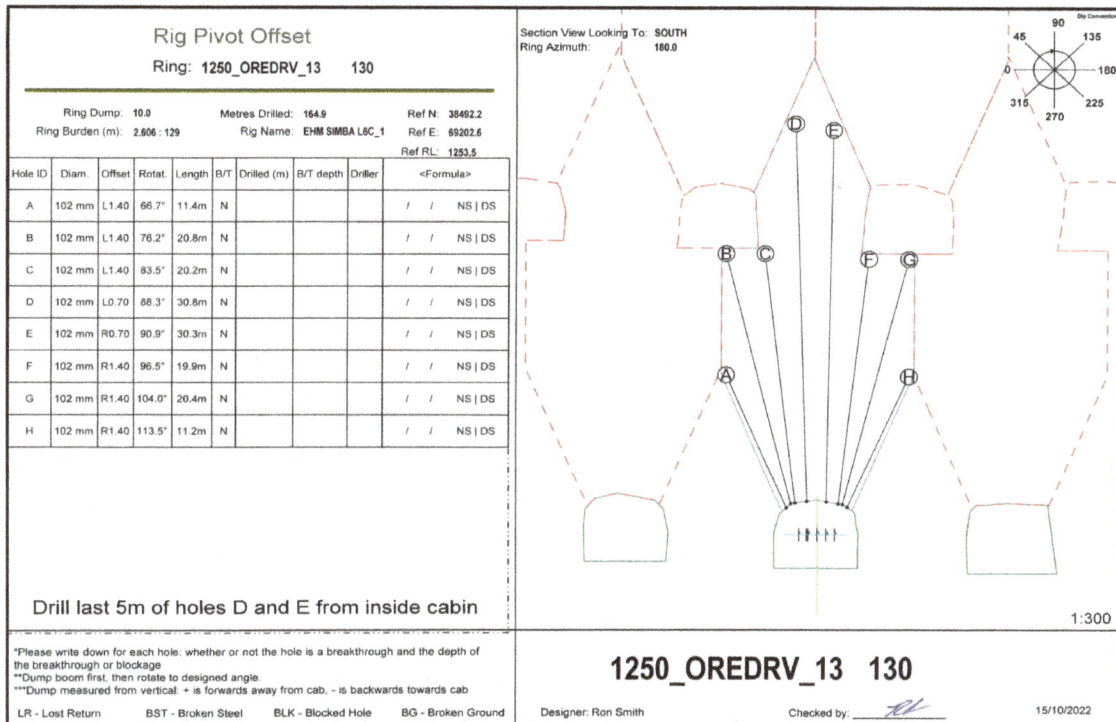

FIG 3 – Standard production ring design at EHO.

PRE-CHARGING RINGS

Pre-charging production rings has been developed through various trials and risk assessments at SLC mines both in Australia and around the world. Unlike stope mining operations where charging and firing rings occur simultaneously, SLC mines are able to pre-charge rings and leave them to 'sleep'. Benchmarking against mines such as Kiruna (Sweden) and Ridgeway (New South Wales, Australia), has proven that pre-charging production rings increases operational efficiency and scheduling flexibility, and is now considered commonplace in SLC mines (Trout, 2002; Lebrocque, 2016). As seen in Figure 4, 16 out of 24 SLC mines worldwide pre-charge production rings. Each production ring has a bulk emulsion explosive, plus boosters and detonators loaded into each hole. As demonstrated at Ridgeway Gold Mine (RGM), best practice is to have three pre-charged rings in each ore drive before firing the next ring (Wiggin, Trout and Macaulay, 2005).

Additional benefits of pre-charging rings include:

- Increase operational efficiency and productivity.
- Improves safety around drawpoints.
- Decreases brow exposure and provides ergonomic positioning for the operator.
- Increase charge stocks and provide flexibility for drill and loader movements.
- Eliminates/reduces redrilling risks caused by seismic events.

- Aids in potential lost production from hole blockages.

Precharging

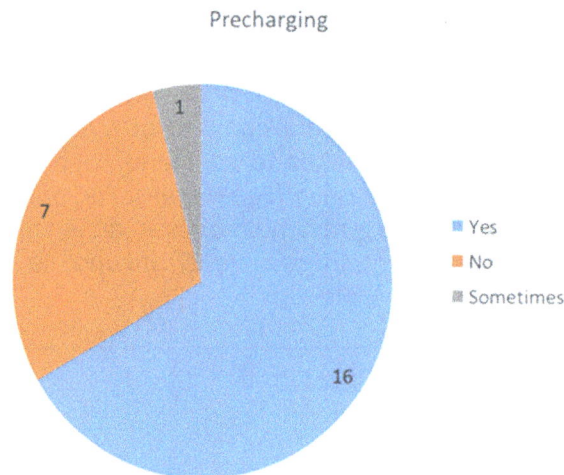

FIG 4 – Pie chart indicating 16 out of 24 SLC mines worldwide utilise pre-charging of production rings (Campbell, 2022).

Sleep time

Sleep time can be defined as the time from charging until blasting. EHO's Management of Overslept Rings document provides guidance and instructions on the acceptable controls of overslept production rings. As part of the production engineer's daily duties, pre-charged rings approaching 28-days sleep time are prioritised for firing. Based on learnings from other SLC mines, and the explosives manufacturer's recommendations, the maximum sleep time before a ring is to be fired at EHO is 28 days. Should a ring 'expire' beyond this time frame, the Management of Overslept Ring Trigger Action Response Plan (TARP), is activated and the necessary steps followed to ensure any hazards are controlled and the firing is conducted promptly and safely.

Reactive ground

As part of the overslept rings TARP, temperature testing is conducted by a mine geologist to ensure hole collars of the overslept rings have not exceeded 40°C. The Reactive Ground Management Plan outlines that the on-site explosives supplier's recommended sleep time is 28 days in areas of reactive ground. Should the holes exceed 40°C, but measure below 55°C, additional steps are followed to minimise risks and hazards associated with explosives in areas of reactive ground. Historically, EHO is classed as having mildly reactive ground conditions. Since 2007, over 3000 Potentially Reactive Ground (PRG) samples have been tested by the site lab as well as the explosives supplier. Results have shown the samples reacting at 80°C, with only 0.15 per cent reacting at 55°C. Historic and recent ambient temperature testing from multiple locations underground range between 29.6°C and 42.4°C.

Blasting products

Since winning the contract for production charging at EHO in January 2021, Orica has utilised both conventional non-electric detonating systems (nonel) as well as WebGen™ 100 Wireless Electronic Blasting System (WEBS). Both nonel and WEBS rings are charged with Subtek bulk emulsion explosive at a density of 1.2 g/cm³. Nonel rings are primed with 250 g Pentex Stopeprime Boosters and Exel Millisecond (MS) detonators and use an electronic i-kon starter detonator and detonating cord to initiate the firing.

WEBS technology relies on a receiver that is sent up the holes at time of charging. This receiver has a three-month battery life (Table 1). Signals are sent wirelessly, and rings are fired remotely from the surface, along with nonel rings. The split between the two types over the last two years is approximately 60/40 per cent biased towards nonel products. The main advantage of using nonel is the cost per unit, whilst WEBS, being more expensive, is ultimately safer due to not exposing charge-up operators to ground conditions at the brow of the drawpoint. WEBS also allows rings to initiate even if a brow snaps back due to deterioration of ground conditions. This is important as it does not hold up production while brow conditions and other issues are rectified, which can be a costly exercise.

TABLE 1
Explosives products used at EHO in 2023.

Product	Maximum sleep time	Temperature range (°C)
Subtek emulsion	30 days	0–55*
Pentex Stopeprime booster	N/A	0–70
Exel MS detonators	Two weeks	0–70
i-kon electronic detonator	Three months	5–45
Pentex WebGen™ booster	Three months	5–55
DRX100 WebGen™ receiver	Three months	5–45

* dependent on reactive ground.

From July 2016 to April 2017, Orica conducted WEBS trials at EHO which included several large-scale signal surveys to test the reliability of WebGen™ transmission equipment. Test holes were originally trialled with one ring fired shortly after (using a safety net of redundant nonel systems). Over the next two and a half months, 30 rings were charged and fired successfully (Orica, 2017). Since Orica became the permanent supplier of production charging products in January 2021, EHO has successfully used WebGen™ 100 detonators in hundreds of rings and is the only SLC worldwide presently using the technology. Figure 5 shows that EHO is the only SLC mine in the world to use wireless production blasting technology on a permanent basis.

Detonator type

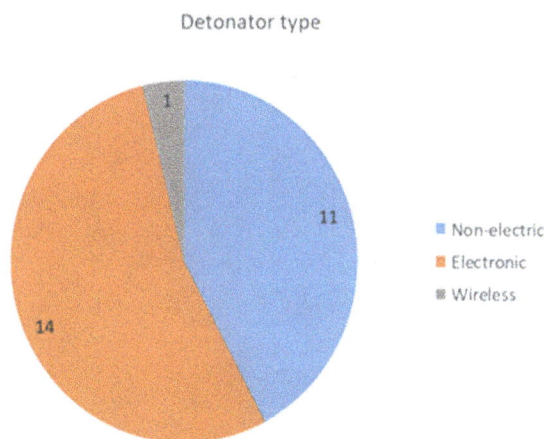

FIG 5 – Pie chart indicating only one SLC mine (EHO) worldwide using wireless blasting technology on a permanent basis (Campbell, 2022).

The criteria for selecting WEBS to charge rings instead of nonel is based on geotechnical advice in areas where high seismicity may be present due to existing faults and shear zones. Along with the standard ground support systems used at EHO, these areas consist of additional ground support in the form of plated and tensioned 6 m or 9 m cables. Charge plans are then slightly modified to allow the engineer the flexibility to control the brow condition post-blasting, as outlined below.

As seen in Figures 6 and 7, the collars of the two apex holes in WEBS designs are brought down to 3 m in length. This reduces hang-up risks due to the existing cables holding the ground up, with brow snapback less of an issue in these areas. Apex hole designs in nonel rings remain at the standard 4 m to avoid any snapback concerns. Furthermore, to reduce costs of WEBS units, successful trials were completed on-site to assess the function of only using one WEBS unit in the shoulder holes, thus reducing primer totals to 12, instead of 16 in a standard ring design.

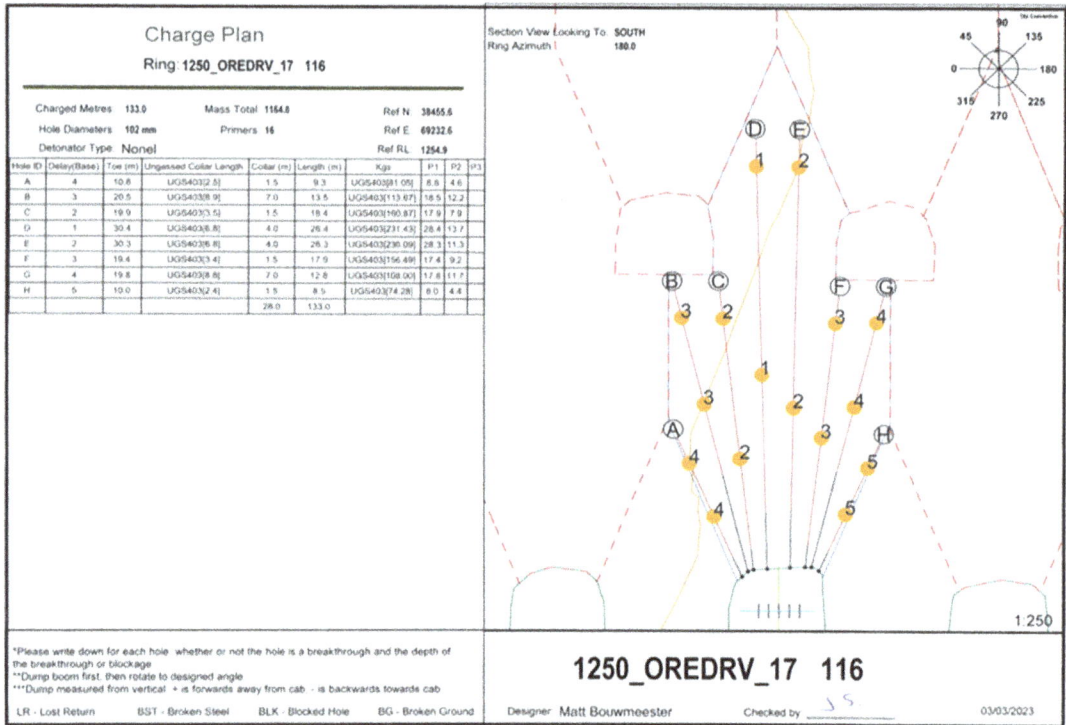

FIG 6 – Standard nonel charge plan with collars of the apex holes at four metres and 16 primers.

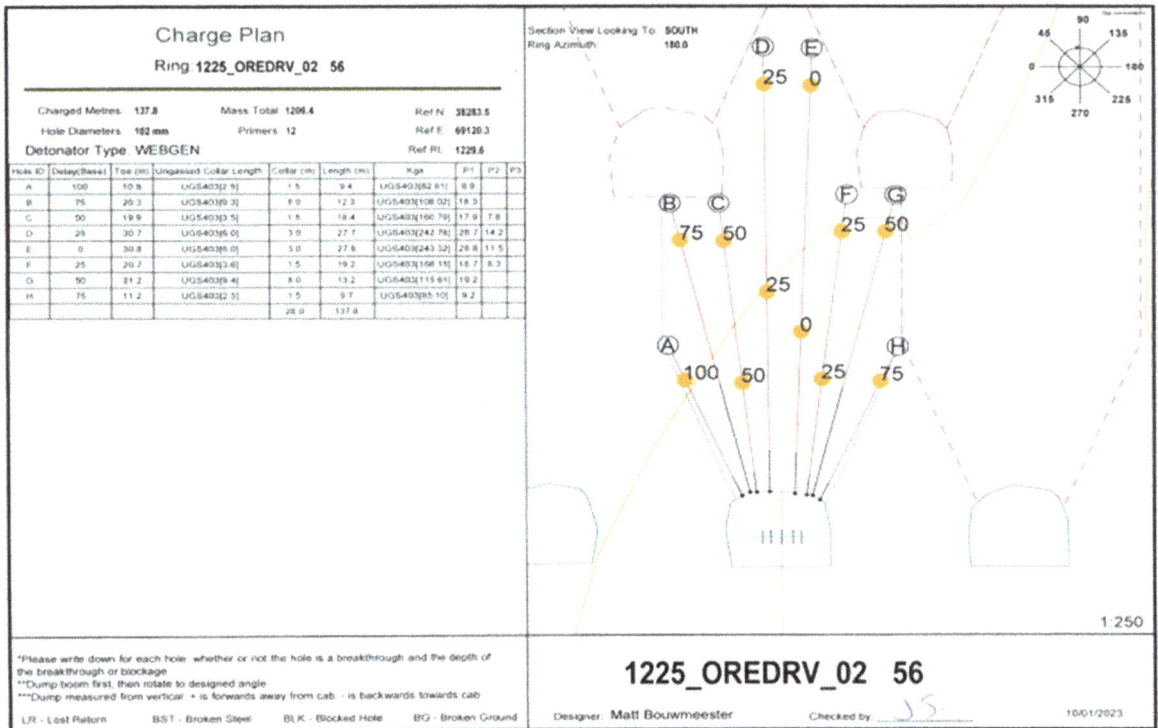

FIG 7 – Standard WEBS charge plan with reduced primers and amended collars.

BLAST VIBRATION MONITORING ANALYSIS

Previous studies conducted at both EHO and RGM have both shown a common denominator of longer sleep time of explosives potentially impacting the blast performance. Ideally, sleep time is analysed and quantified by controlled studies measuring Velocity of Detonation (VOD). This is optimally done using in-hole technology, however, due to the large-scale nature of SLC mines, this type of analysis can present significant challenges. In a 2015 study at EHO, 67 production blasts were captured by locating four geophones in various locations at specific distances. Particle vibration data was acquired by applying a vector sum formula from x, y, z coordinates after each blast. From here, a vector peak particle velocity (VPPV) was determined. Subsequent changes in site constants, k and alpha, could be interpreted to explain changes in intensity and attenuation levels of particle vibrations and their direct associations with sleep times. Furthermore, blast signatures could be predicted based on the timings of each blasthole detonation, providing a baseline for blasthole timing analysis. By correlating the two sets of predicted and actual data, successful blasts were established, while low correlation suggested the blast had not been initiated to its full capacity (Lebrocque, 2016).

The Lebrocque (2016) study found that based on two different sleep time ranges, 0–21 days, and 22 days plus, the k value, as represented by vibration particle intensity, decreased by more than half in the rings fired with longer sleep time. The rings with the shortest sleep time showed evidence of greater blast signature predictability and had the best blast performance overall, whilst blast signatures became harder to predict the longer the ring slept. Oversize rock was also studied and found that minimal correlation existed between the creation of oversize rock and sleep time. During the study, no ring was fired at greater than 28 days.

RGM also conducted studies on the blast performance of their Dyno T6100 emulsion product over sleep time up to 28 days. This product was used at EHO from 2011 until 2020, and the Subtek emulsion currently used has very similar properties. Results from the RGM study indicated 1.1 g/cm^3 emulsion density levels measured a lower VOD as the rings slept longer. These results provided valuable understanding of the product and through this knowledge the higher density levels of 1.2 g/cm^3 were used moving forward. Blast performance improved and the operation became more productive (Wiggin, Trout, and Macaulay, 2005). EHO has maintained a density level of 1.2 g/cm^3 consistently over many years now.

Application

Since the scenario at EHO was more of a response to a crisis then a controlled study, in hole testing, blast curve predictions and other *'luxuries'* of studying each ring were not afforded. The main objective was firing off each ring as promptly and safely as possible to ensure the mine returned to full production. Applying this knowledge, consultations with Orica then followed as they agreed to supply some blast vibration monitors to assist with the analysis and performance of overslept rings.

The ShotTrack ViB 500G vibration monitors (Figure 8) are specifically designed for near field testing where high frequencies are required. In addition, Orica provided information that the products used on-site went through rigorous laboratory tests. These tests indicated the Subtek emulsion product was the most robust explosives product that Orica supply, hence its use in SLC mines. Not all rings received blast monitoring, therefore, not all data was captured for analysis. This was universally accepted at the time; however, it opens the door for possible further controlled studies to verify the results achieved.

FIG 8 – ShotTrack ViB 500G blast vibration monitor.

RISK ASSESSMENT

As an overarching view of the entire operation, many hours were spent by multiple personnel completing risk assessments to recover and restart the operation. Both conventional and dynamic risk assessments were conducted, and it is testament to the personnel at EHO for their valuable input and experience provided. One of the most important risk assessments conducted was centred on production recommencement. We had 18 participants contributing, including the mine manger, superintendents, mining engineers, geotechnical engineers, geologists, safety advisors, shift supervisors, charge-up and loader operators. Several risks and hazards were identified, and controls were put in place. Notably, many risks revolved around the overslept rings, of which the following were identified:

- Unplanned initiation of overslept rings due to reactive ground causing injury to personnel or damage to equipment.

- Unplanned initiation of hole drop-outs in rill during bogging activities resulting in injury to personnel and/or damage to equipment.

- Poor blast performance of overslept rings (degradation of emulsion) resulting in oversize/ hangups causing production delays.

- Poor blast performance of WebGen™ rings due to reduced battery life resulting in oversize/ hangups causing business delays.

- Poor blast performance due to hole slumping resulting in oversize/hangups causing production delays.

Additional hazards outside of the impact of explosives included bogging activities, secondary ventilation, re-entries, and infrastructure concerns, of which the following were noted:

- Mud-rush released from a cave/orepass drawpoint causing injuries to personnel and/or delays to production.

- Risk of NOx gases produced from potential reaction entering mine/RAR system following re-establishment of secondary ventilation resulting in injury to personnel.

- Rockfall as a result increased seismic risk due to re-commencement of cave front resulting in injury to personnel.

- Hang-up in orepass(s) following extended duration with no movement resulting in injury to personnel or business delays.

- Damaging of infrastructure on production levels when loading/tramming trucks resulting in business risk.

Re-entry QA/QC

Once access was available to production levels, inspections were able to occur under the proviso that the risk assessment be strictly adhered to. Given the lack of secondary ventilation, conditions were understandably humid in the ore drives. Mandatory PPE and other items of use included:

- respirators

- gas monitors

- temperature gun

- barricade tape

- enough water for hydration

- 'Geotechnical Production Inspection – post rainfall event' inspection sheets.

As seen in Tables 2 to 4, and Figures 9 to 11, most ring inspections showed no charging retention issues and maintained dry rills, which was encouraging. As expected in a SLC environment, the drawpoints that contained moisture were noticeable on the periphery of the cave front, whilst the centre drives on each level largely remained dry. Nearly all slumping of emulsion and detonator drop-outs occurred on the eastern and western edges of the cave. Temperatures remained steady mostly in the range of 32–38°C, with the exceptions of 1250 transfer drive (TRNDRV) 29 and the unfinished rise in 1225 ore drive (OREDRV) 18 (*NB. these tables only outline next ring to be fired, not all pre-charged rings*).

TABLE 2

1275 Level – Post-weather event inspections 04/04/2023.

OREDRV	Ring	Temperature	Charge retention from heat/general comments
14	159	32.9°C	No charging retention issues. Dry rill.
16	156	34.0°C	No charging retention issues. Dry rill.
18	156	33.4°C	No charging retention issues. Dry rill.
20	154	35.8°C	No charging retention issues. Water trickling from base of drawpoint.
22	153	32.8°C	No charging retention issues. Signs of moisture at drawpoint.

FIG 9 – 1275 OREDRV 22 drawpoint, 04/04/2023.

TABLE 3

1250 Level – Post-weather event inspections 28/03/2023.

OREDRV	Ring	Temperature	Charge retention from heat/general comments
01	116	33.5°C	R118 holes F and G explosives slumped at brow Signs of moisture at drawpoint
03	118	32.8°C	R120 Hole G slumping in backs Signs of moisture at drawpoint
05	121	32.3°C	No charging retention issues. Dry rill.
07	119	32.3°C	Snapback on right shoulder, no charging retention issues. Dry rill.
09	118	33.1°C	Brow open, no charging retention issues. Dry rill.
11	116	34.2°C	No charging retention issues. Dry rill.
13	115	34.7°C	No charging retention issues. Dry rill.
15	115	35.7°C	No charging retention issues. Dry rill.
17	113	35.5°C	R115 hole D slight slumping in the backs. Dry rill.
19	114	37.6°C	R114 hole B bottle brush and tails hanging. R115 hole B explosives and detonators on rill. Signs of moisture in the rill, drawpoint open
TRNDRV 29	18	42.4°C	No charging retention issues. Dry rill.

NB. Transfer drive 29 in the south-east lens inspected 18/04/2023.

FIG 10 – 1250 OREDRV 19 drawpoint, 28/03/2023.

TABLE 4

1225 Level – Post-weather event inspections 27/03/2023.

OREDRV	Ring	Temperature	Charge retention from heat/general comments
98	60	37.8°C	Ring 60 holes F slight slumping of explosives. Signs of moisture at drawpoint
00	62	36.0°C	Rings 62 and 63 explosives slumping in several holes and detonators on the rill. Signs of moisture at drawpoint
02	62	34.5°C	Rings 62, 64 and 65 slight slumping of explosives. Dry rill.
04	62	34.3°C	No charging retention issues. Dry rill.
06	64	36.0°C	No charging retention issues. Dry rill.
08	64	34.7°C	No charging retention issues. Dry rill.
10	64	35.7°C	No charging retention issues. Dry rill.
12	66	36.1°C	No charging retention issues. Dry rill.
14	69	38.1°C	R66 hole D slight slumping in the backs. Dry rill.
16	69	39.8°C	No charging retention issues. Dry rill.
18	59 (rise)	41.3°C	No charging retention issues. Rise charged only. No rill.

NB. OREDRV 18 had only the widenings and forwards slashing rings charged at time of weather event.

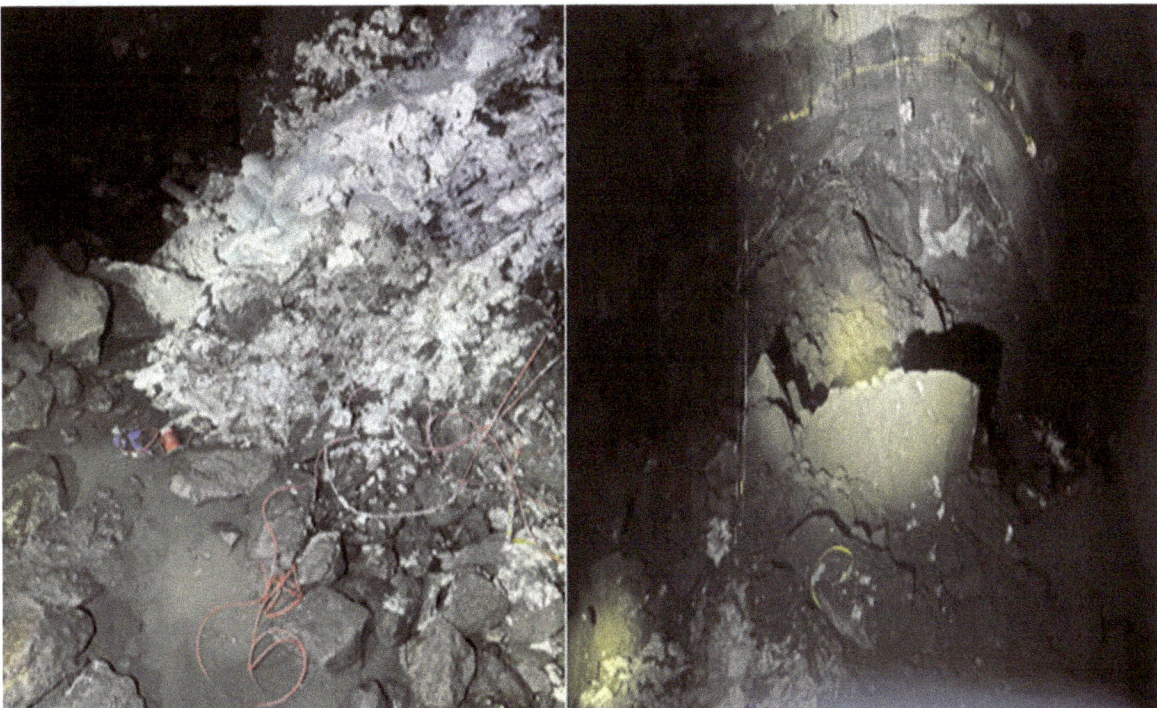

FIG 11 – 1225 OREDRV 00 drawpoint, 27/03/2023.

The initial inspections provided an insight into the behaviour of the explosive products, the nature of the cave material and ambient environment throughout the production halt. Confidence was taken in the fact that only a small number of rings had charging retention issues, and the ore drives themselves were mostly dry and not subjected to excessive heat, especially given the lack of secondary ventilation. NOx and other irritant gases were also not present; therefore, it was feasible to now move towards a potential restart for production.

Mud-rush risks

Prior to firing off the first overslept ring, geotechnical assessments coupled with strategic risk-assessed re-commencement of bogging the cave front were used to mitigate potential mud-rush. As full power and ventilation was restored, bogging to a maximum draw of 250 t (approximately 15 buckets) per drawpoint was permitted evenly across the cave front. Not only did this mitigate a mud-rush release, but it also alleviated geotechnical stress as the cave front had been static for nearly two months. In addition to these points, porosity levels of the rocks were discussed. Historically, the orebody at EHO is considered to have low porosity levels, therefore, rocks do not hold vast volumes of water. Furthermore, given the nature of SLC mining, water tends to move to the periphery of the cave, as was evident in the drawpoint inspections.

RESULTS

The 1250 level was the first level to have full power and ventilation restored, therefore, was ready to fire overslept rings first. Firings would be slow to begin with as the hoist was still out of operation and ore was being transported via trucks to the surface. At this stage all tasks associated with the risk assessment were actioned. A blast monitor was installed conservatively 50 m back from the brow. OREDRV 17, ring 113 (nonel) was the oldest sleeper on the level. Technically considered 49 days overslept by the management guidelines, it had been sleeping 77 days since the ring was charged. Once the remaining tonnes in the draw from the previously fired ring had been bogged, firing of the first post-weather event overslept ring commenced.

Upon re-entry of the blasting, the following steps outlined in the risk assessment were taken:

- verification of initiation from remote blasting box
- secondary ventilation observations
- gas monitoring
- geotechnical-production inspection forms
- blast monitor retrieval
- visual observations of the drawpoint.

Initial visual observations of the first fired ring were not very encouraging, however, this was due to existing brow collars leftover from the previous ring that was fired in early March 2023. Blast monitoring results confirmed that the ring had in fact detonated reasonably well, which was extremely encouraging. Furthermore, reports from the loader operator were reasonably promising, with only some oversize present, and the majority of the ore fragmentation moving freely through the orepass grizzly system. This provided enough evidence and confidence to fire off additional rings.

Figure 12 shows the nonel charging plan and acceleration data for the first fired ring, 1250 OREDRV 17 ring 113. Beginning at detonator number one (25 ms), subsequent detonator timings are sequenced at 25 ms apart. Clear spikes indicate the first detonator initiated at 25 ms, whilst the second and third detonators are delayed about 10 ms. Two further spikes in acceleration occur between 100 and 110 ms, followed by small spikes shortly thereafter. Whilst the data does not perfectly align with the manufacturer's specifications, there is enough satisfactory evidence to conclude that five MS detonators were initiated, which corresponds to the charge plan.

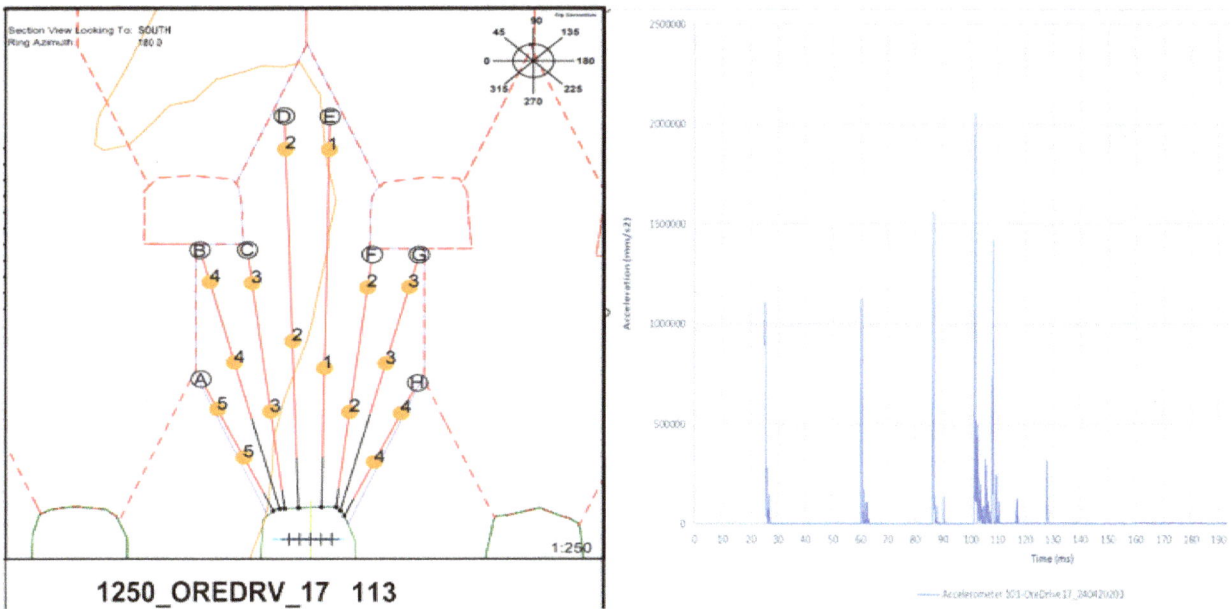

FIG 12 – First ring fired post rain event: charge plan and blast vibration results.

For the purposes of this paper, it is impossible to detail each firing and subsequent blast vibration data analysis. In any case, as mentioned earlier, it was not the environment to be performing a controlled study. Multiple rings needed to be fired in a 24-hour period, often on the same level. In addition, the availability of blast monitors and personnel was insufficient for these purposes, as the prerogative was to ensure a safe return to full production in the most reasonable time frame. Nonetheless, around two-thirds of the rings were monitored for blasting vibration data analysis, and post-firing photographic evidence captured.

Throughout 37 days of the operation, most of the 74 overslept rings initiated successfully with all available tonnes bogged out. Post-firing drawpoint inspections typically revealed a choked brow of dry and reasonably finely fragmented ore with no damage to the ground support as is expected from normal post-firing conditions. Blast monitoring was somewhat revealing, with some reports indicating fairly accurate data whilst others suggested small inaccuracies in timing intervals and anomalies around the data. This could perhaps be explained by the fact that some shifts saw two ore drives in each level being fired with one monitor picking up vibrations from the ore drive it was stationed in, as well as additional vibrations from different ore drives either on the same level or those merged and fired with a level 25 m above or below.

A typical ring is seen in Figure 13. 1225 OREDRV 16 ring 70, was a WebGen™ firing completed 19 May 2023 at 81 days slept. Clear spikes are visible approximately 25 ms apart, albeit slightly lagging. Holes D and H at 50 ms delays appear to have potentially misfired due to a lack of visible acceleration data. Figure 14 shows an image taken upon re-entry of the firing and is indicative of most of the successful firings during the production restart.

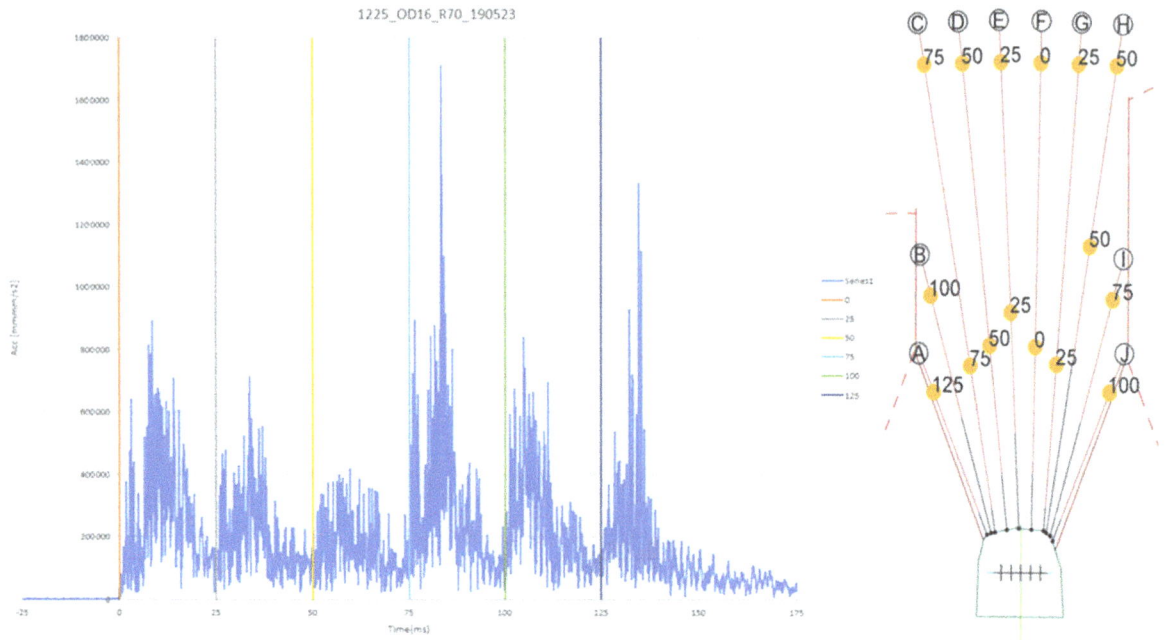

FIG 13 – Blast monitoring vibration data and charge plan for 1225 OREDRV 16 ring 70.

FIG 14 – Re-entry inspection. 1225 OREDRV 16 ring 70. WEBS firing 81 days slept.

In other instances, the data and visual inspections did not present cohesively. 1225 OREDRV 00 ring 63, for example, indicates clear spikes of acceleration at 25, 100, and 150 ms, whilst there are weak signals around 50, 75, and 125 ms. Additionally, the largest spike is seen around 215 ms, well past the last delay of 150 ms. This nonel ring was fired on 22 May 2023 at 79 days slept. The photographic evidence in Figure 15 shows a vastly different picture to how most of the rings initiated. The ground appears to have frozen, and detonators have rifled out of the holes of the next ring to be fired, with explosives laden on the rill. Ground support shows bagged and torn mesh, whilst cables can be seen holding up ground in the brow of the ring that was fired. There were a few occasions in which overslept rings fired did not present preferably, however, this instance provided the most challenges.

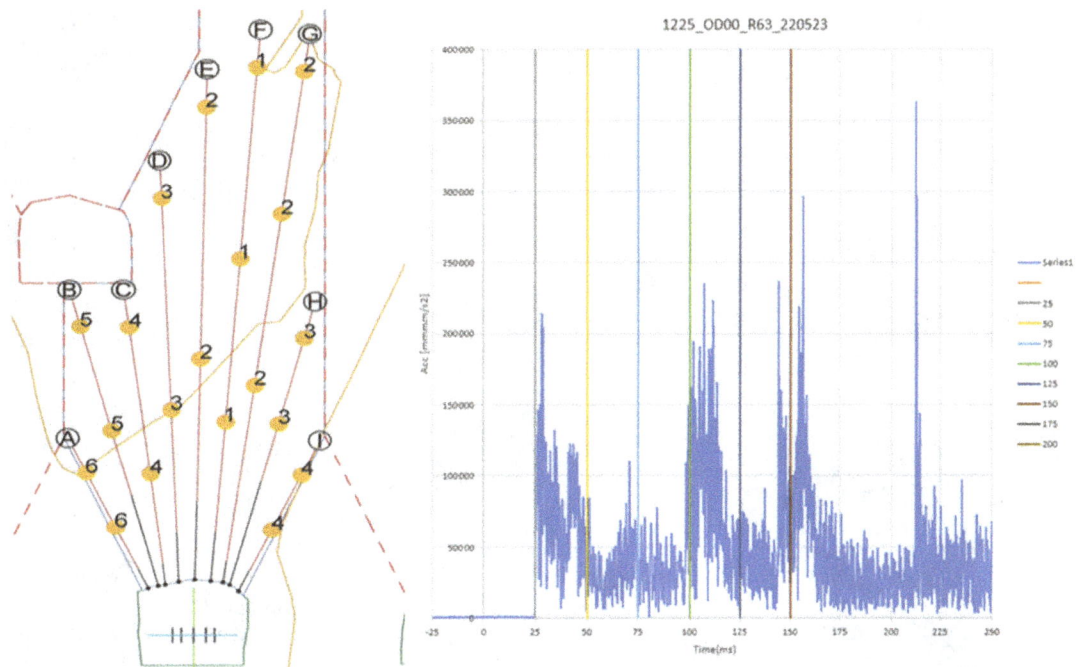

FIG 15 – 1225 OREDRV 00 ring 63 – nonel firing 79 days slept, fired 22/05/2023.

An additional risk assessment was required for this scope of work, and after a clean-up of the explosives and ground support corrected, the decision was made to bog out a small amount of the broken ore to see if that may have any impact on the brow. Historically, the ground at EHO has been reasonably *'forgiving'*, especially since the cave became fully established. During the formative years of SLC production, the initial levels at EHO were incredibly difficult to work, due to the open pit connection. Extreme cases of oversize reported into drawpoints from the open pit and hang-ups occurred during most shifts, causing great difficulties for underground crews. As the cave depth became more established, ground behaviour changed, and along with tweaking drill and blast designs through years of QA/QC, operations have been carried relatively smoothly.

Mining crews and engineers at EHO have come to accept hang-ups and oversize as a constant in SLC mining. Should a hang-up occur during bogging practices, it is bunded for secondary blasting at the end of shift. Should this fail to bring down the hang-up, subsequent bogging of adjacent ore drives is undertaken to try and disturb enough cave material and assist in rectifying the hang-up in the initial drive. Failing this, adjacent ore drives are bogged to completion and the next ring(s) fired. These measures, on nearly all occasions, solve cave-flow issues, and in the above scenario these steps were followed, and the ground conditions returned to normal.

Of note in Figure 16 are the cables holding up the ground at the brow. Installed by a cable bolter, these 6 m cables are grouted, plated and tensioned as an additional ground support standard at EHO. As mentioned earlier, in areas deemed necessary, geotechnical engineers issue cable bolting plans, meaning rings in these areas are usually charged with WEBS. 1225 OD 00 ring 63, however, was a nonel design. Whilst it is encouraging to know the cables are working as they have been designed for, this ring should have been designed using a WEBS plan. As mentioned earlier, WEBS charge plans ensure apex holes are charged lower to the collars to fire off extra ground support. They also safeguard operators against conditions seen in Figure 15. The cave management strategies outlined above ensured that this ring, along with the handful of other poor post-weather event ring firings, eventually returned to normal operating conditions.

FIG 16 – 1225 OREDRV 00 ring 63 – nonel firing 79 days slept, fired 22/05/2023.

Oversize

Oversize rock has long been undesired in SLC mining and has been the subject of studies both at EHO, RGM and LKAB's SLC mine in Malmberget, Sweden. Essentially, oversize rock is a by-product of the fragmentation of rocks during blasting. The size of fragmentation is determined by a series of factors including drill and blast design, drill and blast practices, and the rock mass characteristics, all of which can affect the productivity of the mining operation (Manzoor, Gustafson and Schunnesson, 2023).

Oversize material at EHO is classed as anything too large to fit through the engineered grizzly system at the top of an orepass. Blocked and hung-up orepasses can result in lost production. The grizzlies are constructed from reinforced steel and contain four apertures at a diameter of 1.2 m. During the feasibility study at RGM, fragmentation modelling was undertaken to evaluate drill and blast designs. Particle size distribution and its relationship with oversize material created was studied with oversize categorised as anything too large to fit through the grizzly aperture (Trout, 2002). Results suggested approximately 3 per cent of all material fragmented from blasting standard SLC rings was classed as oversize.

Like RGM, oversize rocks at EHO are transported to a stockpile, or rock bay, where they are added to the amount of material that has been drawn from the cave. Secondary breaking is performed mechanically by a rockbreaker, which is either a loader or excavator with a moil (jackhammer) fixed at the front of the machine. Previous studies at EHO assessed the relationship between oversize rocks and the firing and bogging of 67 overslept production rings. Statistical analysis found there was only marginal correlation between extended sleep times and oversize material created (Lebrocque, 2016).

Similarly, the overslept rings during the 2023 weather event showed the same marginal relationship to oversize rocks. As seen in Figure 17, total oversize reported per month indicates around 12 000 t per month leading into the event. February shows a 22 per cent decrease due to a scheduled week-long maintenance shutdown at the mine. Eight days of data before the weather event in March reported 3600 t of oversize created before 0 t recorded during the downtime in April. Three rings were fired at the end of April, however, due to the orepasses being unavailable for use in this period, ore was trucked out of the mine.

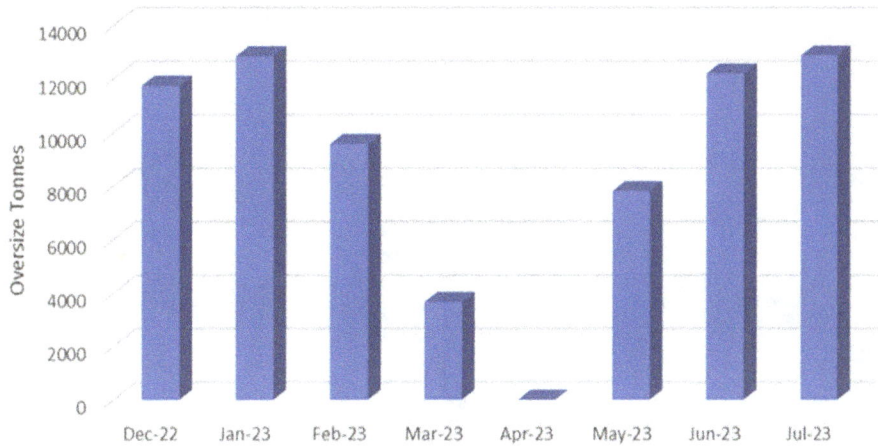

FIG 17 – Monthly oversize tonnes at EHO from December 2022 to July 2023.

Upon recommencement of full-scale production, oversize material recorded per month appears to remain consistent with those months leading up to the event. During the three weeks of production restart, 7800 t was reported. In terms of the percentage of total material moved from the cave, further analysis can be seen in Figure 18. During May 2023, all remaining sleeper rings were fired off, the final overslept ring on 31 May. This data shows an increase of total oversize from around 2.5 per cent up to 2.77 per cent as all overslept rings were fired and bogged out. June saw the operation return to normal percentages whilst July recorded the lowest percentage of the period examined. Using a baseline average of 2.5 per cent oversize, the jump to 2.77 per cent in May is roughly a 10 per cent increase in oversize stocks reported during the firing and bogging of most of the overslept rings. At no stage did the percentage of oversize material exceed the predicted 3 per cent outlined in feasibility studies conducted at RGM.

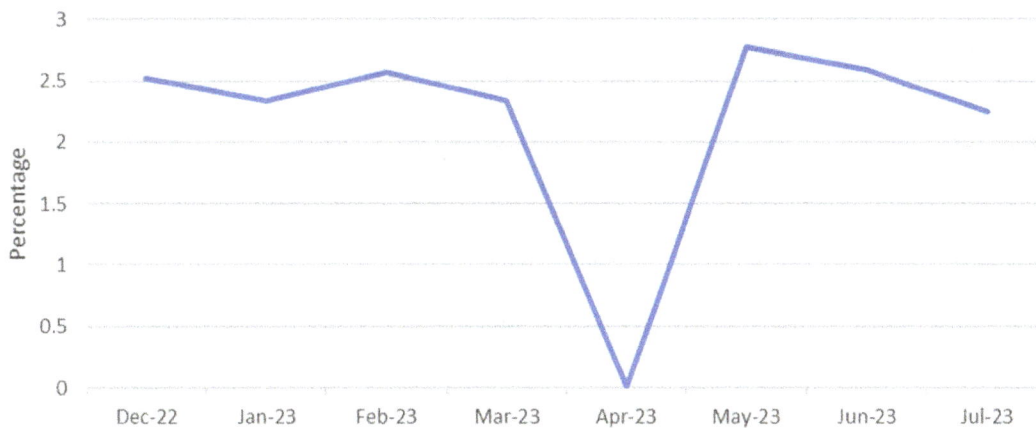

FIG 18 – Percentage of oversize to total cave material moved.

The histogram in Figure 19 indicates all rings fired were in the range of 68 days to 104 days overslept. 16 rings were fired in the time range of 80–83 days and three WEBS firing occurred over 90 days, indicating WebGen™ batteries could potentially maintain their charge after this period. Many rings were pre-charged and over 300 000 t of ore drawn from the cave to ensure the rings could be fired.

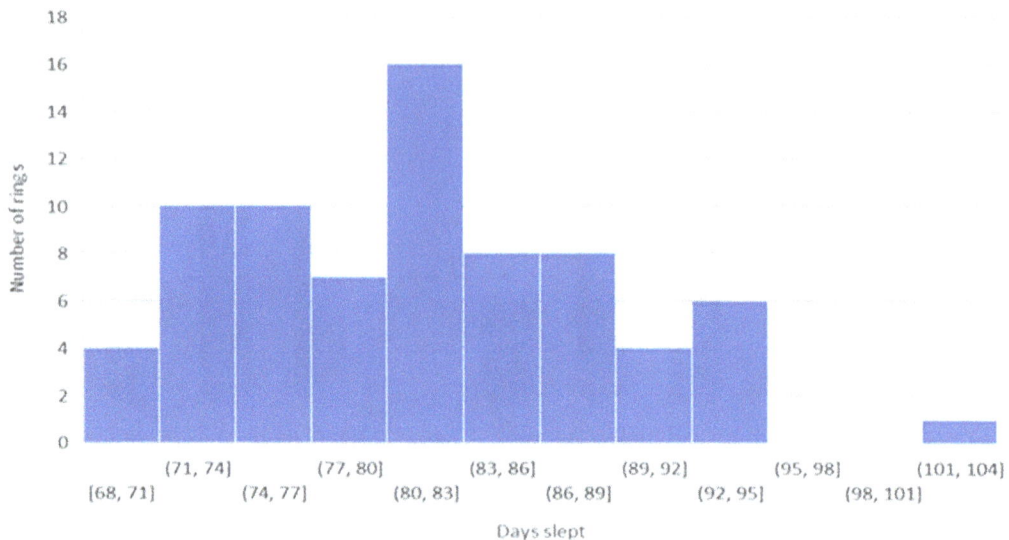

FIG 19 – Summary of overslept rings fired.

CONCLUSION

A review of academic literature was conducted with the purpose of investigating sleep times and their effect on blasting performance of production rings in a large-scale sublevel cave operation. Though there is some evidence to suggest longer sleep time affects blasting results through vibration analysis and tracking of oversize material, no information could be found to analyse rings that had slept a minimum of 68 days and a maximum of 104 days. The main aim of this paper is to provide a detailed review of the approach, controls, and results of initiating overslept rings following an unplanned weather event which had a two-month impact on production on 8 March 2023 at Ernest Henry Operations, Queensland, Australia.

As dewatering the mine progressed, discussions around resuming production occurred as a detailed risk-assessment outlined hazards and suitable controls to begin firing overslept rings. Along with a review of the explosives used for production, several geological, geotechnical, and environmental factors were assessed, and conditions monitored leading up to a restart of production. Throughout a 37-day campaign, dozens of rings were pre-charged and over 300 000 t of cave material bogged to ensure 74 overslept rings were fired. When available, blast monitors were fixed in ore drives and vibration data analysed. Along with post-firing visual inspections, this data confirmed that in all cases, detonators initiated in each ring and all available tonnes were bogged to full capacity. In further analysis, oversize rocks saw an increase of approximately 10 per cent during this period, before returning to the standard operational baseline.

RECOMMENDATIONS

Whilst it has now been proven that a large-scale SLC mining operation can withstand extreme cases of overslept rings, it is not recommended that any operation attempt to initiate overslept rings as a standard procedure on a regular basis. The risks that can impact a mining operation far outweigh any benefit this can achieve. Due to the fact this was a response to a production halt, more than a controlled study, it is recommended that EHO conduct further controlled studies on VOD and/or blast vibration monitoring in areas of the SLC that will not see the operation impacted environmentally or economically. Extended sleep times beyond 28 days, where possible, would be of great benefit to the study. Any results that favour a hypothesis that states the operation is not seriously impacted by the study may result in a review of internal documents, as well as potentially amend manufacturer specifications and industry standards across sublevel cave mining operations worldwide.

ACKNOWLEDGEMENTS

I would like to thank all those from EHO's mine management and technical services teams who recognised the potential for this event to be documented as a conference paper. Your encouragement and assistance made it possible for me to research and write on behalf of our operation. I have learnt a great amount during this process and for this I am thankful. I would also like to thank the underground workers and supervisors, some of whom I have known for well over a decade, both as a fellow crew member and/or work colleague. The dedication and team-spirit displayed by the crews during the mine recovery emphasised the work culture that has existed at EHO since the production mining began in 2012. This paper is dedicated to you.

REFERENCES

Campbell, A, 2022. Global review of sublevel caving, Benchmarking and current best practice, Ernest Henry Operations (EHO) internal document.

Evolution Mining, 2024. Our assets, Evolution Mining. Available from: <https://evolutionmining.com.au/our-assets/>

Lebrocque, T, 2016. The effects of Pre-charging on Sub Level Cave Mining Production Ring Blasting. Available from: <www.espace.library.uq.edu.au>

Manzoor, S, Gustafson, A and Schunnesson, H, 2023. Dumping oversize rock fragments in orepasses: the impact on the production cycle of a sublevel caving operation, *Mining Technology: Transactions of the Institutions of Mining and Metallurgy*, 132(3):215–224, https://doi.org/10.1080/25726668.2023.2215560

Orica, 2017. WebGen™ 100 Introduction, Ernest Henry Mine, Ernest Henry Operations (EHO) internal document.

Trout, P, 2002. Production Drill and Blast Practices at Ridgeway Gold Mine, in *Proceedings Eighth Underground Operators Conference 2002* (The Australasian Institute of Mining and Metallurgy: Melbourne).

Wiggin, M, Trout, P and Macaulay, B, 2005. Solving the problems of precharging sublevel caving rings at Ridgeway Gold Mine, in *Proceedings Ninth Underground Operators' Conference 2005*. pp 81–90 (The Australasian Institute of Mining and Metallurgy: Melbourne).

Mine-scale subsidence at Cannington Mine

A Clarkson[1], C Hall[2] and M Sandy[3]

1. Senior Project Geotechnical Engineer, South32, Cannington Mine, Qld 4823.
   Email: andrew.clarkson@south32.net
2. Geotechnical Superintendent, South32, Cannington Mine, Qld 4823.
   Email: christopher.hall@south32.net
3. Principal Geotechnical Engineer, AMC Consultants, Perth WA 6005.
   Email: msandy@amcconsultants.com

ABSTRACT

The Cannington mine is an underground silver-lead-zinc operation located in North-West Queensland, which has been in production since 1997. The ore is extracted using the longhole open stoping method. Backfilling usually incorporates paste to confine voids with a specific strength to facilitate adjacent stoping. The intent of this approach is to achieve 'full extraction' of the orebody, minimising the incorporation of rock pillars in the mine design.

Feasibility studies for Cannington recognised that the mine could be prone to surface subsidence given the synclinal geometry and moderate depth of the orebody. The potential for occasional but severe flooding of the Trepell and Hamilton creeks required that underground mining should not significantly disturb the surface. Avoiding disturbance to key underground infrastructure including the decline and the rock hoisting shaft was also critical.

To achieve this objective the proposed mining method involved open stoping with paste fill. A systematic layout of regional pillars was proposed to further reduce the potential for settlement, or at least delay it until late in the mine life (Struthers, Lee and Bailey, 1994).

As Cannington enters the final phase of mine operations; understanding the historical response to mining is key to forecasting the surface response to the removal of the remaining pillars. As Cannington enters the final phase of mine operations, this hypothesis is being tested.

Measured changes on surface infrastructure which were focused on the shaft provided the first indication of subsidence on the surface, however the location and scale of the subsidence went largely un-noticed until InSAR technology was introduced to monitor the site's tailings storage facility.

The path to understand the scale and scope of mine subsidence has been a complex process that has evolved over several years. A combination of newer technologies and traditional methods has been employed to measure, correlate, and forecast movement. The outcome is a site-wide awareness of the phenomenon, with stability safety cases developed to manage the risk of mine-scale instability.

INTRODUCTION

As described by Brady and Brown (1985) 'subsidence is the lowering of the ground surface following underground extraction of an orebody. Subsidence is produced, to a greater or a lesser degree, by almost all types of underground mining.'

Regional subsidence is usually associated with pillar-less or non-fill mining such as underground longwall coal mining, however there are cases where regional subsidence has occurred in response to extracting ore from large scale metalliferous, underground operations. Synclinal orebodies are particularly vulnerable to large scale subsidence associated with removal of support of the core as mining proceeds on the limbs and eventually under the 'keel'. Mines such as Nifty in Western Australia (Niessener, 2014), and Luanshya (Roan Antelope) in Zambia (Broome, 1981) have a documented history of large-scale subsidence.

As part of the Cannington feasibility study, Struthers, Lee and Bailey (1994) referred to the concept of designing regional pillars with the intent to minimise surface disturbance and to prevent connection with the main water table. At the time the mining production schedule catered for a 21 year underground mine life in the Southern Zone of the deposit (BHP Minerals Pty Ltd, 1994).

Thirty years later, observations on the surface and underground have confirmed the occurrence of subsidence on a mine-scale at Cannington.

Subsidence is widely understood to be a response to the volume of material mined. Specifically, in its 27 years of mine production to date, Cannington Mine has extracted:

- Over 1200 stopes, designed to range in volume from 1500 m³ to 140 000 m³. Stope design volume is largely dependent on geotechnical design constraints, and ore quality and distribution.

- Volume of hoisted and hauled stope material to surface has been relatively consistent throughout the mine's life with a stope-dirt-to-surface rate average of 750 000 m³ per annum, culminating to a total stope volume extraction of approximately 21 000 000 m³.

- Pastefill with a designed Unconfined Compressive Strength (UCS) of 500–1000 kPa has been used to confine approximately 80 per cent of the stope voids, with the remaining 20 per cent comprised of waste rock fill or failed stopes that have choked off due to swell factor.

Figure 1 shows the location of the underground workings relative to surface infrastructure.

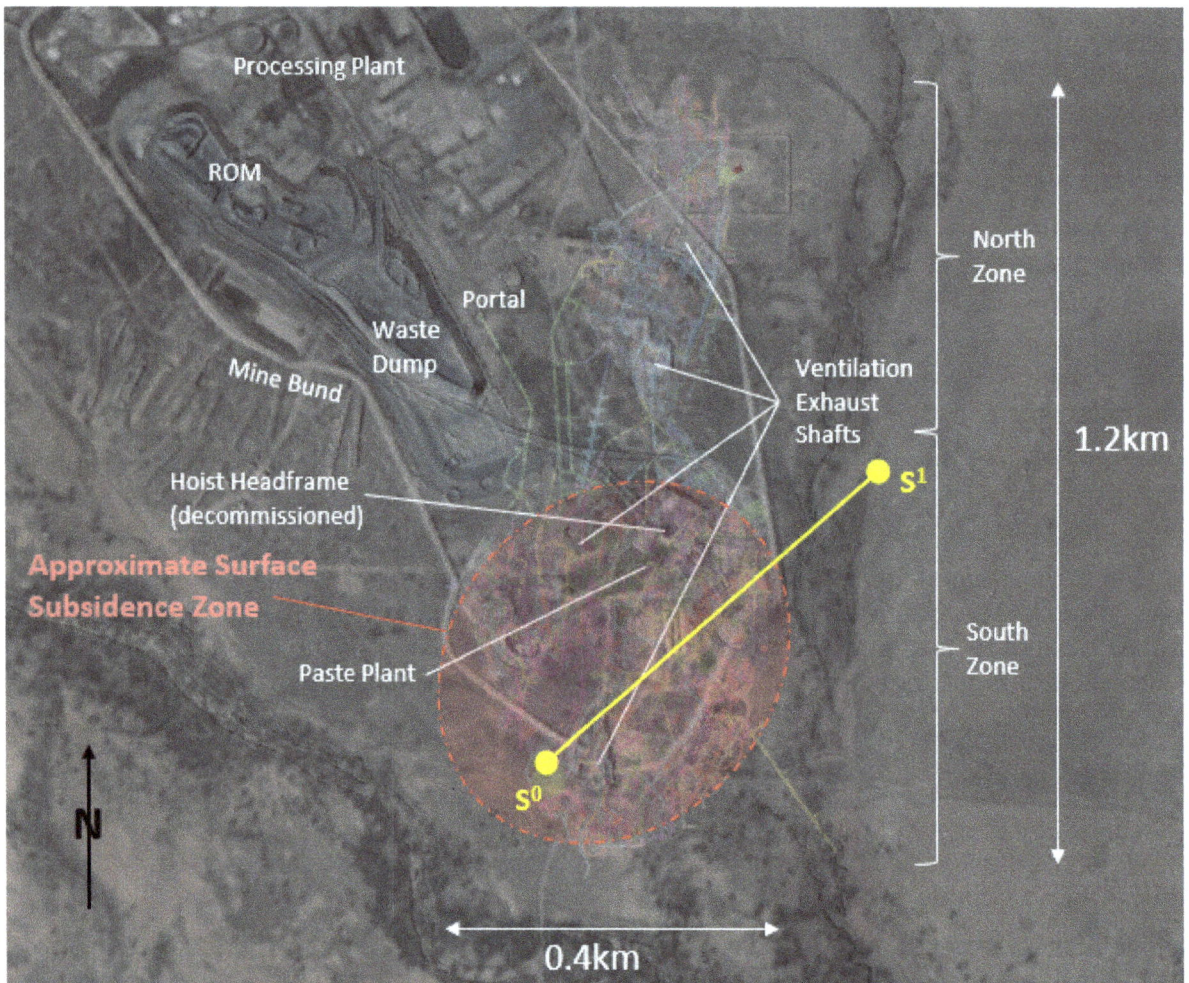

FIG 1 – Plan view of Cannington Mine footprint relative to surface infrastructure and subsidence zone.

OREBODY GENESIS, GEOMETRY AND GEOLOGY

The Cannington deposit formed 1675 Ma during a time of tectonic rifting and post-rift thermal subsidence (Bodon, 2002). The Cannington area was then subject to multiple deformation events as part of the Isan Orogeny (Bodon, 1998; Giles and Nutman, 2003), including coaxial folding and later-stage faulting that has produced the limbs of the synform that dip 40–70° to the east (Walters *et al*, 2002).

Although Cannington has complex geological conditions with significant variability between lithologies, the host rock surrounding the shaft can be broadly summarised by the following rock types; Cretaceous sediments which includes weathered mudstones and clays, gneiss which is hard and foliated, quartzites which are competent and strong, schist which is soft and jointed and amphibolite which is blocky and jointed.

The geometry of the Cannington orebody and its extraction to date is the root cause of subsidence. Figure 2 shows a cross-section through the middle of the South Zone, where the greatest surface subsidence has been measured with Interferometric Synthetic Aperture Radar (InSAR) technology.

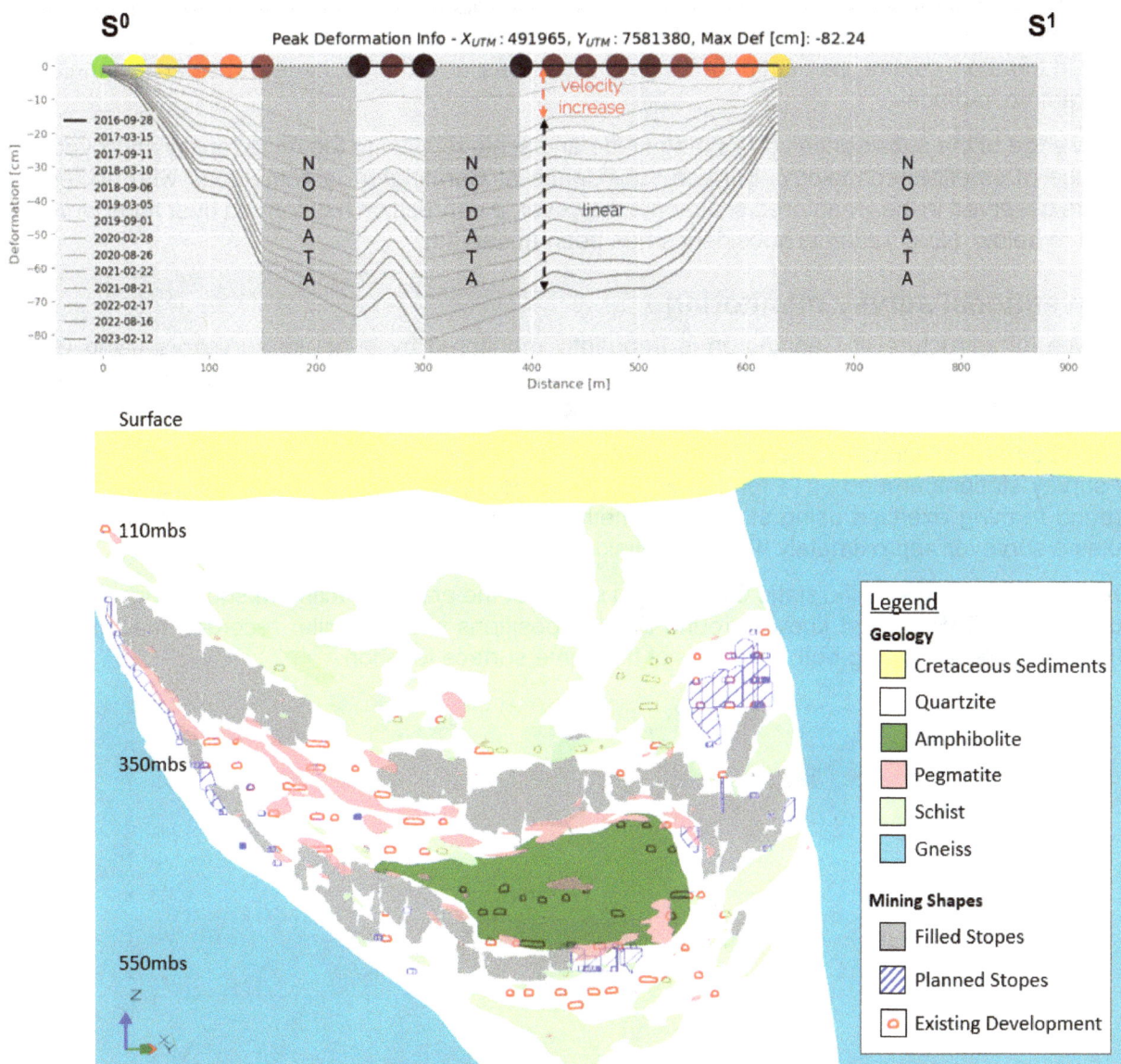

FIG 2 – Cross-section (S^0 to S^1 from Figure 1) of underground mine workings relative to dominant geological interpretations, transposed below equivalent surface subsidence measured by InSAR monitoring (DARES, 2023) looking north-west.

Several key observations can be made from this section view, including:

- The intensity of stoping around the perimeter of the amphibolite 'core'.

- The prevalence of sillimanite-muscovite schist between the ore zone and the surface.

- The subsidence zone is slightly elongate in the South-West/North-East strike –this is the same orientation as the mine-scale faults that are prevalent across the mine.

- Accessible development drives are widely distributed across the mine.

Collectively, these points contribute to the regional, geotechnical change and require effort to ensure a controlled work environment.

SURFACE OBSERVATIONS

The full scope of surface subsidence at Cannington was identified relatively late in the mine life due to the assumption that the use of systematic, tight backfilling and deferred mining of certain pillars in line with the feasibility study recommendations would result in negligible impact to the surface profile.

The occurrence of mine-related subsidence at Cannington was first identified during a routine survey exercise to verify the site's contamination containment bund, and openings to underground were adequately elevated to control the risk of surface water inundation during extreme rain events. Aside from inundation control verification activities, there was no other surface subsidence monitoring programme in place.

Knowledge of the subsidence zone's magnitude and scale relative to the mine rapidly improved with the use of satellite technology. Surface subsidence at Cannington is continuous with no tension cracks observed in the weathered sediments, however, a depression has formed over mine workings ~300 m below. No differential subsidence has been observed.

Conventional survey monitoring

Surface infrastructure at Cannington is frequently monitored by in-house surveyors using Global Navigation Satellite System (GNSS), Global Positioning System (GPS), total station and monitoring prisms. These methods are used to measure fixed plant movement including subsidence and tilt.

In addition to monitoring fixed plant infrastructure, Cannington's Survey Department set-up a series of survey stations around 2014 for a site traverse project which have subsequently picked up for a ground truthing exercise using static GPS methods (Jeavons, 2020). A single static GPS pick up takes a surveyor approximately 40 mins in-field.

There has been no specific static GPS station set-up at the area of maximum subsidence, however, comparing InSAR data at known ground truthing positions shows similar recorded magnitudes as seen in Figure 3 showing both data sets at the same surface location.

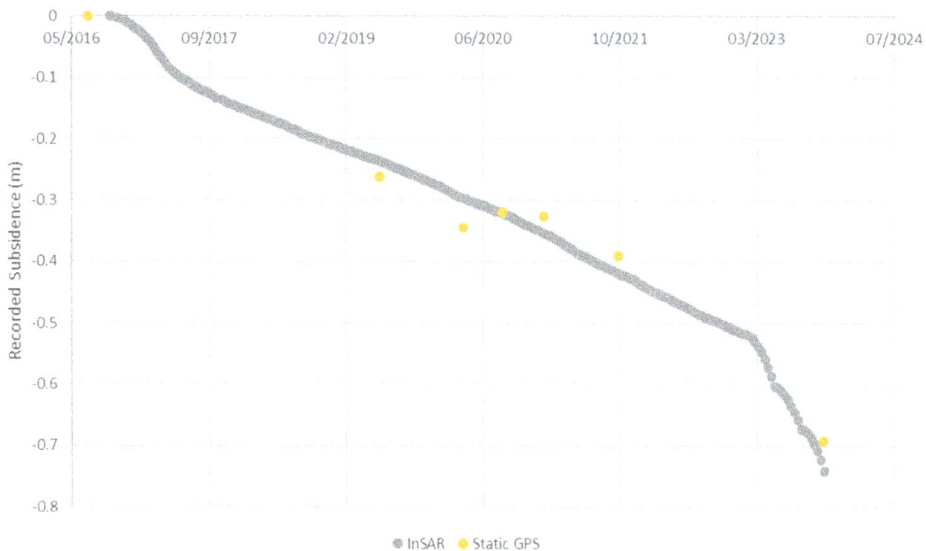

FIG 3 – InSAR and Static GPS comparison data set.

InSAR monitoring

A review of InSAR monitoring data was commissioned in 2014 at Cannington to monitor for changes on the tailings storage facility (TSF) embankments and water volumes. Within six months the opportunity arose to increase the InSAR scope to include the surface footprint above the underground mine. It was at this time that the true picture of the South Zone subsidence zone was realised.

The limitations of InSAR monitoring include line of sight (LOS) issues and calculation methods. LOS issues at Cannington have been problematic in getting reliable InSAR observations due to cloud cover and vegetation which block line of sight. The other LOS limitations found have occurred when the angle of the ground being monitored is too acute to the viewpoint of the satellite. This limitation means some infrastructure such as stockpiles and dam walls are not seen by each orbit of a satellite.

Significant known movements at Cannington stockpiles and TSF have not been detected by InSAR. It is assumed that these large movements are not included in reporting and alerts due to the scale of filtering required to disregard all planned surface material movements.

Alternative InSAR providers using different satellites have calculated slightly different movement rates. Cannington's InSAR providers have supplied data primarily from RadarSat-2 and TerraSAR-X satellites (Jackson and Vakilli, 2023).

Cannington relies on external InSAR providers to process and report on collected data. Static GPS methods are advantageous regarding being able to get an accurate pick up anytime, whereas TerraSAR-X satellite re-visit orbits occur every 11 days. For Cannington's slow rate subsidence monitoring, the emergence of InSAR technology has shown to be highly effective.

Fixed plant infrastructure

Evidence of surface movement has been observed at various surface infrastructure locations proximal to the zone of maximum recorded subsidence. Infrastructure impacted includes the paste plant, shaft headframe, paste conveyors and mine bund. Figure 4 shows the layout of impacted infrastructure and zone of maximum recorded subsidence.

FIG 4 – Aerial view of fixed plant infrastructure and their sense of movement within the surface subsidence footprint, looking south-east.

Specific observations include:

- Changes to the paste plant and connecting conveyor structure where the paste conveyor was designed and built with pin-joint connections to footings to allow the structure to absorb horizontal movement due to conveyor operations. The support legs of this structure began to 'domino' towards the paste plant and centroid of the surface subsidence zone.

- The headframe is observed to be tilting toward the centre of the subsidence zone. Hoisting operations ended in June 2022 due to damage throughout the hoisting shaft (Clarkson, Hall and Sandy, 2023). Whilst the primary concern at the time was skip clearance to the shaft liner, the headframe's skip guiderail alignment was also an emerging issue as the skip emptying mechanism became less reliable over time due to the tilt of the headframe. The surface subsidence is geometrically correlated to the mine-wide movement which lead to decommissioning of the shaft.

- Occurrences where the mine contamination containment bund and the collar of an underground-to-surface ladderway had failed to meet the tolerable elevation from a mine inundation perspective. This observation prompted actions to restore flood inundation controls, but also drew attention to the possibility of subsidence above the mine footprint.

UNDERGROUND OBSERVATIONS

Evidence of mine-scale subsidence has been observed in various forms within the underground mine. The type of rock mass deformation is dependent on the depth below surface, geology, and location relative to mined-out areas. Figure 5 shows some examples of rock mass response to the full extraction approach to production at Cannington.

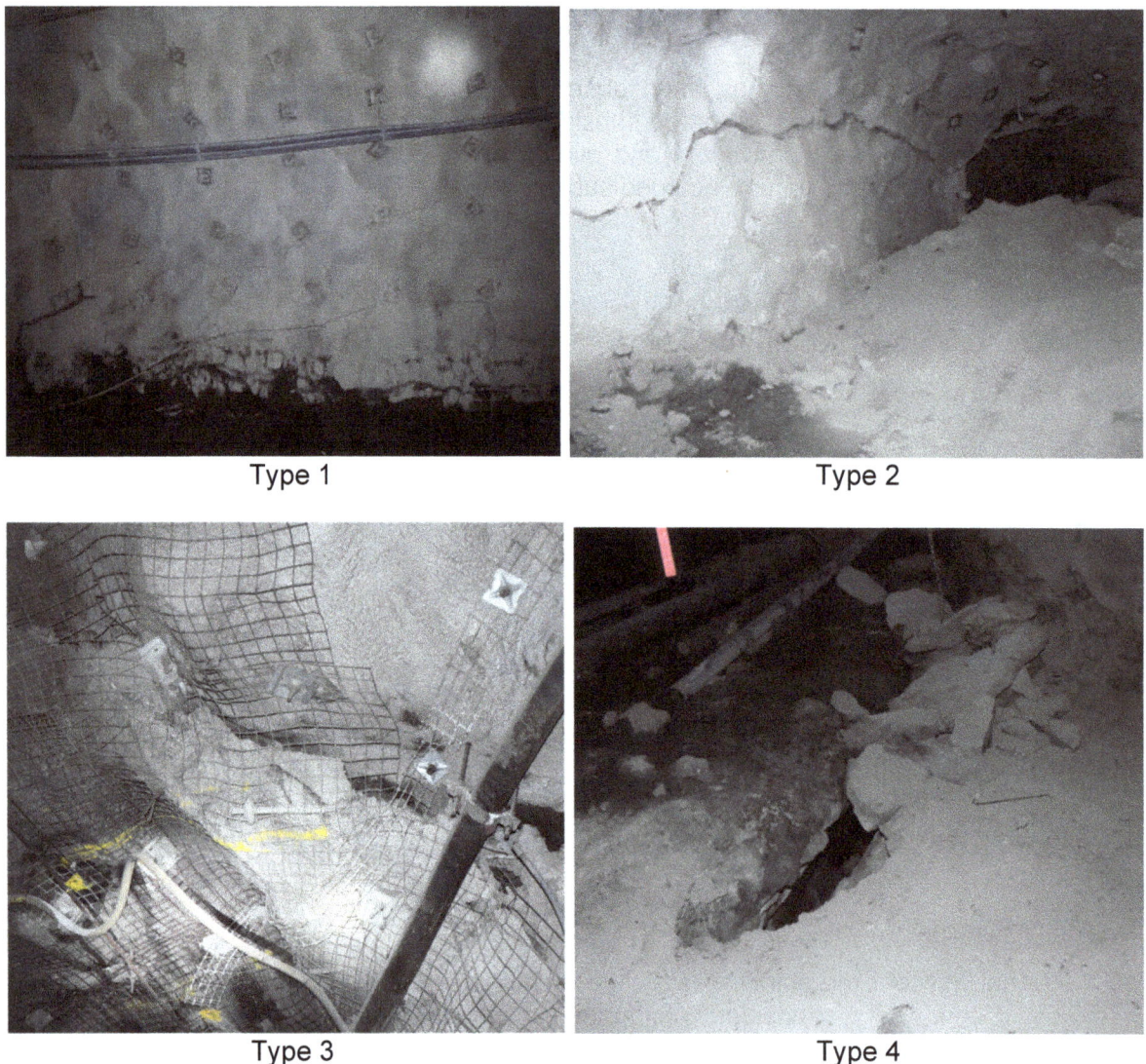

Type 1

Type 2

Type 3

Type 4

FIG 5 – Typical types of rock mass deformation related to mine-scale subsidence.
Type 1 – 165 mbs: Minor, sub-horizontal joint dilation in South Zone decline.
Type 2 – 100 mbs: horizontal tension cracks in the North Zone stoping block crown pillar.
Type 3 – 475 mbs: shear zone fault slip at the eastern abutment of South Zone.
Type 4 – 375 mbs: tension crack with 400 mm aperture in the footwall of the South Zone.

Sub-horizontal tension cracks in walls

Dilation of cracks in the walls of shallower (<200 mbs) mining areas can range from subtle hairline cracks in fibrecrete, to more substantial dilation immediately above the crown of a mined-out ore block (see Figure 5). In these scenarios the rock mass quality is excellent and deformation is a result of the tensile capacity of the rock mass being exceeded in response to the subsidence mechanism.

These areas are monitored for change over time with Light Detection and Ranging (LiDAR) point cloud comparison methods, which inform trigger action response plans to trigger rehabilitation when movement exceeds the capacity of the ground support installed.

Sub-vertical tension cracks in floor

Significant, tension cracks in certain underground workings were detected several years prior to the measurement of surface subsidence using InSAR technology. The most obvious example is the structure located at the southern end of the footwall orebody, and extends between 325 and 400 mbs. On levels 350 and 375 mbs the aperture of the tension crack is measured at approximately 400 mm as shown in Figure 5.

The aperture of the southern footwall tension crack has progressively grown over the years despite the nearby stopes being paste-filled to within 1–3 m of the crown, and there being no recent stope activity in the mining block on these levels. This observation speaks to the reality that subsidence is occurring, similar to tension cracks forming behind the final crest of an open pit excavation, despite the lack of a substantial void for the rock mass to move towards.

Tension crack structures are usually sub-vertical and more obvious in the floor than the backs. The risk of localised rockfall in these areas is controlled with cablebolt patterns similar to that used for stope brow control on-site.

Tunnel-scale, fault-slip movement

Fault movement due to regional subsidence has been observed around the eastern extents of the mine workings where abutment stresses are present, regional faults exist and the subsidence is 'differential' over small horizontal areas. This differential subsidence is observed above the interface between undercutting mined out areas (backfilled) and regional pillars.

The occurrence of fault structure mobilisation is more widespread than tension crack precincts. At times it can be difficult to differentiate between fault movement caused by general subsidence and fault movement caused by specific stoping activities that change local stress conditions intersecting the structure. Mining induced seismicity trends indicates that it is possible to have both occurring simultaneously.

In terms of controlling fault-slip movement, point-cloud comparison using LiDAR data has proved to be an effective tool to identify localised rockfall hazard and respond in a timely fashion.

Re-survey traverse

In late 2022 the Cannington Survey Department commenced a project to re-survey historic survey stations on several levels, which traverse the suspected subsidence zone. The process involved working from stations outside the production areas and using them as datum points to measure change across the mine footprint in an east-to-west direction. The intent of this task was to quantify the change over time in the coordinates of survey stations within the suspected subsidence zone (Price, 2023).

The re-survey project scope included six mine levels and re-surveyed 195 survey stations, some of which had not been re-surveyed since their initial installation, 25 years prior. Table 1 provides a summary of results for the underground subsidence survey project which suggests:

- Underground subsidence data is consistent with the scale of change measured by surface InSAR methods.

- The subsidence rate measured at 325 mbs is significantly higher than at the other surveyed levels. A key difference is that the survey station that recorded the maximum subsidence on the 325 mbs was installed in 2016, and an average subsidence calculation over seven years, compared to over 16 years for the other levels. This finding supports the hypothesis that subsidence rates increase with time and the extent of excavation.

- Less subsidence measured at the 520 mLv, possibly due to this level representing the base of the subsidence zone (ie majority of subsidence is occurring above this level.

TABLE 1

Summary of survey station data collection.

Level (metres below surface)	Date of survey station installation	Date of re-survey to inform change	Days between survey station installation and re-survey	Maximum change in survey station elevation (m)	Average subsidence rate (mm/a)
325	01/05/2016	25/06/2023	2611	-1.09	-152
350	11/11/2007	10/8/2023	5751	-1.40	-89
425	01/07/1998	16/05/2023	9085	-1.55	-62
450	20/08/2003	12/04/2023	7175	-1.57	-80
500	01/06/2007	01/04/2023	5783	-1.32	-83
520	09/07/1999	30/10/2022	8514	-0.86	-37

Limitations in the re-survey process included:

- Inability to clearly observe changes in subsidence rates, as the source data was restricted to the initial installation, with only the subsequent coordinate pick-up for each survey station being available to represent the rate.

- Little opportunity to revive survey stations had been mined-out, damaged bypassing vehicles, or had ground support re-installed over the top of them.

Figure 6 shows the approximate locations of the maximum subsidence relative to other levels surveyed and the underground mine workings.

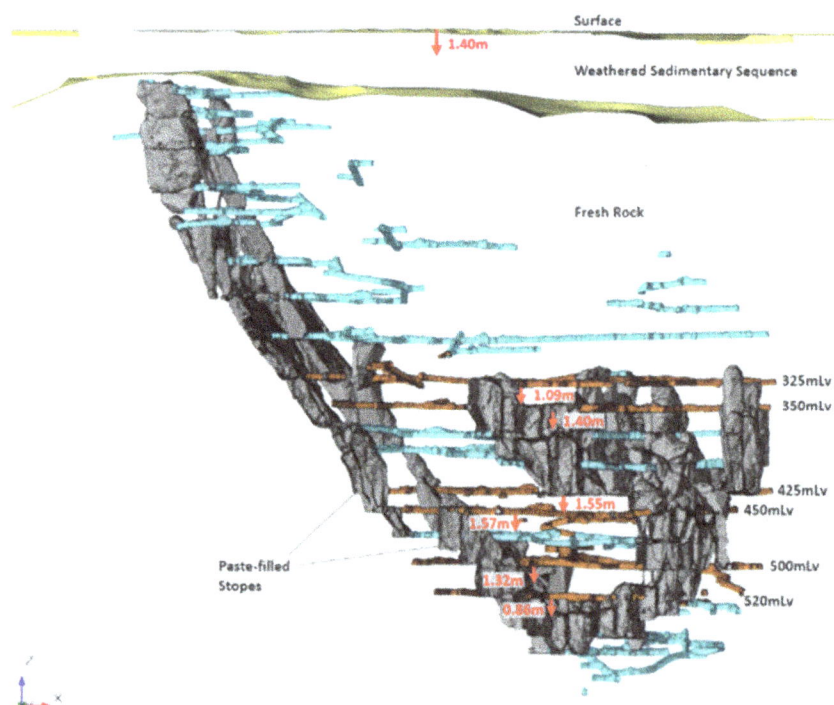

FIG 6 – Cross-section of South Zone relative to measured underground subsidence, looking north.

Seismicity

Mining-induced seismicity is monitored by an array of 26 sensors comprised of accelerometers and geophones. The monitoring system is designed to triangulate accurate event locations to inform changes to mine-scale stability.

A correlation between seismic event frequency and rate of surface subsidence was identified in 2019 following a stope failure event that occurred in the footwall of the mine. Figure 7 shows:

- The seismic response to extracting a footwall stope which was mined within a regional pillar.

- The contemporaneous step change increase in the rate of subsidence measured by the survey station attached to the mine headframe.

- The headframe subsidence rate returned to baseline response at the same time as the regional seismicity in the footwall returned to steady-state.

These observations indicate seismic frequency in certain mining blocks could be considered a monitoring proxy for rate of subsidence.

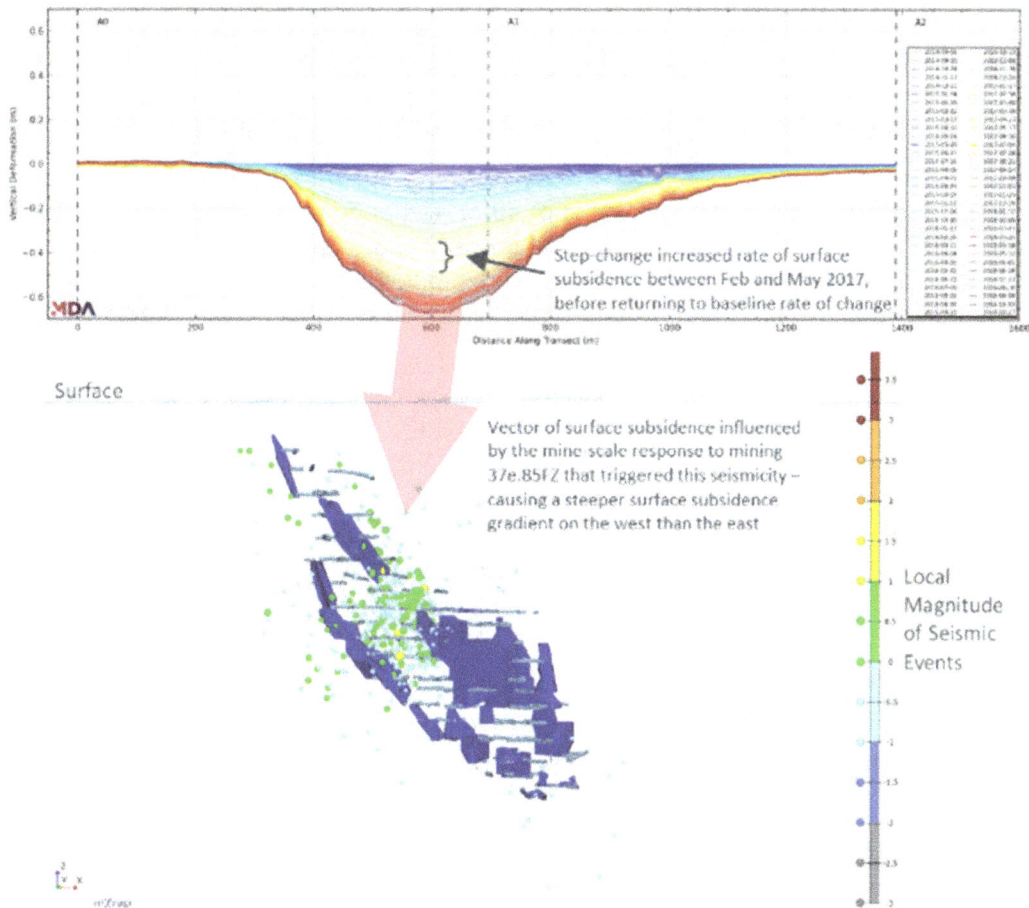

FIG 7 – Cross-section of South Zone showing distribution and magnitude of seismic events for four months following a footwall stope failure in February 2017, relative to measured underground subsidence, Looking North (Hall, 2019).

RISK MANAGEMENT

The approach outlined by Beck (2022) to manage mines with the potential for large-scale seismic events and instability has been adapted and implemented at Cannington for several large-scale failure mechanisms, including mine subsidence. These failure mechanisms are complex and applying a simple trigger action response plan would be inadequate to consider all relevant data sources.

'Mining-induced, regional ground failure' has been identified as a material risk at Cannington. Importantly this risk is independent of 'localised rockfall' risk, due different causal pathways for related events, and their controls.

Surface exclusion fence

When the North Zone stope sequence began to extract stopes with crowns at only 40 mbs, surface barricading of the approximate stope footprint on a stope-by-stope basis was introduced as a control to prevent potential exposure to the underlying open voids. The initial surface barricades were made from flagging tape and iso-containers and proved to be a challenge to set-out and manage their removal.

As a measure to address the issues with stope-by-stope surface barricading, a permanent fence was established around the approximate North Zone footprint in 2023. The gates at the north and south ends of the exclusion zone improved control of vehicle and pedestrian traffic over the North Zone crown pillar and a map of the area was updated quarterly to highlight the areas within the fenced areas where stopes were open below.

Stope extraction sequence

The correlation between mining-induced seismicity and subsidence rate has led to the current assumption that mine sequence is a dominant lever for the control of subsidence and mine-scale geotechnical changes in general. Specifically, the lever is to delay the extraction of remaining regional pillars in the South Zone footwall to as late as possible in the mine life.

The forecast response to footwall mining and other regional pillars and abutments has been investigated with the use of non-linear numerical modelling of life-of-mine extraction completed by consultants Beck Engineering and Mining One. Both models forecast the expected geometry of subsidence areas, however they do not reach the magnitude of subsidence measured to date.

In terms of underground subsidence, numerical modelling results suggest a change whereby ore block scale subsidence zones in the footwall, hanging wall and keel areas eventually converge to form a single halo of damage with an appearance similar to some block cave mines. Figure 8 displays the step-change in subsidence forecast by numerical models.

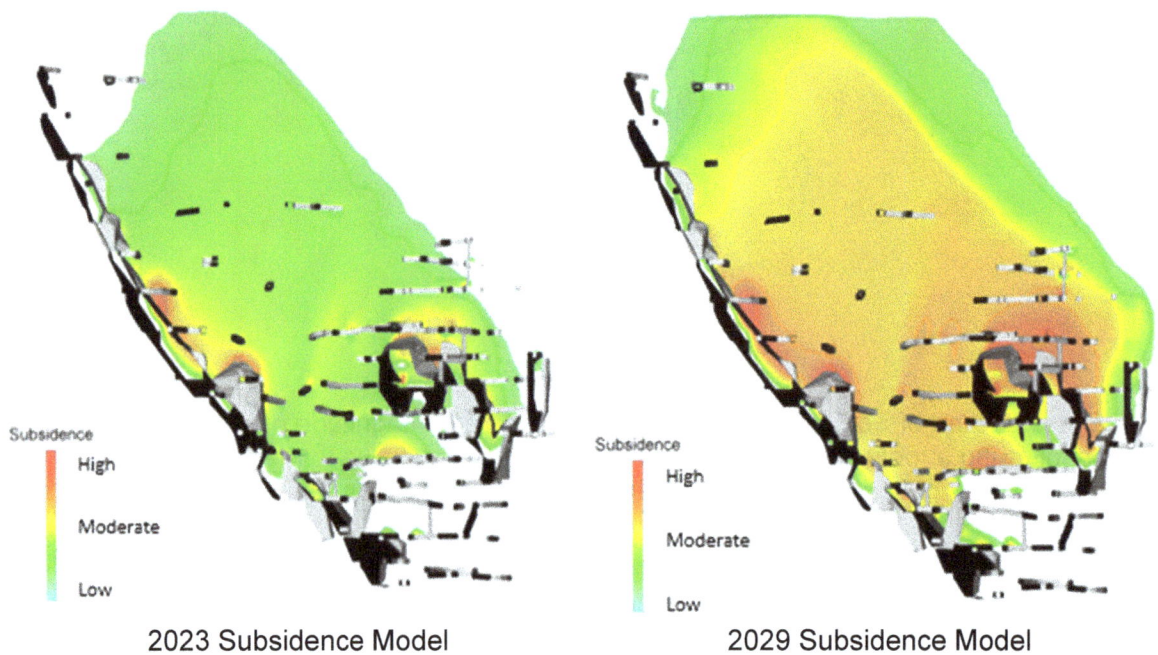

2023 Subsidence Model   2029 Subsidence Model

FIG 8 – Beck (2023) numerical model comparison showing current subsidence conditions with future state.

In-field verification of numerical model results is undertaken in the form of annual rock mass damage mapping. Compliance to the stope sequence set by the annual budget is tracked on a stope-by-stope basis and reported monthly. Any material change to mining sequence is passed through an assessment and escalation process.

Stope backfill

The use of paste with an UCS typically less than 1 MPa and a modulus of 0.35 GPa has been historically used to backfill stope voids and enable the production plan. The effectiveness of paste backfill as a control for mine-scale subsidence is inferred, but in practical terms, outside the scope of backfill design on a stope-by-stope basis.

To replace the extracted ore with a material of similar properties would be impractical (the UCS of intact, mineralised quartzite is approximately 200 MPa and a Young's modulus of 77–97 GPa (South32, 2024)). As a result, the effectiveness of backfill as a control for surface subsidence is questionable.

Regardless of the effectiveness of paste fill, the medium and long-term plans for the remaining mine life include the ongoing standard of tight-filling stope voids with the primary intent to maximise ore recovery and provide adequate confinement to control against stope-cave-to-surface events.

MINE CLOSURE CONSIDERATIONS

The magnitude of surface subsidence has significant relevance to mine closure strategy and final landform plans. Numerical modelling results, supplemented with measured survey data has been used to project subsidence rate and magnitude to forecast surface subsidence beyond the end of mine life. This has informed the scope and budget required to achieve the final landform commitments.

CONCLUSIONS

The mining of underground orebodies will incur a degree of subsidence. The magnitude of subsidence is dependent on host rock, orebody depth and geometry, ore extraction strategy, and backfill quantity and quality.

Subsidence on a mine scale can be safely managed if systems are established to monitor for change and respond with appropriate engineering or exclusion-based controls.

Collection and management of historical survey data is important when striving to quantify subsidence magnitude and rate on the surface and underground. The value of this data is exemplified when numerical models fail to accurately simulate the response to mining.

Backfilling underground voids does not eliminate surface subsidence if the extent of extraction is sufficient. Modelling forecasts of the ultimate subsidence response should be treated with caution as the actual total may be substantially greater. This may be especially important when seeking approval for mining under sensitive locations such as housing, heritage sites, water bodies or national parks.

ACKNOWLEDGEMENTS

The authors would like to thank South32 for allowing this paper to be shared with the mining community. They would also like to thank the past and present members of Cannington's Geotechnical and Survey Departments for their efforts to collect data, contribute to the interpretation and workshop controls for mine subsidence at Cannington.

REFERENCES

Beck, D, 2022. Maintaining a Stability Safety Case in Seismically Active Mines, in RaSiM10: Proceedings of the Tenth International Conference on Rockbursts and Seismicity in Mines (Society for Mining, Metallurgy and Exploration: Englewood).

Beck, D, 2023. CNGTN2023JUL24 Day 1 Quick appreciation, informal presentation notes.

BHP Minerals Pty Ltd, 1994. Cannington Project Feasibility Study: Summary and Financial Evaluation.

Bodon, S B, 1998. Paragenetic relationships and their implications for ore genesis at the Cannington Ag-Pb-Zn deposit, Mount Isa Inlier, Queensland, Australia, *Economic Geology*, 93:1463–1488.

Bodon, S B, 2002. Geodynamic Evolution and Gensis of the Cannington Broken Hill-type Ag-Pb-Zn deposit, Mt Isa Inlier, Queensland, PhD thesis (unpublished), University of Tasmania, Hobart.

Brady, B H G and Brown, E T, 1985. Rock Mechanics for Underground Mining (George, Allen and Unwin: London).

Broome, M T, 1981. The Subsidence of the Roan shaft pillar at Luanshya mine, Zambia, The American Institute of Mining, Metallurgical and Petroleum Engineers.

Clarkson, A, Hall, C and Sandy, M, 2023. Challenges and Solutions for Deformation Management of Cannington's Hoisting Shaft, in Proceedings Underground Operators Conference 2023 (the Australasian Institute of Mining and Metallurgy: Melbourne).

DARES, 2023. Historical Report Cannington Underground Mine Sep 2016 - Feb 2023.

Giles, D and Nutman, A P, 2003. SHRIMP U-Pb zircon dating of the host rocks of the Cannington Ag-Pb-Zn deposit, south-eastern Mount Isa Block, Australia, *Australian Journal of Earth Sciences*, 50:295–309.

Hall, C, 2019. GE-0375 – Surface Subsidence update, Internal South32 Cannington Report.

Jackson, S and Vakilli, A, 2023. Surface Subsidence Assessment Cannington Mine.

Jeavons, C, 2020. Surface Subsidence Ground Truth Report, Internal South32 Cannington Report.

Niessener, J, 2014. Subsidence impacts Nifty production, *Mining News,* 1 April 2014.

Price, D, 2023. Underground Subsidence Report, Internal South32 Cannington Report.

South32, 2024. Ground Control Management Plan, South32 Internal Document.

Struthers, M A, Lee, M F and Bailey, A, 1994. Cannington Mine Feasibility Study Rock Mechanics, AMC Consultants report 192005.

Walters, S, Skrzeczynski, R, Whiting, T, Bunting, F and Arnold, G, 2002. Discovery and geology of the Cannington Ag-Pb-Zn deposit, Mount Isa Eastern Succession, Australia: development and application of an exploration model for Broken Hill-type deposits, in *Integrated Methods for Discovery: Global Exploration in the 21st Century*, special publication 9, pp 95–118 (Society of Economic Geologists: Littleton).

Innovative geotechnical management at the Wira Shaft

R Coad[1], C Scott[2], C Hill[3] and D Lagacé[4]

1. Experienced Geotechnical Engineer, pitt&sherry, Hobart Tas 7000. Email: rcoad@pittsh.com.au
2. Principal Geotechnical Engineer, pitt&sherry, Hobart Tas 7000. Email: cscott@pittsh.com.au
3. Project Manager – Shafts, Byrnecut Australia Pty Ltd, Brisbane Qld 4000.
   Email: christopher.hill@byrnecut.com.au
4. Package Manager – Shaft, BHP, Adelaide SA 5950. Email: daniel.lagace@bhp.com

ABSTRACT

The Wira Shaft is a 1326 m deep, 7.55 m diameter, concrete lined haulage shaft located at BHP's Prominent Hill Mine in South Australia; it is currently under construction. It is being constructed using the strip and line method, with the use of steel fibre reinforced shotcrete (FRS) as the primary means of temporary ground support.

The ground support system for the Wira Shaft has been designed and implemented to consider both safety and efficiency. Permanent support, achieved via a cast *in situ* concrete liner, is placed approximately 12 m behind the advancing shaft bench. Between the liner and bench, temporary ground support is necessary to ensure the safety of personnel. FRS has largely been used as the primary support element with the required thickness and early strength designed and modified as required to achieve varying levels of support pressure. Where possible, the use of rock bolts and embedded support has been eliminated.

Geotechnical mapping of the shaft walls following firings allows for rock mass classification; this determines the required support pressure and subsequent ground support standard. The continual process of mapping the ground conditions allows for specific support recommendations and removes the needs for a conservative blanket pattern bolting approach.

The specification and application process of the FRS has been optimised for ease of use and to produce the quickest possible re-entry times. With a high reliance on the quality of the FRS, a substantial quality assurance and control process was developed and executed. It includes ongoing workability, thickness and early strength testing.

Prior to and during the sinking process, innovative geotechnical management has enabled safety and operational improvements that have enhanced the shaft's advance rates, to the benefit of the overall project cost and schedule.

INTRODUCTION

Prominent Hill Mine is an underground copper mine, owned by BHP, located in the north-western region of South Australia. Construction works for the Wira Shaft began in late 2021 and completion of the sink is expected during 2025. Byrnecut Australia are the principal contractor. Byrnecut engaged pitt&sherry to provide on and off-site geotechnical support for the project.

Current mining horizons at Prominent Hill sit up to ~950 m below the surface (considering surface as shaft 0 m), with the lower extent of the orebody indicated to extend to ~1400 m from surface. The current production rate of the mine, per annum, is ~4 Mt, with the ore trucked from underground to surface. The completion of the shaft and materials handling system will enable a production uplift from the Prominent Hill underground mine to 6.5 Mt/a.

The Wira Shaft will be a 1326 m deep, 7.55 m diameter, concrete lined haulage shaft. The shaft's pre-sink (94 m) was excavated using a Herrenknecht Vertical Sinking Machine (VSM). The remainder of the shaft is being constructed using the strip and line method (cycle outlined in Figure 1). The general strip and line cycle involves drill and blast firing of 3 m rounds to advance the shaft (at an excavated diameter of 8.3 m). The poured *in situ*, concrete liner lags by 9 m to 15 m behind the shaft face; in this lag area, the walls are supported using temporary ground support. The temporary ground support designed for, and used, in the Wira shaft is primarily a standalone FRS support system with no embedded support where possible. A substantial geotechnical mapping, ground support assessment and ground support quality assurance and quality control (QA/QC) process has been implemented to allow specific support recommendations based on the encountered conditions. This removes the need for the implementation of a blanket conservative ground support design.

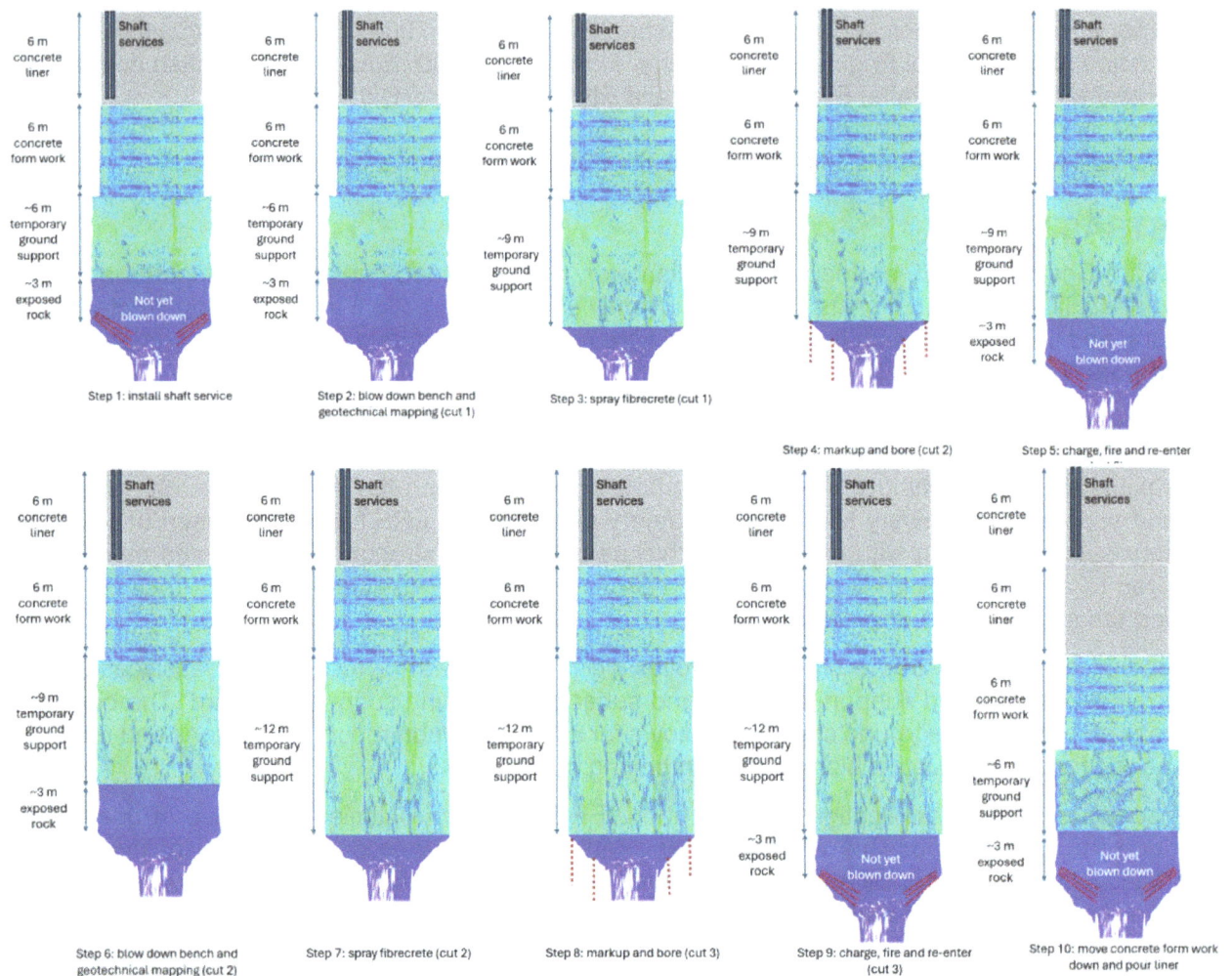

FIG 1 – Strip and line cycle at the Wira Shaft.

This paper focuses on the geotechnical aspects of the cycle, including the mapping and assessment of ground conditions, temporary ground support design, quality assurance, and geotechnical improvements to the overall process. At the time of this paper, the shaft has been excavated to ~770 m below surface.

GEOLOGICAL SETTING

The Wira Shaft sits into the footwall of the orebody, ~500 m away from active production fronts. The lithologies through the area of the shaft are:

- Cover sequence: from surface to a depth of 95 m (surface at 10216.9 mRL) the cover sequences were encountered. This included silcrete, siltstones (Bulldog Shale), Cadna-Owie Formation (very weakly cemented unit, easily washed out and disintegrated) and diamictite. These units are all weak sediments.

- Volcanics (VOLC): the main rock type encountered through the Wira Shaft (and the main footwall sequence at Prominent Hill) are volcanics. The alteration of the volcanics is generally noted to be red rock hematite, earthy hematite and speckled hematite but bands of chlorite and sericite alteration are also noted. Occasional brecciation is noted as well as pseudo brecciation texture. Generally observed to be a strong competent rock mass.

- Steely hematite (HG): steely hematite is a sub classification of the hematite altered volcanics but has distinct properties that separate it from other hematites. It is a very hard and abrasive rock mass that is generally very blocky (three plus joint sets, tight spacing) and micro fractured, with poorer joint properties (smooth/slickenside, planar surfaces). Although it required higher levels of support than the other volcanics, it is generally noted that this unit appeared worse in the drill core than it did when intersected by the shaft.

- Albite altered volcanics (ABVC): this is a poorly understood rock mass unit at Prominent Hill. It sits in the lower footwall and has only been intersected in development on a small number of occasions. It is a very strong, massive rock unit.

- Mafic/dolerite intrusions: occasionally dark volcanic intrusions are noted. When present, there is a sharp contact to the surrounding rock. No issues have been observed.

FIBRECRETE AS TEMPORARY SUPPORT – CASE STUDIES

Two instances, as summarised below (Hills *et al,* 2014; Wu, Dase and Howell, 2022), of previous works conducted around boltless fibrecrete support were used as guidance for the ground support works undertaken at the Wira Shaft.

Exploring the use of fibre reinforced shotcrete in large diameter shafts

Hills *et al* (2014) discussed the following:

- A detailed literature review and limited field testing were undertaken to investigate the use of FRS (both with and without bolts) during excavation of large diameter shafts.

- It is vital to have a strong understanding of the geotechnical conditions encountered and ensuring these conditions are transferred into a usable classification scheme (Hills suggested the use of the Q-system).

- In addition to the overall rock mass quality, wedge formations may cause wall stability concerns. It is important to identify the likelihood of wedge formations and potential size/orientation during the investigation study, and then verify this during geotechnical mapping.

- Only one occurrence of boltless FRS in a conventionally sunk shaft was cited (South Deep Shafts), there were however many instances of boltless FRS being used in lateral development (where personnel access was required). In lateral development, where there are likely to be many headings available, there is the luxury of allowing the fibrecrete to cure for a substantial period to achieve the required re-entry strength. In single heading development (ie a shaft sinks), there becomes a need to ensure a timely re-entry is achieved (where a timely re-entry was defined as re-entry to the fibrecrete significantly sooner than it would take to support the heading with rock bolts).

- The critical parameter for determining the early age strength of the fibrecrete was identified to be shear strength. The required shear strength can be related back to compressive strength which is easier to test in the field.

- A high intensity quality assurance regime is required for FRS if its use is being optimised as the key support element in a ground support scheme. This includes thickness testing, early age testing and adhesion testing, as well as assessment of the required capacity of the FRS.

- As part of the quality control process, there is a need to manage the human aspects that influence the quality of the fibrecrete. Ongoing training of spray operators is vital to ensuring a successful outcome for the fibrecrete, as their influence on the quality cannot be underestimated.

- With detailed cut by cut geotechnical assessments, a specific FRS mix deign and a rigorous QA/QC process, the use of boltless FRS in large diameter, conventionally sunk shafts, is expected to be successful.

Turquoise ridge #3 shaft sink, Nevada

Wu, Dase and Howell (2022) discussed the following:

- The #3 Shaft at Turquoise Ridge Underground Mine is a haulage shaft developed using a blind sinking method. The diameter of the finished shaft was 7.5 m with a depth of ~1000 m.

- A system of boltless temporary support was used. This involved using FRS to replace the traditional bolt and mesh support system.

- The design of the boltless temporary support system was reliant on an accurate understanding of the encountered ground conditions. A geotechnical investigation hole was drilled and covered ~80 per cent of the shaft depth. The information obtained from this hole was used for initial geotechnical assessments, with geotechnical mapping conducted on every fired round to verify the conditions encountered.

- Rock mass quality was logged using the RMR_{76} system. Generally, the ground type seen from site mapping was consistent with the core logging from the shaft investigation hole, with some variation between the very good/good classification. Slightly poorer (still classified as good) rock mass was logged in site mapping than from the investigation hole. This was attributed to blast induced damage, scale between the mined shaft and drill hole, and variations in logging classifications between engineers.

- The maximum temporary support wall height (lag between the bench and liner) was 15 m. This was dependent on encountered ground conditions, the better the conditions, the longer the lag distance. Poor ground had a maximum temporary wall height of 7.5 m with very poor ground being reduced to a 4 m exposure.

- An early age strength of 1.0 MPa was required prior to personnel working adjacent to the sprayed fibrecrete. A special fibrecrete mix was developed, with 1.0 MPa being reached within an hour. This was verified by penetrometer testing for every round.

- The thickness of the fibrecrete was varied to achieve the required ground support capacity for different ground conditions. For good ground 75 mm fibrecrete was required, poor ground required 100 mm fibrecrete and very poor ground required 125 mm fibrecrete. These thickness increases were designed in conjunction with the liner lag distance.

- This project successfully used FRS as the primary means of temporary ground support along the length of the shaft, with no bolting required.

GROUND SUPPORT DESIGN

Several levels of ground support capacity have been designed for use within the Wira Shaft. These are documented in and developed as per the Principal Ground Control Hazard Management Plan (GCMP) written for the Wira Shaft. The level of support required for a particular round is dependent on encountered ground conditions. The mapped ground conditions are recorded in terms of Q-system parameters; these are used in support pressure calculations which dictate the level of support required. The capacity of the fibrecrete is calculated with consideration of the fibrecrete's shear resistance to ground movement (load) and is used to specify the levels of standard ground support.

Q-system

The Q-system is the rock mass classification scheme used within the Wira Shaft. It was developed at the Norwegian Geophysical Institute (NGI) in 1974 and is revised as necessary to remain relevant with advancements in the field. The NGI developed a handbook for the use of the Q-system in rock mass classification and ground support design (NGI, 2022), which is used to assist with the classification of the ground encountered within the Wira Shaft. The Q-value is determined using Equation 1, with the six input parameters as per NGI (2022). The rock quality designation (RQD) was determined within the shaft using the volumetric joint count method as proposed by Palmstrom (2005) (Equation 2); this accounts for the mapping of RQD in three-dimensional space. Equation 3 is used to determine the volumetric joint count (Palmstrom, 2005).

$$Q = \frac{RQD}{J_n} \times \frac{J_r}{J_a} \times \frac{J_w}{SRF} \tag{1}$$

$$RQD = 110 - 3.3J_v \tag{2}$$

Where:

$$J_v = \frac{1}{S_1} + \frac{1}{S_2} + \cdots + \frac{1}{S_n} + \frac{N_r}{5\sqrt{A}} \qquad (3)$$

Where:

$S_x = average\ joint\ set\ spacing$

$N_r = number\ of\ random\ joints$

$A = area\ in\ m^2$

Required support pressure

The required support pressure is the maximum load the rock mass will place onto the ground support scheme. It is calculated from the mapped Q-value parameters and is highly influenced by the joint roughness. When selecting the joint set parameters to use for the Q-value and support pressure, the likely influence on the stability of the excavation is considered. The shaft can be domained into different areas if the dominant/influencing joint set changes depending on the location around the shaft (ie several required support pressures/ground support standards may be required in a single round).

The equations for support pressure are shown in Equations 4 and 5, with the required support pressure taken as an average of the two. Equation 4 is as proposed by Barton, Lien and Lunde (1974). Equation 5 is as proposed by Singh (1993) (as cited by Grimstad and Barton (1993)).

$$P_{arch} = \frac{2J_n^{1/2} \times Q^{-1/3}}{3J_r} \qquad (4)$$

Where wall adjustments are applied equivalent to:

$$Qw = 5Q\ where\ Q > 10$$

and

$$Qw = 2.5Q\ where\ 0.1 < Q < 10$$

$$P_{arch} = \frac{2}{J_r} \times 5Q^{-1/3} \times f \qquad (5)$$

Where:

$$f = 1 + \frac{H - 320}{800}\ for\ f \geq 10$$

Where:

$H\ is\ depth\ below\ surface\ (m)$

Fibrecrete capacity

The primary failure mechanism of the FRS used for design purposes is direct shear failure. To determine the load bearing capacity of the fibrecrete at varying thicknesses and early strengths, the shear strength of the fibrecrete, the perimeter of the design area (taken per base unit of 1 m² for this purpose) and the thickness of the fibrecrete are considered (Figure 2). Using the relationships in Equation 6 (Rocscience, 2024) and Equation 7 (relationship between fibrecrete shear strength and compressive strength determined from Figure 3 (Bernard, 2007)), Equation 8 can be obtained for the available support capacity of fibrecrete (provided in t/m²), where it is equivalent to the resistance of fibrecrete in shear over a surface area of 1 m². Although the primary factor in the capacity of the fibrecrete is the shear strength, the final relationship is stated in compressive strength as this is easier to assess in the field.

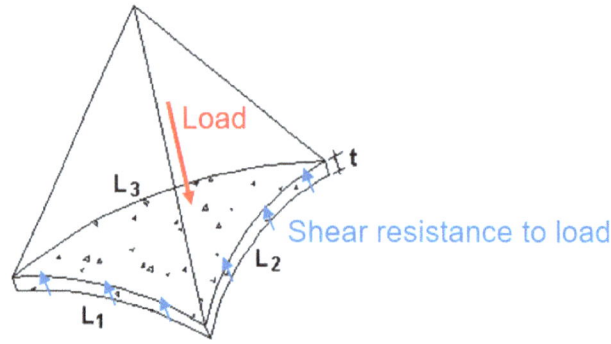

FIG 2 – Parameters considered when assessing the load bearing capacity (in shear) of fibrecrete (Rocscience, 2024).

$$V = vpt \qquad (6)$$

Where:

$V = resistance\ of\ fibrecrete\ in\ shear\ (MN)$

$v = shear\ strength\ of\ fibrecrete\ (MPa)$

$p = perimeter\ length\ (over\ an\ area\ of\ 1\ m^2)\ (m)$

$t = fibrecrete\ thickness\ (m)$

$$v = 0.28f_c^{0.6} - 0.11 \qquad (7)$$

Where:

$f_c = mean\ compressive\ strength\ of\ fibrecrete$

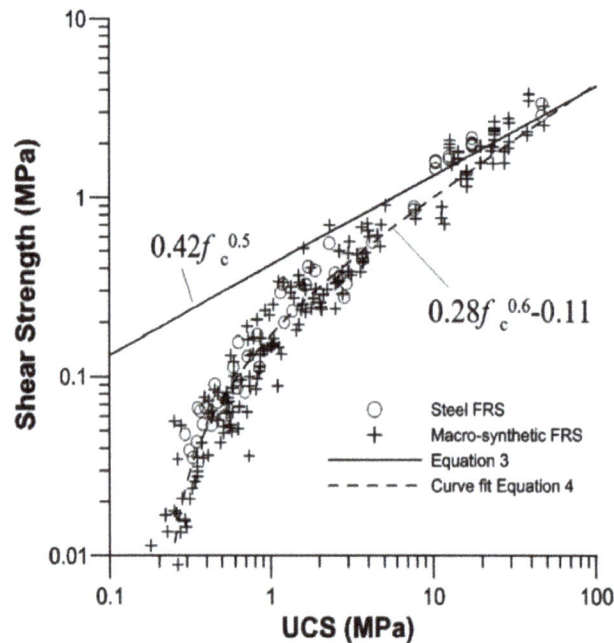

FIG 3 – Relationship between fibrecrete shear strength and compressive strength (Bernard, 2007).

$$Support\ capacity_{FRS} = (112f_c^{0.6} - 44)t \qquad (8)$$

Equation 8 shows the support capacity of fibrecrete (provided in t/m²), where it is equivalent to the resistance of fibrecrete in shear over a surface area of 1 m². There are limitations to this support capacity estimate, refer to the 'Limitations in Design Process' section.

Bolt capacity

The load capacity of bolts (per bolt) for input into the available support pressure calculations were determined considering the tensile strength of the bolts (Table 1). The bolt type and spacing were varied to achieve the required support pressure per m^2 (taking into consideration the support pressure provided by the fibrecrete).

TABLE 1

Bolt capacities for input into support pressure calculations.

Bolt Type	Length (m)	Tensile strength (kN)	Load capacity per bolt (t)
Resin bolt	2.4	254	25
Friction bolt	2.4	-	16.8 (at 7.0 t/m)

Although the majority of the shaft has been sunk in strata where the ground conditions have been assessed as being suitable for fibrecrete only support, it is anticipated there will be situations where bolts are required in addition to the fibrecrete. The main cases where bolting will be required are:

- Any cracking in surrounding fibrecrete indicating movement.
- Seismic or adverse stress conditions are expected to be encountered and deeper-seated reinforcement will be required.
- Geotechnical Engineer's discretion.

The use of self-drilling anchors (SDA) and Falcon Bolts (self-drilling point anchor bolts) were researched as potential alternatives for resin bolts and cables. These would save time in the ground support scheme installation process. These bolting alternatives have not yet been implemented in the shaft, given bolts have not been required for the temporary ground support scheme.

Levels of ground support

Two sets of ground support standards have been developed:

1. A standalone fibrecrete system (spot bolting only when required) (Table 2).
2. A combined fibrecrete and bolting system (Table 3).

The ground support standard required for a round is recommended based on the encountered ground conditions, the recommended solution will be dependent on the above discussed conditions, with respect to bolting requirements. Given the speed and ease of spraying, the operational preference is fibrecrete only profiles, however, there are cases where bolting is required.

TABLE 2

Fibrecrete only ground support standards.

Name	Support pressure (t/m^2)	Fibrecrete thickness (mm)	Fibrecrete re-entry strength (MPa)	Additional
Blue (F)	<5	75	1.0	Spot bolting as required
Green (F)	5 to 10	150	1.0	Spot bolting as required
Yellow (F)	10 to 14	150	1.5	Spot bolting as required
Orange (F)	14 to 19	150	2.2	Spot bolting as required
Red (F)	19 to 24	150	2.9	Spot bolting as required
Dark Red (F)	>24	150	7.6	Spot bolting as required

TABLE 3

Bolting ground support standards.

Name	Support pressure (t/m²)	Bolt type	Ring spacing (m)	Bolts/ ring	Fibrecrete thickness (mm)	Additional
Blue (B)	<5	-	-	-	75	Spot bolting as required
Green (B)	5 to 10	2.4 m friction bolt	1.5	15	75	-
Yellow (B)	10 to 14	2.4 m friction bolt	1.0	15	75	-
Orange (B)	14 to 19	2.4 m resin bolt	1.0	15	75	-
Red (B)	19 to 24	2.4 m resin bolt	1.0	21	75	-
Dark Red (B)	>24	2.4 m resin bolt	1.0	21	75	6 m cable bolts as required

Distance of ground support to the bench

It was assessed that ground support was not required on the bottom 600 mm of the shaft walls. This was determined considering personnel exposure to the lower walls while on the bench (for mark-up and charging) and assessing, similar to grade line in underground mining, what would be considered exposure to unsupported ground. This was implemented to assist with the transition of spraying between rounds and to reduce the blast effects, and subsequent support rehab requirements, on the fibrecrete at the immediate bench. This improved the overall quality of the fibrecrete sprayed. When spraying the next round, the lower 600 mm of the previous round is also captured. Prior to personnel working on the bench, the exposed ground is check scaled. During geotechnical mapping, if any ground with unravelling potential is identified, the recommendation is made to support down to the bench.

Mesh alternative

Although this project is heavily reliant on fibrecrete as the means of primary temporary support, there were limited cases where a mesh alternative profile was required (eg if batch plant issues were experienced). The mesh type used has been Hock PP Biaxial Geogrid mesh. For the support capacity assessments conducted, it was assumed that the mesh carried no support load (ie support load carried by bolts) and was installed for scat control only. The bolt length (0.9 m, 1.8 m or 2.4 m) and type (friction or resin bolt) were varied to achieve the required support pressure. A minimum bolt and ring spacing of 1.0 m was required to ensure the mesh is profiled to the shaft walls. Additional bolts are used where required to ensure the mesh is tight to the rock surface as to avoid excessive scat build-up and loading of the mesh. The use of mesh is only prescribed following geotechnical inspection and mapping of the ground conditions; ground with a high tendency to unravel/deteriorate is not considered suitable for use with the mesh alternative.

Limitations in design process

As discussed previously, the primary failure mechanism of the fibrecrete was assumed to be direct shear failure. This is not the only failure mechanism that may occur, and it is important to be mindful of this. If other failure modes do occur, they may decrease the effective support pressure of the fibrecrete. Adhesion loss between the fibrecrete and rock face can lead to more complex failure modes including flexural failure (Rocscience, 2024). Compressive/tensile failure may also occur if the fibrecrete is put under extreme loading conditions. Assessment of large wedges occurs on a case-by-case basis to determine if the standalone fibrecrete support is suitable. In some cases, wedges may need bolting.

Given the limitations in the design process for the fibrecrete support, controls are required. These include a comprehensive understanding for the rock mass conditions encountered, a rigorous fibrecrete QA/QC regime (including early strength, thickness and adhesion testing), inspection of previously sprayed shaft walls for indications of possible damage and longer-term assessment of potential stress and seismic related concerns.

REAL TIME GEOTECHNICAL ASSESSMENTS

The importance of a thorough understanding of the ground conditions encountered cannot be understated. It provides the basis for the entire approach of optimisation of the ground support scheme. Detailed geotechnical mapping and ground support standard recommendations occur for every fired round. This is achieved with 24-hour geotechnical coverage during the shaft sinking phase.

Geotechnical mapping

A geotechnical inspection is undertaken on every fired round. This occurs during or immediately after blowing down the bench. To optimise time, the following steps are taken regarding the timing of mapping.

- While blowing down of the bench is occurring, the stage (and personnel) sit ~3 m above the top of the exposed ground. During blow down, the bench is cleared off and hydro-scaling of the walls occurs. Once one section has been blown down, the geotechnical inspection can begin; the geotechnical engineer follows the spray operators as they continue the blow down. Generally, the geotechnical inspection is complete within 5 mins of blow down completion, with minimal disturbance to the production cycle time. Important information recorded during the geotechnical inspection:
    - Face sketch and photographs.
    - Bench depth relative to the current liner curb (as to calculate depth along the shaft).
    - Rock type and alteration.
    - Estimate of RQD (volumetric joint count method).
    - Structures – number of sets, roughness, alteration, groundwater.
    - Strike and dip of structure sets.
    - General comments on profile (including roughness, overbreak, undercutting etc).
    - Potential failure modes (wedges, slabbing etc).
    - Rock noise, spalling etc.
    - Additional scaling (manual, hydro or mechanical scaling).
- If insufficient data is obtained by looking down on the rock mass from the blow down position, the stage can be moved down to get a closer look at the rock face once the bench as walls are cleared.
- An inspection of the previous fibrecrete also occurs during the geotechnical inspection, with any issues rectified as necessary.

Work is ongoing to determine how LiDAR scanning may be used for geotechnical mapping purposes.

Ground support recommendations

Once the geotechnical inspection of the rock mass is complete, the required ground support standard can be determined. This occurs by assessing how the joint sets interact with the shaft and whether they have the potential to impact the stability of the excavation. From the sets that may influence that stability of the shaft walls, the most conservative joint parameters are used to determine the Q-value and required support pressure. Where different areas of the shaft may have varying stability concerns, the ground can be split into multiple domains, to be supported accordingly. Discussions are held between the shift supervisor and geotechnical engineer on the ground conditions encountered, any additional scaling requirements, points of geotechnical concern and the required ground support standard. Feedback is obtained from the shift supervisors as to any issues or variations that were noted during drilling of the bench and support of the round. This is also used as an input into the geotechnical considerations.

Actual versus anticipated ground conditions

Prior to shaft construction commencing, a geotechnical investigation drill hole was drilled within/near the shaft footprint along the entire length of the shaft. This hole was geotechnically logged, which provided input into the initial disposal hole raise bore stability assessments and design, and into the expected ground conditions, in terms of Q-value parameters. From the anticipated ground conditions, the expected support pressure and subsequent support standard could be assessed. Up to 75 per cent of the shaft was expected to fall within the <5 t/m² category, with an increase in support class from 950 m along the shaft to account for potential stress effects.

Mapped ground conditions were typically better than expected (Figure 4). To date, there have been no significant adverse ground issues observed. The highest level of support pressure used has been within the 5–10 t/m² range (green support). At the commencement of sinking, there were six cases where spot bolting for wedges was recommended. This was a somewhat conservative approach while details within processes were being resolved and fibrecrete quality assurance was in the early stages. Only two rounds have been fully bolted to date. These were the first two rounds in the transition zone between the cover sequence and the volcanics. Approximately 97 per cent of the stripped shaft (to date) has only required fibrecrete support.

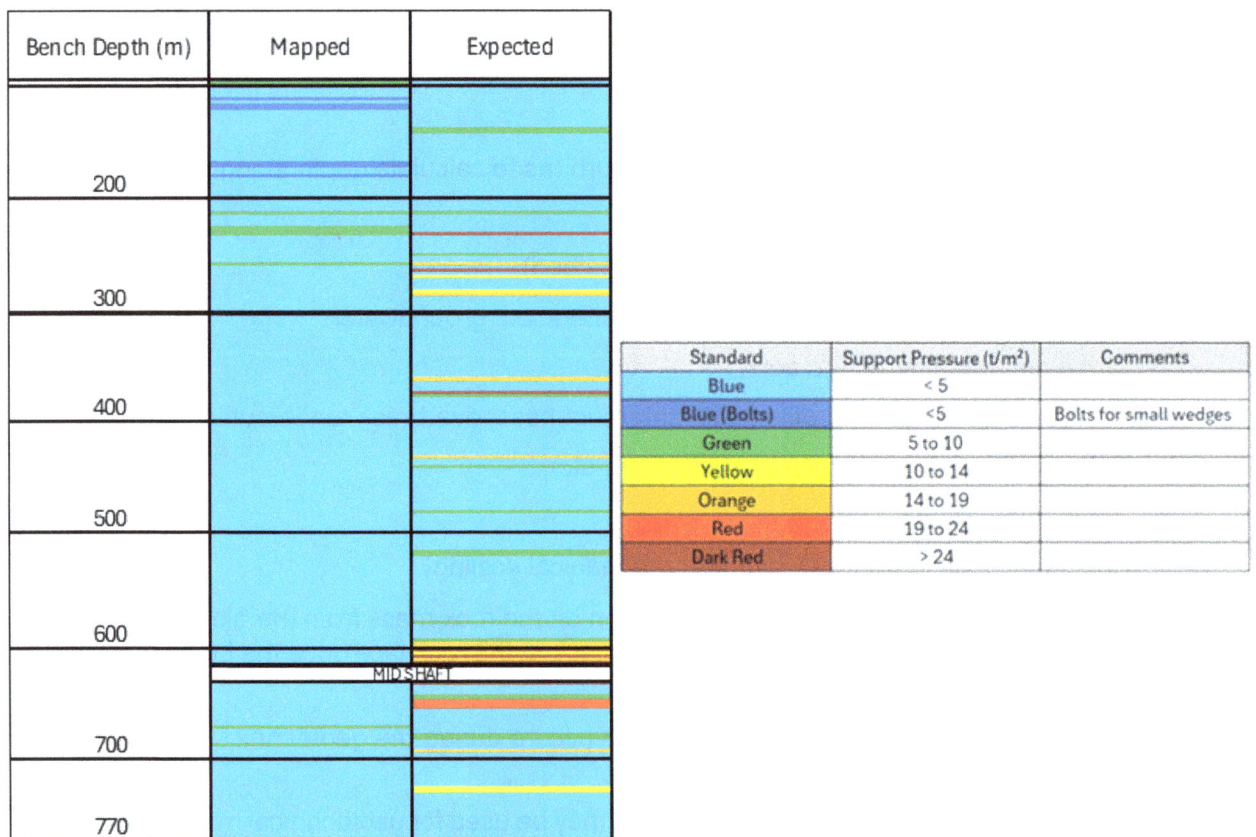

Standard	Support Pressure (t/m²)	Comments
Blue	< 5	
Blue (Bolts)	<5	Bolts for small wedges
Green	5 to 10	
Yellow	10 to 14	
Orange	14 to 19	
Red	19 to 24	
Dark Red	> 24	

FIG 4 – Mapped versus expected ground support classes (down to 770 m).

The expected ground conditions from the geotechnical logging have typically overestimated the anticipated support requirements. This is potentially due to:

- Orientation of the poorer joint friction sets and their actual influence on the shaft stability.

- Damaged core, particularly through areas of higher micro-fractured rock, due to driller induced damage during the core removal process (particularly through steely hematite).

- Intersection of isolated pods of poorer material and/or vertical/semi-vertical structures which indicated greater impact on the diamond drill hole than was seen in the shaft.

When using LiDAR scanning within the shaft, part of the raise bored disposal hole is picked up. This can be used to assist with the expected upcoming conditions.

To date, there have been no observed stress effects or seismicity that have influenced the strip and line excavation of the shaft. The expected Q-values had an elevated stress reduction factor (SRF) based on logged ground conditions and expected stress regime (SRF value of 2 from a depth of ~475 m with 5–7.5 through areas of poor ground/structural influence), however mapped SRF has generally remained at 1 given the ground conditions encountered.

Instrumentation overview

In-ground multipoint borehole extensometers (MPBX, 6 m long) have been installed at locations along the shaft as stipulated by the permanent liner designers. These instruments are required to assess convergence and are installed within 2 m of the bench; they are read following every firing. Two MPBXs are installed per ring, one perpendicular and one parallel to the major principal stress direction. To date, five sets have been installed (at 230 m, 375 m, 575 m. 675 m and 740 m down shaft); in general, less than 2 mm of total movement have been recorded by the instruments.

FIBRECRETE SPECIFICS

A fibrecrete mix was specifically designed for use in the Wira Shaft. This mix design took into consideration the required use within the slickline and the need for the fibrecrete to achieve early strength as quickly as possible.

Slickline considerations

A large factor in the design of the fibrecrete mix was its compatibility with a slickline delivery system. The mix had to be able to be useable with the Normet spray system installed on the stage, once deposited down the slickline, with a maximum drop of ~1330 m. The workability of the mix at the point of discharge (at the brace) varied throughout the project, as the shaft depth increased. Initially recommended as a slump of 240 mm ± 10 mm, the workability was increased to a slump of 260 mm ± 10 mm at ~300 m down shaft, with another increase occurring at ~700 m down shaft to a spread of 550 ± 50 mm. Further variations to workability limits are expected as depth increases further. This high slump mix allowed for losses in the slickline while remaining workable once at the nozzle. Key factors when considering the mix design with the slickline use were:

- Fine sand/crusher dust was incorporated into the mix as fibrecrete requires a high fines content for spraying. The fine sand/crusher dust also provides lubrication to the system. The intention was to reduce the friction, both between the slickline and the FRS mix and the mix's internal friction between aggregates. This was to avoid significant water loss from the mix due to evaporation from frictional effects.

- An allowable water to cement ratio (w:c) of 0.25 to 0.45 was specified. This allowed the water content of the mix to be adjusted, depending on external factors, to provide a more consistent mix. As depth along the shaft increased, the water loss seen in the slickline also increased and the water content of the mix was adjusted accordingly.

- A high range water reducer (Master Glenium SKY 8708) was incorporated into the mix to assist with workability and slump retention through the slickline. The levels of the water reducer were varied, in conjunction with the water, to achieve the required workability.

Early strength considerations

Given the importance of the efficient advancement of the shaft sink, it was critical to optimise all processes to reduce delays and ensure the time lost between productive activities was minimised. The fibrecrete was designed to reach the minimum required strength (as determined by encountered ground conditions) in as little time as was possible. Key considerations for the early strength gain were:

- The cement content used in the fibrecrete mix was 550 kg/m^3. This high cement content assisted with the early strength gain of the fibrecrete.

- A high-performance accelerator (MasterRoc SA167) was used with the fibrecrete mix. The allowable dosage of accelerator was from 7.0–12.0 per cent (calculated as a percent of the cementitious material). During the early stages of the project, varying dosages were trialled, with 9.0 per cent being the optimal dose rate. This dose rate considered initiation point of the hydration process as well as impacts on application quality and ease of spraying.

- When considering slickline concrete/fibrecrete mixes, the substitution of some cementitious material for fly ash may occur as the rounded shape of the fly ash granules can assist with workability and flowability. Substitution of fly ash for cement to assist with workability, within the fibrecrete mix designed for the Wira Shaft, did not occur as, due to the lower reactivity of the fly ash, a delayed early strength gain would have been observed.

- The fibrecrete mix incorporated a hydration control admixture (HCA 20) to allow sufficient stand time of the mix in case of delays. The HCA 20 was modified depending on the ambient conditions (ie cold weather HCA 20 was taken to lower limit (1.1 L/m^3), hot weather it was increased (4.0 L/m^3)) as to avoid overdosing the mix and impacting the initiation of the hydration process.

Fibre selection

The addition of fibres to a shotcrete mix to form FRS is a typical practice for ground support. The fibres improve the mechanical properties of the FRS including the flexural capacity and crack resistance. Polypropylene (PP, synthetic) fibres are the most typical type of fibre used. This is due to their cost-effectiveness and availability. It was determined however, that steel fibres would be more appropriate for use with the slickline. Key points regarding the comparison between fibres were as follows.

- Based on testing conducted by Sika (Taylor, Rieger and Atkinson, 2022), steel fibres (at a dose rate of 25 kg/m^3) should have minimal impact on the workability of the FRS mix. PP fibres (at a dose rate of 6 kg/m^3) can have a reduction in spread of the FRS mix of 50–70 mm, when compared to concrete with no fibres. To increase the workability to that of the reference concrete additional admixtures and mix design considerations are required.

- Both steel and PP fibres do not have an adverse impact on the 1-day and 28-day compressive strengths of the FRS.

- Steel fibres have a lower risk of balling up and causing blockages in the slickline than the PP fibres. Given the importance of the slickline operation to the project, this was a key deciding factor in the fibre type used at the Wira Shaft.

- The steel fibres used (Dramix 4D65/35 or SikaFibre Novocon HE5535 HT) have hooked ends, which allows greater interlocking capacity and increases their pull-out strength in the final product. They are 35 mm in length and a diameter of 0.90 mm (for Dramix) and 0.55 mm (for Sika).

- Considering a dosage of steel fibres of 25 kg/m^3, an energy absorption class of E500 – E700 is anticipated and for the PP fibres with a dose rate of 6 kg/m^3, an energy absorption class of E700 would be expected (Taylor, Rieger and Atkinson, 2022).

QA/QC regime

The ground support system used within the Wira Shaft has a high reliance on fibrecrete. Given the role ground support plays in ensuring the safety of personnel, the quality of the fibrecrete used was of utmost importance. A significant QA/QC regime was developed and undertaken, this process involved the following testing.

Workability testing

The workability of the fibrecrete is critical to enable the fibrecrete to be pumped, sprayed and compacted, while achieving required early age strength without segregation or bleeding. Workability testing occurred on every load of fibrecrete at both the batch plant and the brace (prior to discharge) to ensure it meets the requirements set out in the specification. Workability testing occurred on the stage when deemed necessary as to assess potential concerns and losses in the system.

The main method of workability testing to begin with was a standard slump test, moving to a spread test when required workability of the mix increased. Alternatives such as the K-slump tester are also being trialled.

Early strength testing

Given the importance of achieving a specific re-entry strength for the fibrecrete (determined based on encountered ground conditions), in a timely manner, the accurate collection of re-entry strength data is of high importance. Previous works (Bernard, 2005) have been undertaken on the methods of early age fibrecrete re-entry, with this used as a guide for the work undertaken in the Wira Shaft. Several early strength re-entry methods have been trialled within the Wira Shaft, with varying degrees of success. As previously mentioned, when considering the direct shear failure mechanism of the fibrecrete, the shear strength is the driving strength behind the fibrecrete capacity. Given the in-field difficulties of testing shear strength, compressive strength is tested instead; this is then related back to shear strength for capacity purposes.

- Beam testing: testing of fibrecrete beams using a beam end tester is the only readily available means of directly testing the compressive strength of fibrecrete at a young age. Unlike other methods trialled, the beam tester directly measures the compressive strength of a fibrecrete sample (other methods convert to compressive strength based on force required to penetrate). The sample preparation for the beam testing is more complex than *in situ* methods of early age strength testing. For the beam testing, moulds (75 × 75 × 400 mm) are sprayed; once set, they are de-moulded and crushed using the beam end tester. An efficient system of beam spraying and retrieval was developed for use in the Wira Shaft and beam testing caused minimal delays to the cycle. The beam end testing, as it provides a direct measure of compressive strength, is the most reliable method of fibrecrete early age testing and was determined to be the most suitable method for use in the Wira Shaft.

- Hilti gun: the Hilti BX 3-SCT (with X-M6-8-87 DP7 SCT B3 studs) was trialled as an *in situ* method of determining the early strength of the fibrecrete. Given the very early age testing (1 MPa), this method was deemed unsuitable for application within the Wira Shaft. The results obtained tended to indicate the Hilti method was underestimating the compressive strength of the fibrecrete at a very young age.

- Soil penetrometer: it is a gauged plunger type device. It is pressed into the fibrecrete (or soil), and an estimate of the compressive strength is obtained. Given the end of this device is flat, the strength of fibrecrete can be overestimated if a piece of aggregate is intersected during testing. This was determined to be unsuitable for use within the Wira Shaft.

- Shotcrete Penetrometers: both needle and electronic penetrometers were trialled within the Wira Shaft. The basic principle behind both is the same; the compressive strength of the fibrecrete is calculated based on the force required to embed a needle a certain distance into the fibrecrete (on the needle penetrometer results are read from a graduated scale and the electronic penetrometer is a digital device). A similar issue to the soil penetrometer may be seen, where results are overestimated by intersection of aggerates, it is however less likely with the shotcrete penetrometers, due to the needle end. Several results are taken at each location to determine an average of the results, with significant outliers removed. The use of these devices to test *in situ* fibrecrete was determined to be unsuitable due to the distance between the stage and the wall (hazard of personnel having to overextend to conduct the testing). Use of the shotcrete penetrometers with a test panel would be suitable.

Thickness testing

The thickness of the fibrecrete is key to achieving the required support capacities for the encountered ground conditions. It is important that fibrecrete be applied as per the standard and checked to ensure it has passed. Two methods of testing have been used at the Wira Shaft; these are:

- Drill holes: holes were drilled using a handheld drill and the depth recorded. Limitations with this method were only obtaining data from set points (8 points per cut). This does not always provide an accurate representation of the overall thickness.

- Point cloud comparison: using LiDAR and CloudCompare (point cloud software), the thickness of the fibrecrete was able to be assessed. This involved obtaining a LiDAR scan of the fired round before and after spraying. Painted control points were used as common points between the scans to allow them to be overlaid. A heat map of fibrecrete thickness was generated, with thick and thin patches easily identified (Figure 5). This method was validated using drill hole depth testing to ensure its suitability.

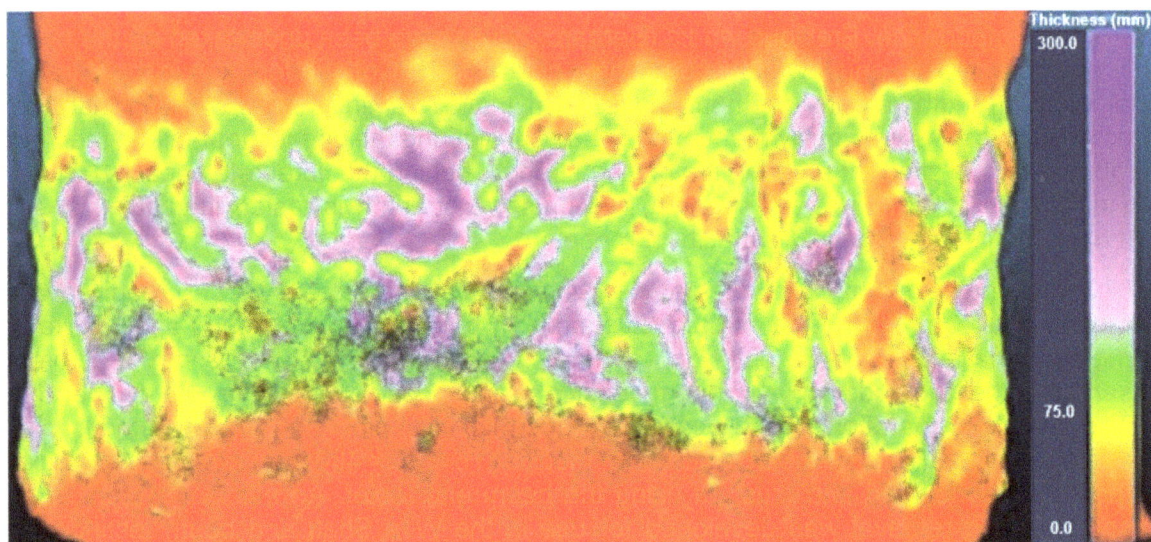

FIG 5 – Fibrecrete thickness assessment using LiDAR scanning.

Adhesion testing

Adhesion testing was conducted to assess the competency of the connection between the fibrecrete and the rock and ensure there was no risk of delamination. This testing was conducted using a pull off tester, which is partially destructive test, whereby the tensile strength of the fibrecrete/rock interface is checked.

28-day cylinder testing

Test cylinders were taken from every third round of fibrecrete sprayed to be tested at 28-days to determine the compressive strength. The testing of these cylinders had no bearing on the early age testing and re-entry and were simply a means of testing the consistency between mixes and for long-term assessment to back analyse any early strength issues that may have been encountered. Generally, 28-day strengths of 70–80 MPa were achieved.

Aggregate checks

Workability retention issues have been encountered on a number of occasions. A common cause of this was excessive deleterious fines (swelling/reactive clays etc) in the aggregates. The clay absorbs water and admixtures making them 'unavailable' in the mix. This impacts the workability and overall quality of the mix. These fines were typically observed in the crusher dust and sand, with minimal fines noted on the 10 mm aggregate. When this issue was observed it typically led to a requirement to max out the water in the mix to allow sufficient workability while travelling down the slickline; this in turn reduced the early age strength gain of the mix, increasing the re-entry time. Given the significance of this issue on the strip and line cycle, it was important to understand the influence of the deleterious fines and determine when they may cause concerns. Testing for fines was conducted using sand equivalent tests and sieve analysis, with specific deleterious fines testing undertaken using a methylene blue method. This testing assisted with determining when workability retention issues were anticipated and to what magnitude they may impact the shaft's fibrecrete mix.

Temperature plays a significant role in the quality of the fibrecrete. Aggregates have a major impact on the temperature of the mix, particularly if not properly conditioned prior to use. Shading the aggregates and the use of a sprinkler system can assist to reduce the temperature of the aggregates in particularly hot climates and can assist with the quality of the fibrecrete mix.

Human factors

Given the importance of the spray operator's influence on the quality of the final fibrecrete product, it was important to provide ongoing training and quality control around the processes involving human influence. These factors included nozzle angle to the rock surface (should be as close to perpendicular as is possible), nozzle stand off from the wall (should be 1–2 m away), correct priming and cleaning of the chemical lines (to ensure correct dosing of accelerator), appropriate techniques for layering of thicker fibrecrete, and where possible minimising rebound. This was achieved through a verification of competency process for all spray operators.

Fibrecrete re-entry

As discussed, the fibrecrete mix was specially designed to achieve the required re-entry strength in as little time as possible. The requirement was to achieve 1.0 MPa within one hour. Figure 6 shows the re-entry results to date. For all data points shown on this graph the required re-entry strength was 1.0 MPa.

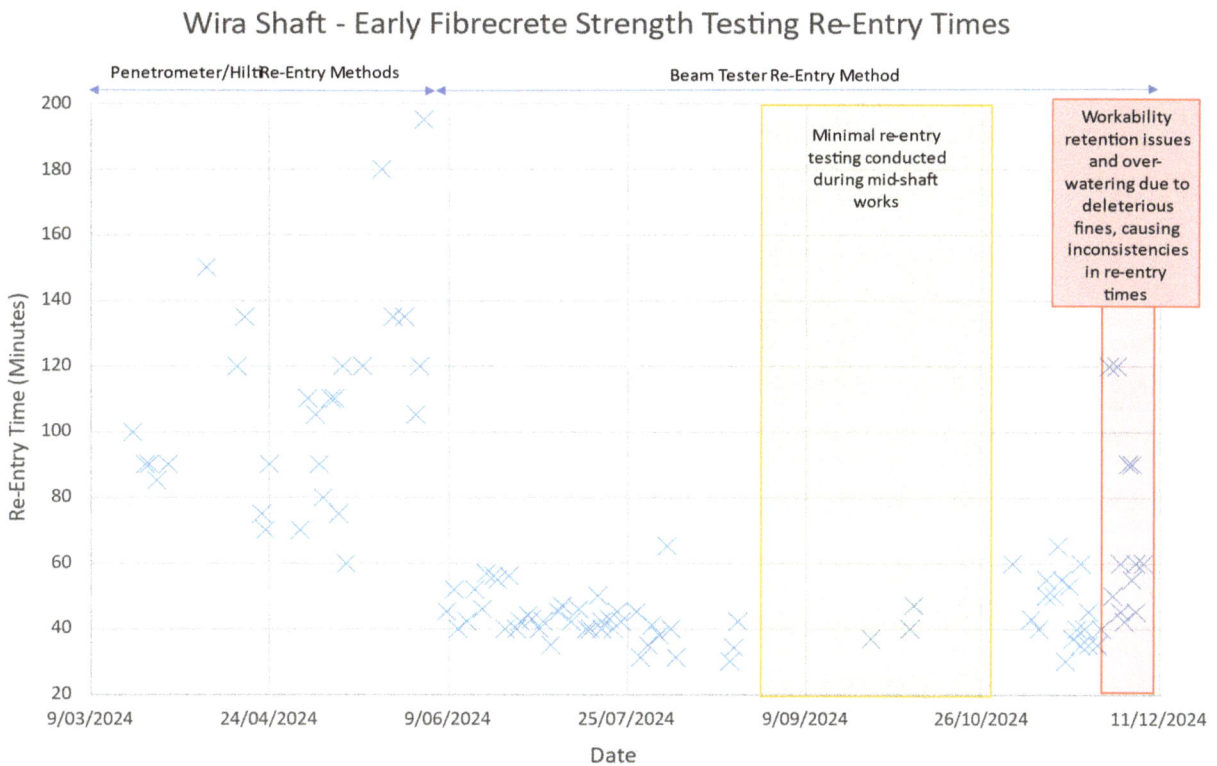

FIG 6 – Fibrecrete early strength testing re-entry time versus date and re-entry method.

The data between March 2024 and May 2024 indicates initial issues in establishing efficient fibrecrete processes; this was towards the start of the strip and line process and was in part due to the use of early strength testing techniques that were under calling the actual strengths achieved. Towards the start of June 2024, a focus was placed on improvements to fibrecrete related processes, with significant improvement seen in the results obtained. Around this time, the early strength testing method was changed from the Hilti gun and penetrometers to the beam tester. This has provided consistent testing conditions and may be a contributing factor to the improvement in results and data observed.

Towards the start of December 2024, it was identified that there were deleterious fines present in the crusher dust. This caused inconsistencies in the mix, and in some cases, the mix required more water than standard (increase to max w:c of 0.45) to achieve the required workability. This caused some extended re-entry times of up to 120 mins.

Not considering the period of higher re-entry strengths due to subpar aggregates, the average re-entry time, from June 2024 onwards, for the fibrecrete to reach 1.0 MPa is ~40 mins.

IMPACT ON CYCLE TIME

The use of the boltless fibrecrete ground support system implemented in the Wira Shaft positively influenced the cycle times and efficiency of the shaft sink. Three cases were considered when assessing the likely influence of ground support installation on cycle times.

1. Boltless fibrecrete – includes set-up, spraying, re-entry and cleanup time.

2. Fibrecrete and bolting – not as per the previously discussed fibrecrete and bolting case, this scenario considers a traditional support scheme where the bolts are considered to provide all required capacity and fibrecrete is surface support only.

3. Mesh and bolting – installation of poly mesh for scat protection, required to be profiled to the walls, with a maximum bolt spacing of 1.0 m × 1.0 m.

Although bolting is conducted using the sinking stage's three jumbo booms in unison, it is estimated that a fibrecrete only support will save at least 4 hrs per 6 m liner advance (estimated variance in cycle times shown in Figure 7). Across the life of the strip and line process, the use of boltless fibrecrete support is expected to save in excess of 800 hrs.

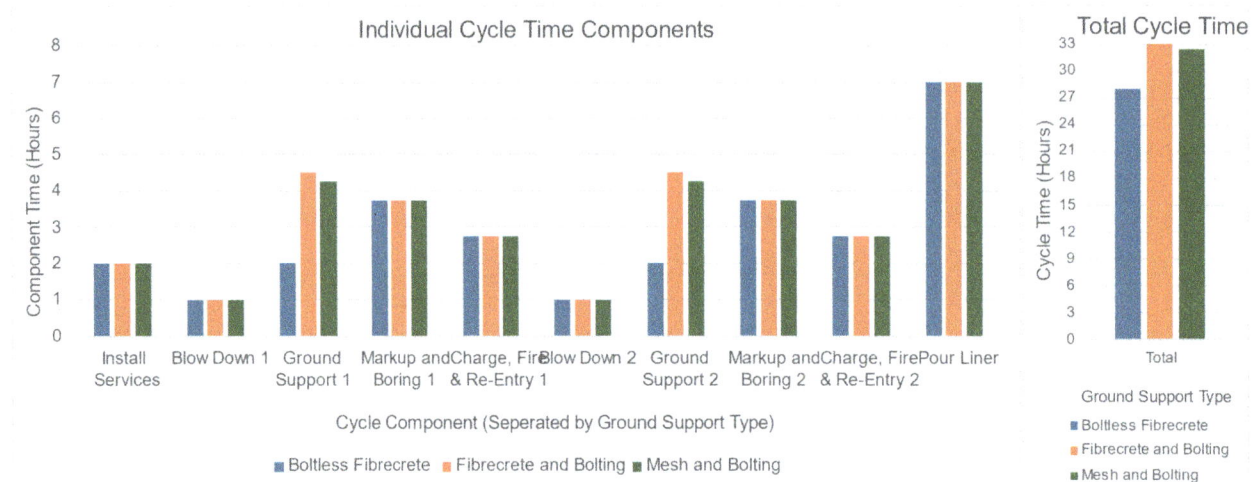

FIG 7 – Component times and full cycle times considering a 6 m liner advance, comparing ground support methods.

DISCUSSION AND CONCLUSION

A boltless temporary ground support system has been successfully implemented at the Wira Shaft, down to a depth of 770 m. The success of this work is highly reliant on a thorough understanding of the expected and actual ground conditions and specific recommendations of ground support. Alterations to fibrecrete thickness and strength allow for variable ground support capacities to account for a range of ground conditions encountered. For an in-depth understanding of the rock mass conditions, full time geotechnical coverage on the project was required.

Given the importance of the fibrecrete, the assurance of its quality was of utmost importance. A substantial QA/QC regime was implemented to ensure a high standard was maintained. The key components of this QA/QC regime were workability, thickness, adhesion and early strength re-entry testing, as well as ongoing operator training. Through the trialling of various early re-entry strength methods, it was determined that the most reliable and accurate method was the use of the beam tester. This system is the only early strength method that provides a direct measure of fibrecrete's compressive strength. Although it is more complex than the penetration type methods, the accuracy and confidence in results obtained outweighed the additional testing requirements. A process was developed for the beam tester sample collection and testing whereby minimal impact was had on productive activities, and it became a concurrent aspect of the cycle.

For the remainder of the shaft, and for future works, the limits of a boltless support system need to be understood. To date, at the Wira Shaft, this limit has not been reached. Understanding the impact of depth and stress on the rock mass and fibrecrete failure mechanisms will assist in providing guidance on the potential limit of the boltless temporary support system.

The systematic geotechnical management at the Wira Shaft has allowed for the replacement of an often-conservative pattern bolting approach with a fibrecrete only temporary support system. The key components of this were full time geotechnical coverage and a substantial fibrecrete QA/QC regime.

REFERENCES

Barton, N, Lien, R and Lunde, J, 1974. Engineering Classification of Rock Masses for the Design of Tunnel Support, *Rock Mechanics*, 6.

Bernard, E, 2007. Early-age load resistance of fibre reinforced shotcrete linings, Penrith, *Tunnelling and Underground Space Technology*, 23 .

Bernard, S, 2005. Early-Age Test Methods for Fiber-Reinforced Shotcrete, *Shotcrete*, pp 16–20.

Grimstad, G and Barton, N, 1993. Updating the Q-System for NMT, Norway, Norwegian Geotechnical Institute.

Hills, P, O'Toole, D, Kilkenny, D and Seah, K, 2014. Exploring the Use of Fibre Reinforced Shotcrete in Large Diameter Shafts, *Ausrock 2014: Third Australasian Ground Control In Mining Conference*.

NGI, 2022. Using the Q-System; Rock Mass Classification and Support Design, 2nd ed, Oslo: Norwegian Geophysical Institute (NGI).

Palmstrom, A, 2005. Measurements of and Correlations between Block Size and Rock Quality Designation (RQD), Sandvika, *Tunnels and Underground Space Technology,* 20.

Rocscience, 2024. Shotcrete Support Force [online]. Available from: <https://www.rocscience.com/help/unwedge/documentation/theory/support/shotcrete-support-force> [Accessed: 31 August 2024].

Taylor, D, Rieger, C and Atkinson, T, 2022. *SikaFiber® Reinforced Concrete Handbook* (Sika Services AG: Tüffenwies).

Wu, R, Dase, J and Howell, R, 2022. Using Fiber Shotcrete As Temporary Wall Support At Barrick Turquoise Ridge #3 Shaft Sinking, Salt Lake City: SME Annual Meeting.

Expanding Prominent Hill

A M Ebbels[1]

1. Superintendent Mine Planning, BHP, Adelaide SA 5950. Email: annemarie.ebbels@bhp.com

ABSTRACT

Prominent Hill is owned and operated by BHP Group Ltd and is located on the Antakirinja Matu-Yankunytjatjara (AMY) lands in northern South Australia. Prominent Hill is currently expanding its open stoping operation from 4.5 Mt/a to 6.5 Mt/a.

The current operation is truck haulage to the surface while the mine transitions to shaft hoisting once the expansion is complete to facilitate the production rate increase. The mine expansion requires additional mine infrastructure to be developed including an underground crusher, hoisting shaft, decline extension and haulage level as part of the ore handling system and an upgrade to the ventilation system including refrigeration.

This paper discusses the planned expansion project and the challenges in implementing the project while maintaining the mines production rate and preparing for the ramp-up in the production rate.

INTRODUCTION

Overview

Prominent Hill is located 650 kilometres north-west of Adelaide, South Australia, 130 km south-east of Coober Pedy and is located on the traditional lands of the Antakirinja Matu-Yankunytjatjara people.

The Prominent Hill copper-gold deposit (Figure 1) was discovered in 2001 by Adelaide-based Minotaur Exploration Limited, and OZ Minerals (then Oxiana) moved to 100 per cent ownership in in early 2005 (Sandery and Light, 2014). In May 2023, BHP acquired OZ Minerals to form BHP's Copper South Australia (CuSA) mining province within the Gawler Craton.

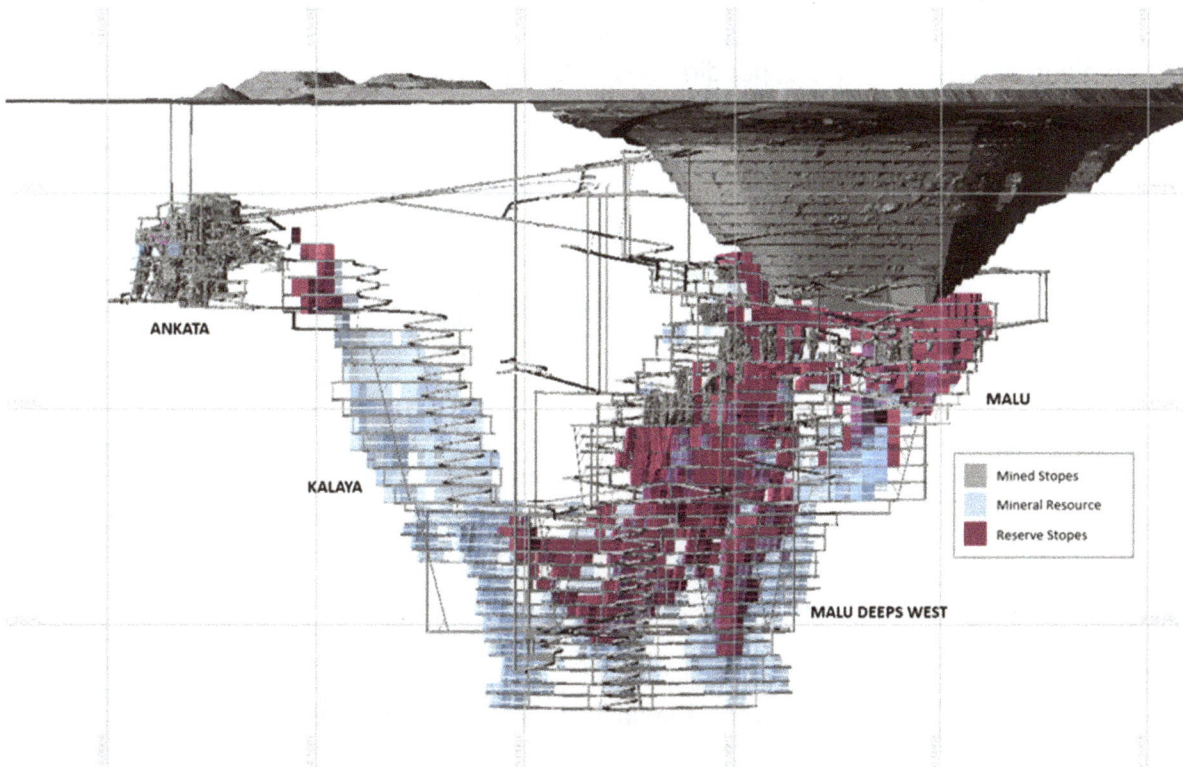

FIG 1 – Long Section of Prominent Hill Mine (OZ Minerals, 2022).

Open pit mining commenced at Prominent Hill in 2009 in the Malu orebody and transitioned to underground mining in 2012 in the Ankata orebody with the last open pit ore mined in 2018. The current mining operation is sub-level open stoping with the ore trucked to the open pit where the ore is then hauled with surface trucks to the run-of-mine (ROM) pad.

Prior to the shaft being approved, Prominent Hill's mine life as a trucking mine was limited to 2033 with ventilation, congestion at depth and long haulage distances constraining the production rates. Studies for expansion of the mine commenced in 2018 considering various material handling options for the mine.

In August 2021, OZ Minerals approved the constructions of a hoisting shaft at Prominent Hill copper gold mine which will extend the mine life to 2036 at 6 million tonnes per annum (Mt/a; OZ Minerals, 2021a). In addition, the shaft would also open up options to access additional ore and enable the potential growth of the resource at depth.

Geology

Prominent Hill iron-oxide copper gold silver deposit is located on the southern margin of the Mount Woods Domain in South Australia's Gawler Craton (Freeman and Tomkinson, 2010). The copper-gold-silver mineralisation is mostly hosted within the hematite-matrix breccia. Copper mineralisation also occurs as disseminations of chalcocite, bornite and chalcopyrite the breccia matrix.

The orebody dips at approximately 70 degrees to the north and is approximately 1000 m along strike.

As part of the Prominent Hill Operation expansion study (PHOX) infill drilling was carried out in the Malu inferred resource below the 2020 Ore reserve to confirm the continuity, thickness and grade to support PHOX.

GENERAL MINE DESIGN

Production

Sub-level open stoping (SLOS) with paste backfill is the method currently employed at Prominent Hill in both bottom-up and top-down mining sequences. The SLOS is also applied in both the transverse and longitudinal direction within the sequences. The stopes sizes are 40–60 m high and 20–30 m along and across strike depending on the location with the mine. Figure 2 shows a typical level layout in the mine.

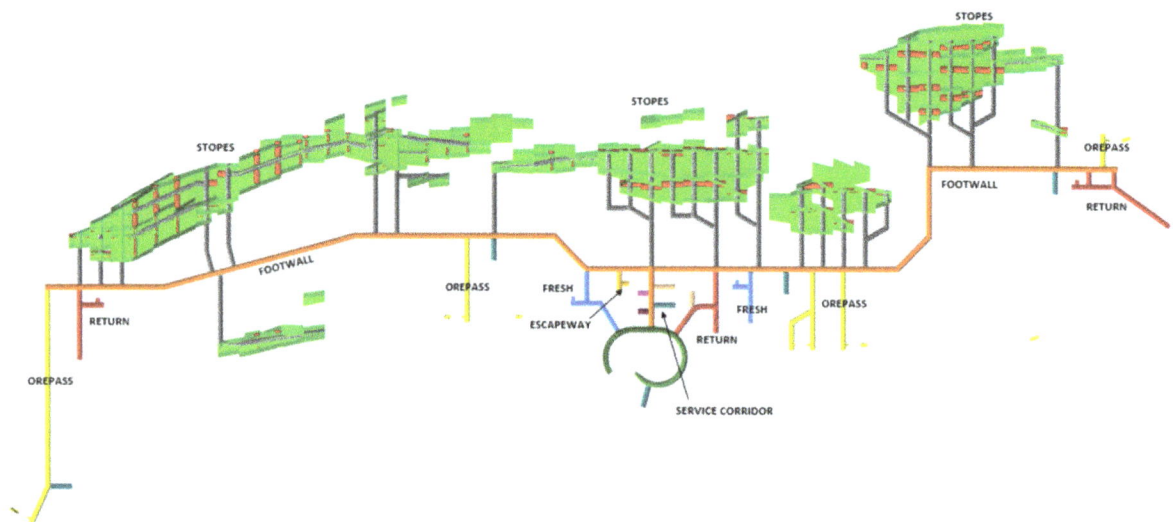

FIG 2 – Typical level layout within Prominent Hill.

Ore handling

The proposed material handling system include bogging to orepasses on production levels to a dedicated truck haulage level to transfer the ore to the crusher and hoisting system. Figure 3 provides an overview of the planned materials handling system. This paper concentrates on the underground portion of the system.

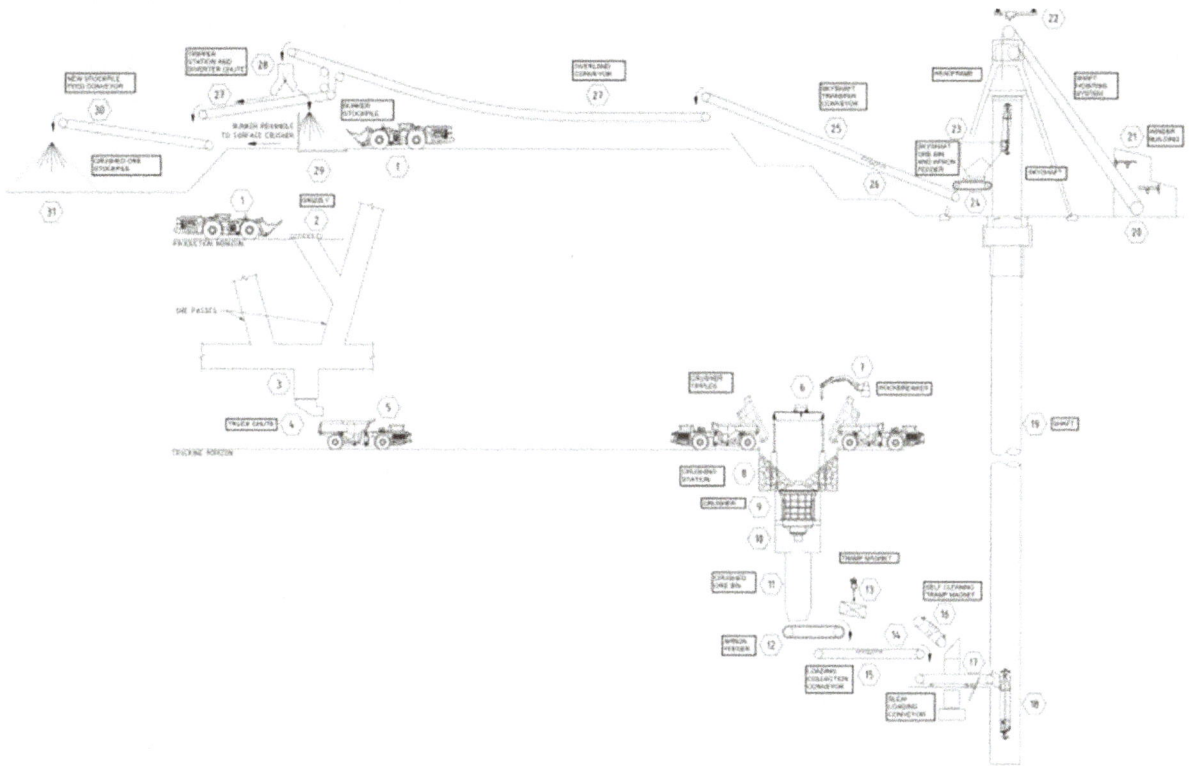

FIG 3 – Schematic of material handling system (AECOM, 2021).

Orepasses

The orepasses required placement to avoid certain lithologies for both stability and drillability. The steely hematite in particular is very difficult to drill through although geotechnically favourable. The initial criteria used for the orepass in the PHOX study were that the preferred dip of the orepasses was greater than 80 degrees with a minimum of 75 degrees. The PHOX design had two orepasses converging into a single ore bin and feeding a reverse-in truck chute (Figure 4).

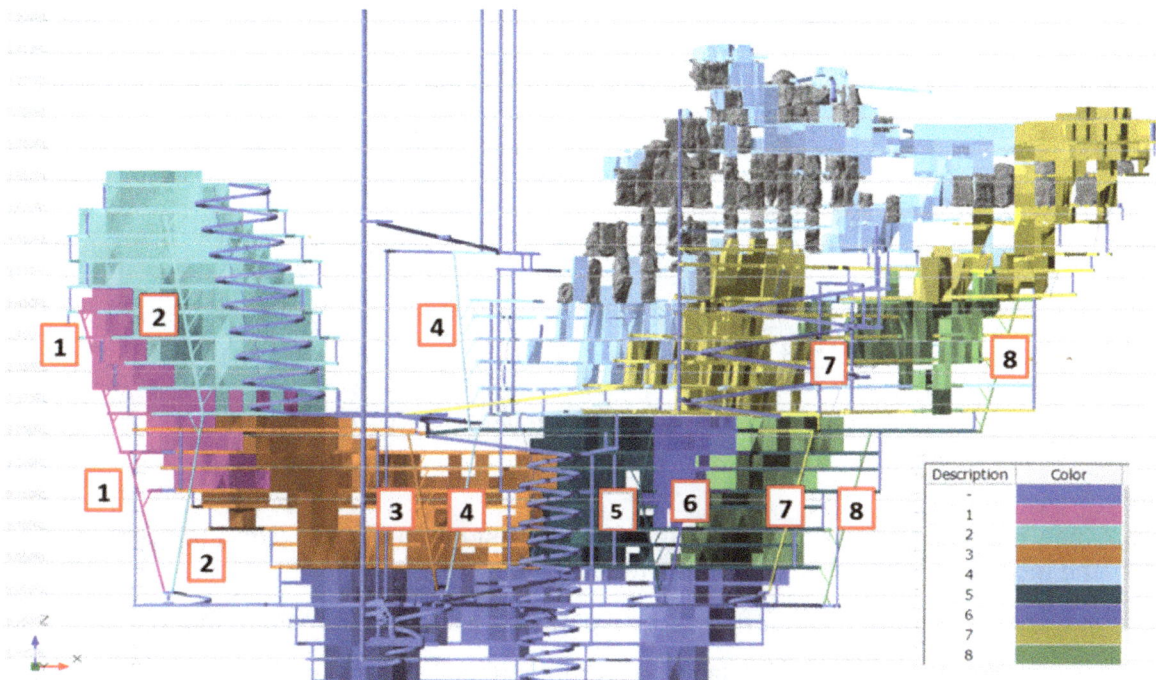

FIG 4 – PHOX orepass layout (OZ Minerals, 2021b).

Subsequent geotechnical work after the completion of the feasibility study was undertaken and the orepasses were moved closer to the orebody and a single orepass was design to feed a single truck chute, as shown in Figure 5, because there were concerns regarding the tramming distances to the orepass on the production levels were considerable and the potential for preferential feed into the single ore bin from two orepasses. Addition to the revised geotechnical consideration collaboration with the operations team identified some key functional requirements to be incorporated into the design including:

- the ability to monitor orepass level
- ability to monitor and manage orepass and finger pass overbreak
- allow parallel tipping into a single orepass
- maximise the design life
- minimise air blast risk
- minimise truck or loader transfers
- tipples require grizzlies
- can deal with orepass blockages
- can cope with 5 per cent paste dilution
- honour the geotechnical stand-off and exclusion zones
- adequate connection to the return air system.

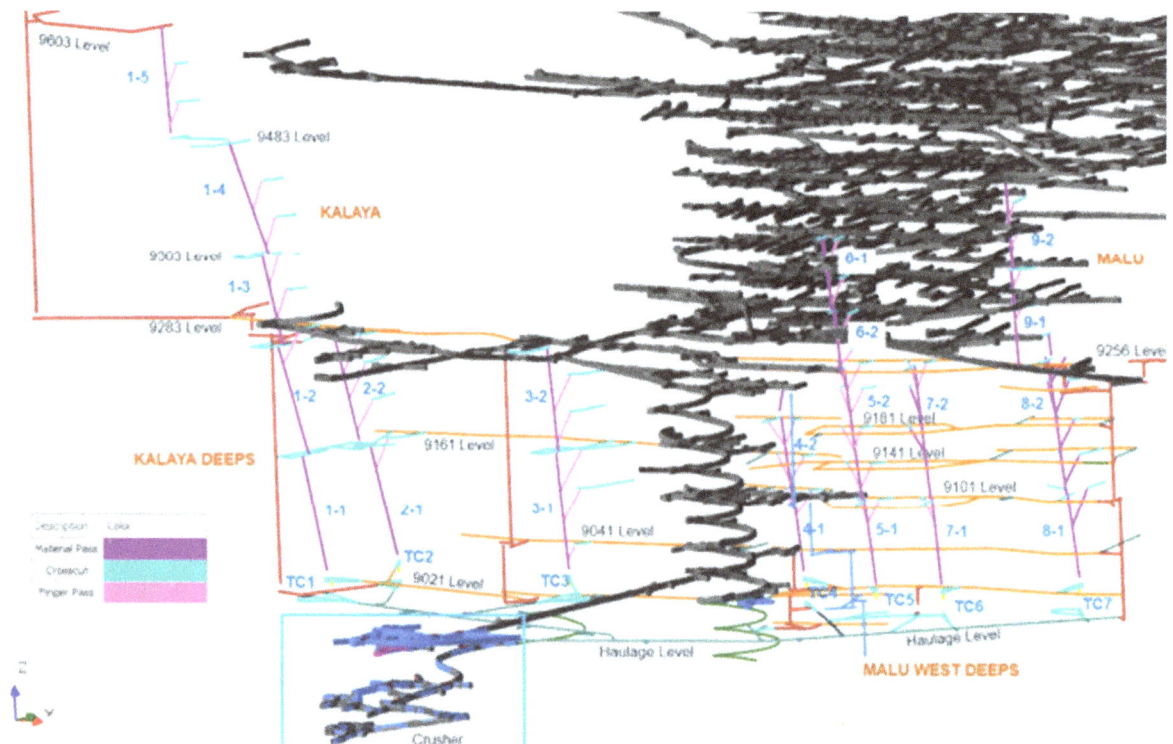

FIG 5 – Update orepass layout post-PHOX.

Haulage level

The haulage level was re-designed post the feasibility study to aligned with the updated orepass design and included the following functional requirements:

- Have adequate ventilation for heat, diesel particulate and radon management.
- Good drainage to provide adequate road maintenance requirements.

- Accommodate daily truck servicing.

- Deliver a 10 per cent higher instantaneous production rate than the annual production rate.

- Consider the potential for future automation.

- Potential for parallel construction/operation activities.

- Accommodate the ability to have trucking operating for 24 hrs on the level, ie no downtime for shift change/firings.

Crusher

As part of the PHOX study trade-off studies were undertaken to determine type of primary crusher to be installed, the options considered were:

- gyratory crusher

- jaw-gyratory crusher

- jaw crusher

- eccentric roll crusher

- roll sizer.

Some of the selection considerations when selecting the suitable crusher included:

- feed to product reduction ratio

- product particle size distribution (PSD)

- maximum feed lump size second dimension

- quantity and type of tramp materials that are likely to be feed into the crusher

- required throughput

- geotechnical considerations for the size of the underground excavations required.

The outcome of the trade-off study was to proceed with a gyratory crusher based on the high percentage of fines experienced at the operation in the surface crusher and the modification that could be made to the gyratory crusher to mitigate the risk of bogging the crusher.

The design of the crusher station includes (Figure 6):

- one 54 × 67 primary gyratory crusher

- a fixed hydraulic rock breaker

- two truck tipple approaches

- lined crusher dump pocket with spider access

- crusher workshop and laydown provisions.

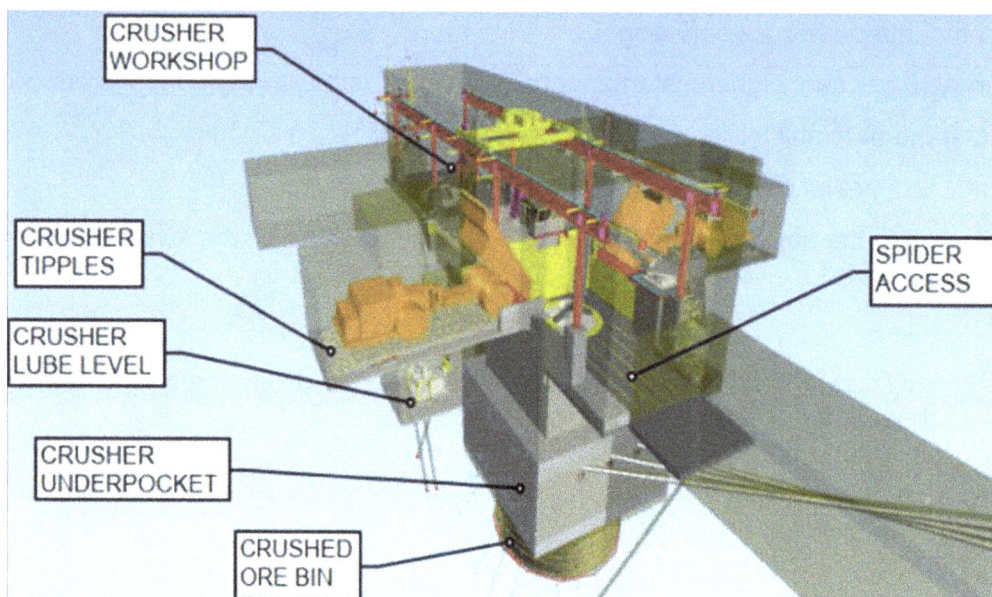

FIG 6 – Crushing Station Arrangement (AECOM, 2021).

Shaft

The shaft has been named 'the Wira Shaft' in recognition of our traditional owner, the Antakirinja Matu-Yankunytjatjara people and their cultural history is important to Prominent Hill. The shaft was named by representatives of the Antakirinja Matu-Yankunytjatjara people. Wira (pronounced wi-rah) is small bowl traditionally used by the indigenous women as bowl and a digging tool.

The location of the shaft was a compromise between the available area on the surface, stand-off from the orebody, the resource centroid, estimated ground conditions and the proximity to the site services and processing plant.

The construction method of the shaft is strip and line with the initial 93 m pre-sink conducted using the Herrenknecht Vertical Shaft Sinking Machine as described by Kilkenny in 2023.

The shaft is 1318 m deep and excavated to an 8.5 m diameter with a final diameter of 7.5 m. An initial 3.5 m raise bore is mined for each leg of the shaft to be stripped into.

The shaft has been designed including the criteria shown in Table 1.

TABLE 1

Shaft performance criteria.

Parameter	Value
Design throughput (wet)	1010 t/h
Maximum hoisting speed	18 m/s
Shaft guides	Rope guided
Winder type	Koepe (friction) hoist
Shaft ventilation velocity	m/s
Skyshaft orebin capacity	>5 skips
Skip payload	36 t
Skip design	Bottom dumping

Ventilation

The expanded mine design required the upgrade of the ventilation system to increase the ventilation capacity and install refrigeration to the lower portions of the mine (Figure 7). The ventilation upgrade required the following to be installed:

- New primary fan installation – Installed underground in adits connecting to the open pit.
- Two twin 5 m diameter ventilation raises for exhaust and one 5 m diameter fresh air raise from the 9993 mRL to 9536 mRL and another leg for each raise to 8981 mRL.
- Two bulk air coolers – one on the Wira shaft, the other on an existing raise.
- Approximately 1000 m of lateral development for the Ankata link.

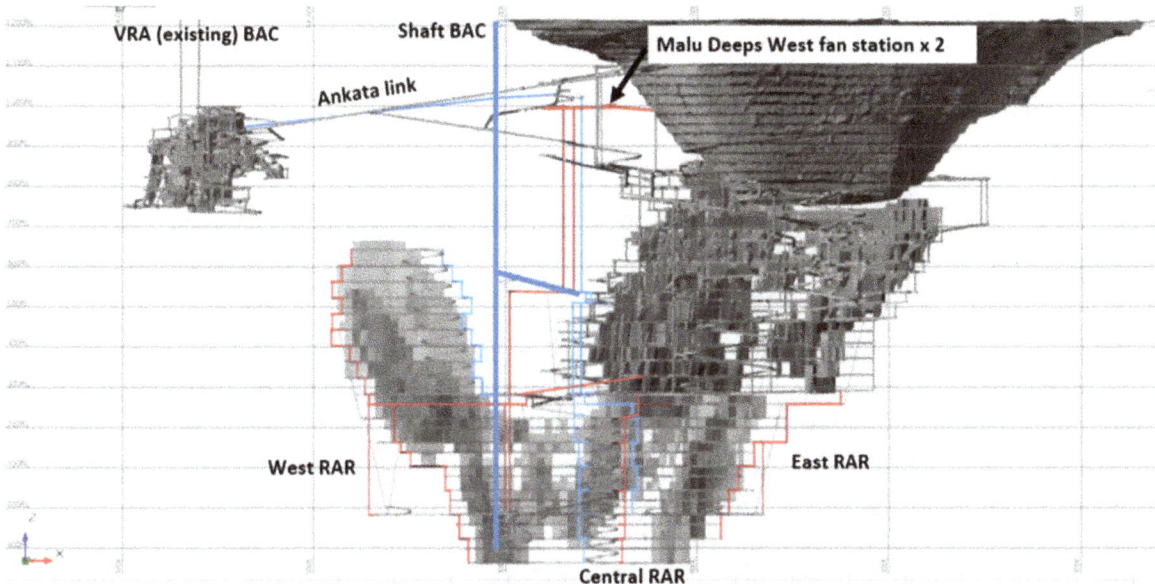

FIG 7 – Ventilation overview (OZ Minerals, 2021b).

INTERACTION WITH CURRENT OPERATION

An early decision was made prior to the OZ Minerals board approving the shaft expansion to extend the decline towards the bottom of the shaft with the development being on critical path for the project. The footwall decline was also extended in the upper section of the mine to create a dual decline system to reduce congestion in the section of the mine where the development waste and production ore interacted with each other.

The additional lateral development requirements increased the development targets by 25 per cent in year one and another 25 per cent in year two to reach the ambitious timelines of the feasibility study. Concurrently, production was planned to increase by 10 per cent each year over the same period.

While the project is still underway, the operation has found that interactions between the work associated with the project and maintaining the mines production challenging at times particularly with the setting up of the ventilation system and the interactions with the raise bore rigs and when the project work is located in the same areas as production.

The operation is now moving into a new phase where the civil and mechanical construction work for the crusher and haulage level will commence requiring a large quantity of material to transported through the mine and a higher demand on the concrete supplies which will require good communication between the operation and project teams to ensure minimal impact on both plans.

RAMPING UP TO 6.5 MT/A

The ramp-up to 6.5 Mt/a per annum requires the mine to increase the rate of all the underground mining activities prior to the shaft being commissioned. The feasibility study was based on approximately 30 per cent Inferred resource (OZ Minerals, 2021a) and therefore diamond drilling is ongoing to ensure that the highest confidence is in the resource estimation is available prior to mining. This has required additional diamond drill locations to be developed and where possible these locations are part of the future development requirements and serve a double purpose.

Ongoing reviews have been undertaken of the mine stoping sequences to optimise the development and backfill requirements. To achieve the higher production rate Prominent Hill needs to increase the number of stopes being mined by 30 per cent to over 100 stopes per annum.

In the background a significant amount of work has and is being undertaken to prepare for the shaft future with procedures being review, management plan being put in place, assessments of workforce requirements, reviewing and changing flights schedules to accommodate the different phases of the project and updating the training requirements. The logistics team have an important role to play as the amount of equipment and components are brought into the operation for the construction of the shaft, crusher and load station as well as the changes that will occur once the shaft is online with changes in maintenance spares required. The maintenance engineering team will be required to implement difference maintenance plans with the new fixed equipment being commissioned.

CONCLUSIONS

The PHOX project is ambitious and challenging particularly in the initial staging with the major ventilation upgrades required to mine to the bottom of the shaft. Communication between the teams involved from the project team, operations team, the mining contractor and other contractors has been key to progressing the project. The communication is required to ensure that every team knows their roles and responsibilities in the project and when they need to lean in and help. As with all projects there are times when the plan gets off track, but good communication and collaboration has allowed the problems to be solved and the project to continue.

Changing the way an operation works is not just about the changes in the physical aspects of the operation but requires a significant amount of work in the background that is not always acknowledged. The preparation to be ready for Prominent Hill shaft future requires a lot of planning and work being undertaken beyond just the physical construction work with the development of management plans, workforce planning including flights and accommodation requirements, updating of procedures, change to logistics and office spaces and developing new maintenance plans.

ACKNOWLEDGEMENTS

This paper summarises the extensive work that has been completed by a large team over the last four years and the author would like to acknowledge the contribution that each and every person has made along the way. The author would also like to thank BHP for the support in publishing the paper.

REFERENCES

AECOM, 2021. Feasibility Report, unpublished.

Freeman, H and Tomkinson, M, 2010. Geological setting of iron oxide related mineralisation in the Southern Mount Woods Domain, South Australia, in *Hydrothermal Iron Oxide Copper-Gold and Related Deposits: A Global Perspective* (ed: T M Porter), vol 3, pp 171–190 (PGC Publishing: Adelaide).

OZ Minerals, 2021a. Green Light for Prominent Hill Wira Shaft expansion, ASX announcement.

OZ Minerals, 2021b. Prominent Hill Expansion Project Study Report, unpublished.

OZ Minerals, 2022. Mineral Resource and Ore Reserves Statement and Explanatory Notes, ASX announcement.

Sandery, L and Light, S, 2014. Managing Diamond drilling interactions in commissioning the Ankata underground mine in *Proceedings Twelfth Underground Operators' Conference 2014,* pp 76–79 (The Australasian Institute of Mining and Metallurgy: Melbourne).

The Site Operations Centre – a Carrapateena story

B Edwards[1], T Webley[2], A Schmidt[3] and E Law[4]

1. Superintendent – Mine Projects, BHP, Adelaide SA 5950. Email: ben.edwards@bhp.com
2. Superintendent – Site Operations Centre, BHP, Adelaide SA 5950.
   Email: troy.webley@bhp.com
3. Senior Technician, Systems – Site Operations Centre, BHP, Adelaide SA 5950.
   Email: amelia.schmidt@bhp.com
4. Senior Engineer, Data – Site Operations Centre, BHP, Adelaide SA 5950.
   Email: eric.law@bhp.com

ABSTRACT

In 2016 Carrapateena began collecting operational data on a PC in a demountable building in the middle of the desert. In 2024 Carrapateena has a world-class Site Operations Centre (SOC), employing 28 personnel and providing full serviceability to the Carrapateena operation, from underground development to concentrate production. The story of Carrapateena's SOC stems from big ideas and then a journey of discovering what a Site Operations Centre is and what it is not.

The SOC at Carrapateena has held many different personas and for it to truly become effective it had to strike the right balance between being a control room, and a service provider. Developing the SOC into a function with the right balance to support the operation requires subtle changes ranging from calling staff members technicians not controllers, accessibility to the room and presence at key stakeholder meetings, through larger modifications such as organisational structure and reporting lines.

Detail is what can make or break a SOC. If starting fresh some key considerations should be location, room layout, functions, resourcing, and well-defined deliverables.

The SOC should be central to the operational areas and departments. A single SOC serves all functions of a site and needs to be easily accessible by all areas.

The room needs to flow in the order of the material, from development, production, materials handling then processing.

Understanding the role and workload of each functional area of the SOC allows it to be set-up for success. This leads to resourcing and must be adequate to maintain operations. A SOC must have 100 per cent technician availability.

Documenting the Carrapateena SOC journey from infancy to maturity will hopefully provide some insight for new mines looking to set one up from the start and those established organisations wanting to take the next step.

INTRODUCTION

Carrapateena

The Carrapateena copper-gold mine is located 472 km north-west of Adelaide, and 160 km north of Port Augusta in South Australia's highly prospective Gawler Craton (see Figure 1). The asset is located on Pernatty Station, and its supporting infrastructure is located within Oakden Hills Station. The Kokatha people are the traditional owners of the land.

First saleable concentrate production for the Carrapateena copper-gold mine in South Australia was achieved in December 2019, and it ramped up to 4.25 Mt/a over the following 12 months. Ore is processed on-site to produce a copper concentrate containing copper and gold minerals.

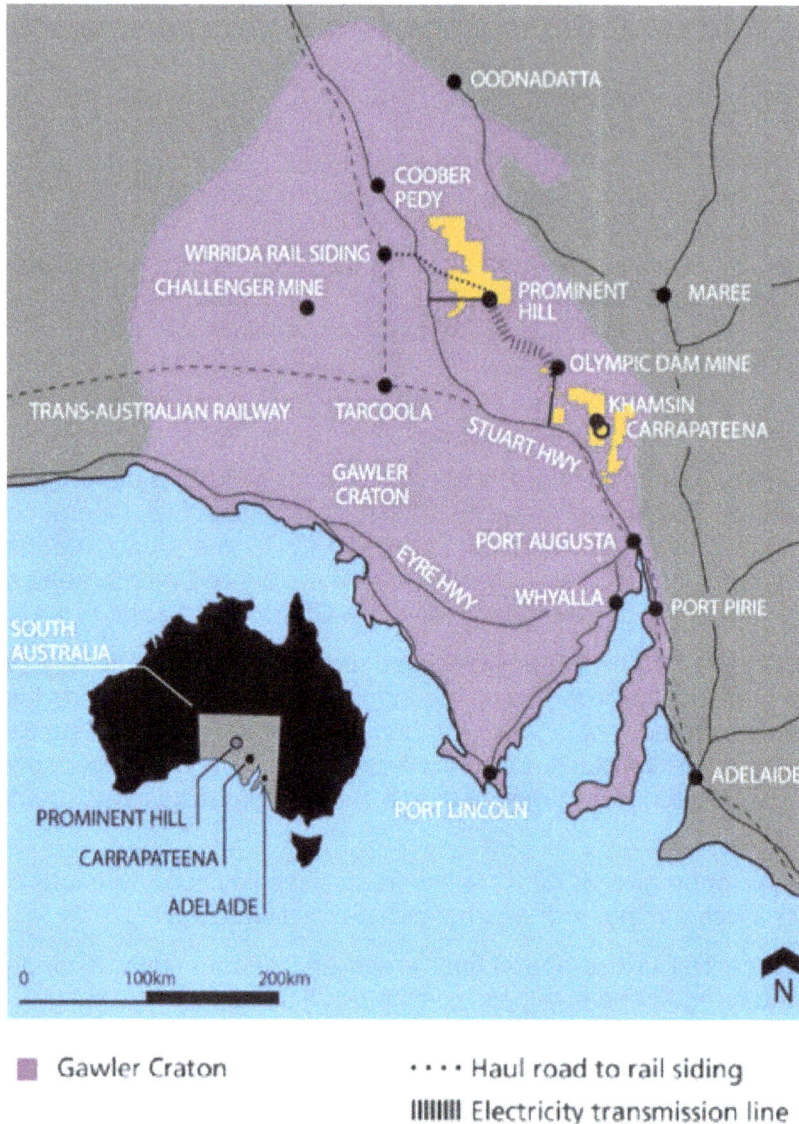

FIG 1 – Carrapateena location.

THE EARLY DAYS

In the early days the Site Operations Centre (SOC) team was located alongside all other teams on a site referred to as The Expo Camp. This was the initial camp built during the exploration days of Carrapateena. The camp accommodation, mess, shared offices and infrastructure were in close proximity to each other. The shared office arrangement brought minor inconveniences to the teams but became a contributing factor to the collaborative and agile culture at Carrapateena and inspired big dreams with the initial design of the permanent SOC.

The first SOC was a 20-foot hire container with a single desk in it (Figure 2). The initial team was five people, four desk operators and a coordinator. During those early days the focus was on face utilisation to achieve rapid development of the two declines. Accurate data capture enabled an increase in productivity as a clear picture was able to be built of where the constraints were and then action improvements to those areas.

The implantation of a digital production management system was welcomed by the small underground contractor teams because it started to do their timesheets for them. The teams just had to stop by the SOC and were able to collect their printed, pre-populated time sheets for their review. This was the beginning of the 'single source of truth' philosophy that is still alive today at Carrapateena. This has a lot of advantages in client/contractor relationships and stops the confusion of having multiple sets of data.

FIG 2 – SOC shared Expo container office.

From the shipping container, the SOC grew into an office space located beside the underground shift supervisor and so began the understanding of the power of good professional relationships between underground frontline leadership and the SOC.

The office was a shared space with the underground supervisors allowing for close conversation and connection between the mining departments and SOC personnel. Relationships built during this time fostered and assisted growth and understanding for the aspirations of the SOC long-term.

The SOC was occupied by a team including one dispatcher for a 12-hour shift, and overall a total team of five dispatchers supported by two Supervisors, a data analyst and a superintendent.

The initial radio system was basic 2-way radio communication for all information, which didn't allow for any recall or replay of recordings of information received via the radio system. This became frustrating when the system was overloaded with critical information required for reporting or escalation. Investigation into a digital radio system that could be hosted within the PC's began.

The office walls were covered with physical print outs of mine level plans and other crucial paperwork, which triggered brainstorming of transforming these items to digital version and online logging procedures (Figure 3).

FIG 3 – Inside the SOC Expo container office.

In the lead up to the commissioning of the processing infrastructure, the SOC expanded to include two additional mill control operators on day shift only. The mill control operators were embedded with the Process Control System (PCS) engineering team and the processing team. They were tasked with contributing towards the development of the mill control philosophy prior to the plant commissioning stage. This participation helped the development of the first operating procedures for mill control in manual mode as control programs were not fully developed at the time of commissioning for various start-up and shutdown sequences.

Non-Processing Infrastructure – construction

The construction of SOC was completed as a part of the Non-Processing Infrastructure (NPI) package. The NPI consists of administrative offices, ablution facilities, muster rooms and the SOC area as shown in Figure 4. The initial layout of the SOC underwent multiple iterations. Weekly meetings were conducted as a part of the Operational Readiness process to coordinate construction, projects, ICT and contractor teams. The initial layout of the SOC changed from the draft plan in Figure 5 to as built in Figures 6 and 7. This later changed to the layout shown in Figure 8 as operational requirements changed with the SOC for Crusher 02 (CS02) in mining operations.

FIG 4 – NPI draft plans.

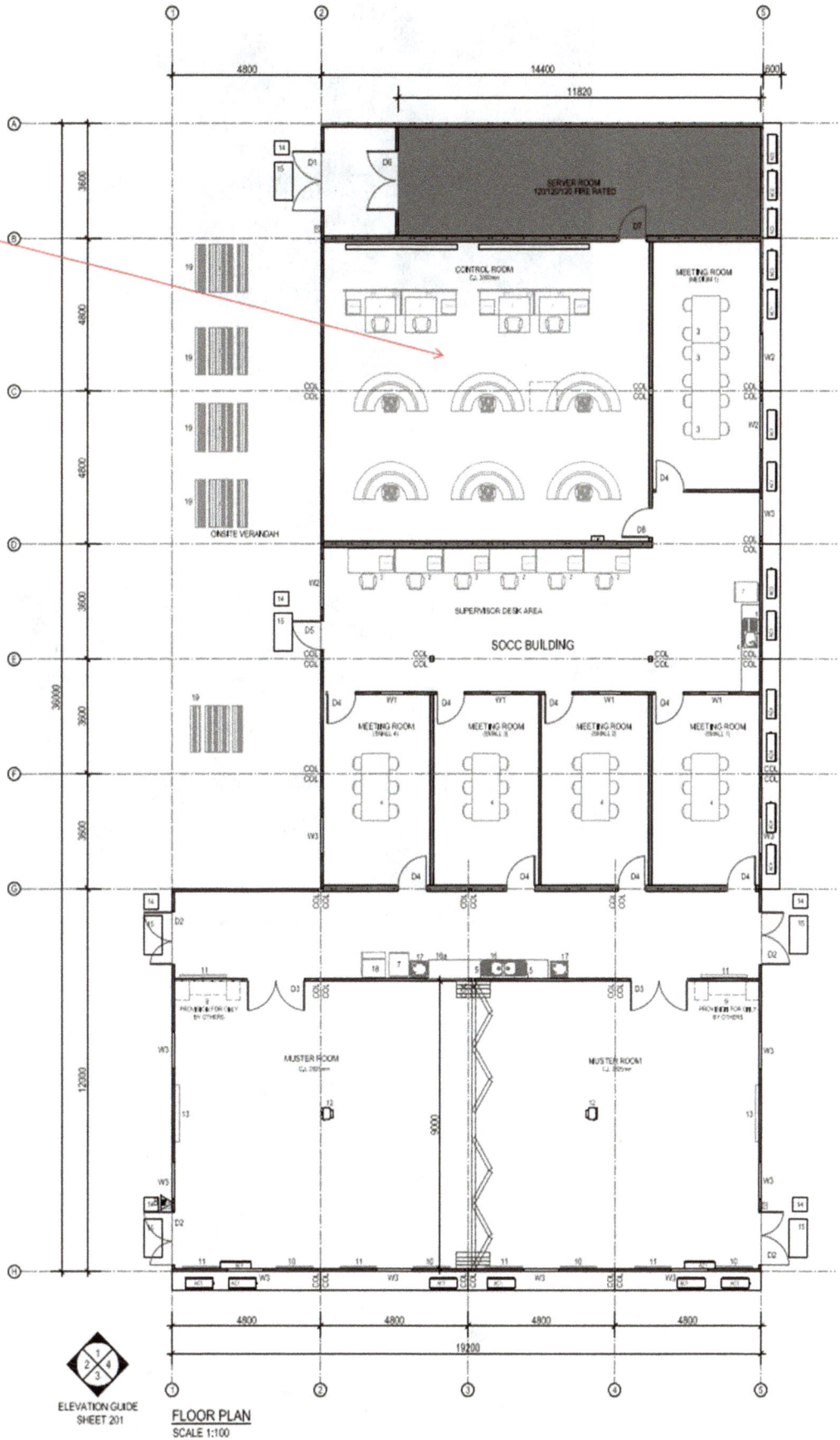

FIG 5 – SOC, IMT, muster area and supervisor office draft plans.

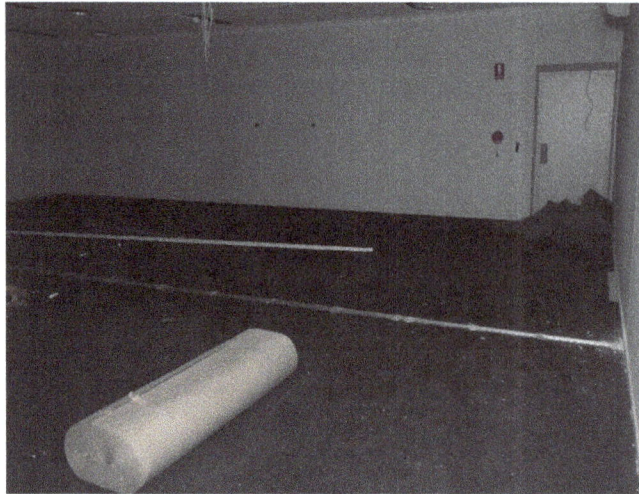

FIG 6 – Initial build of SOC NPI office.

FIG 7 – Installation of SOC NPI offices with updated layout.

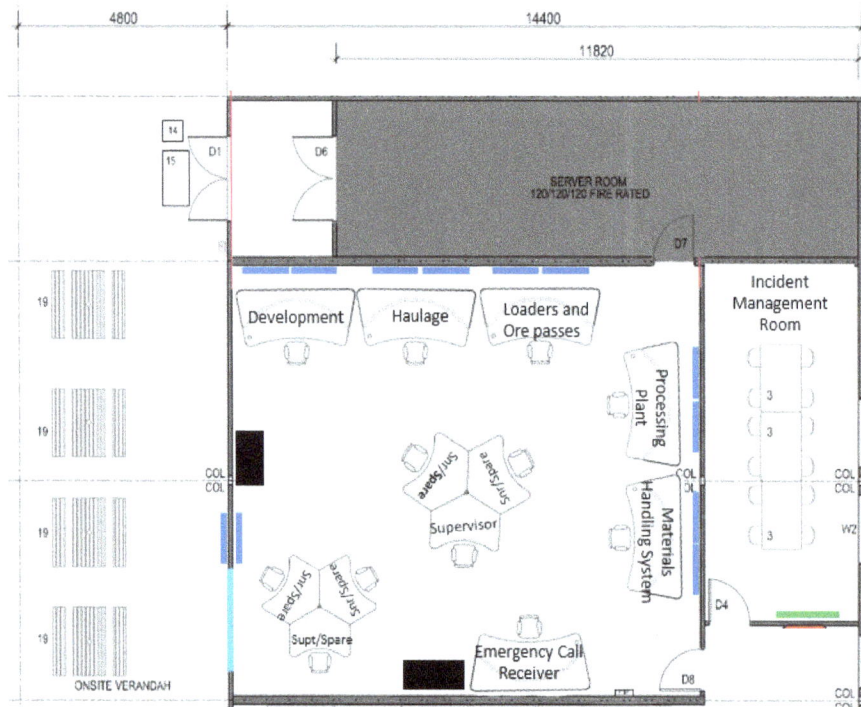

FIG 8 – Current layout of SOC NPI office post-CS02 commissioning.

THE FACILITY AND ITS LOCATION

The Carrapateena SOC covers approximately 210 m² of floor space. Of which, 100 m² is the primary operations centre, 52 m² is a server room, 32 m² for the Incident Management Room and approximately 26 m² of entrance/kitchenette space. This layout of the facility is important and its original design, for the most part, considered the functionality of the facility and its interactions with key stakeholders well. For example, the location of the sites primary data centre right next to the SOC, enables easier access to a complete UPS system, shorter data cable runs and allows for easier upgrades as the site changes and the SOC needs to adjust. Likewise, the Incident Management Room, directly adjacent to the SOC is essential in incident response, as the SOC are the first call receivers and manage the incident before handing over to the Incident Management Team once they are formed. Due to the criticality and timeliness of the work the SOC technicians do, having a dedicated kitchen for the SOC is an imperative.

Some additional considerations for a new facility would be to build with the future in mind. Pre-installing additional cables through the firewall between the SOC and the data centre for data and power enable simpler upgrades and expansion of the room. A dedicated window between the SOC room and the Incident Management Room would aid in improved direct communications between the SOC and the Incident Management Team during an event. Natural light via windows and or skylights is critical and should be accompanied by blinds to manage glare. Further considerations of glare on screens would be fully manageable lighting. Depending on where desks are located, the lights above them can cause glare on the screens and having the ability to turn off or dim each light in the room would be a large advantage. Floor coverings can be a controversial topic if not managed well. Having carpet in the room makes for a nicer environment and aids in sound absorption however if not managed well via use of boot coverings can easily become filthy during winter on a mine site. Sound absorbing material clad to the ceilings would also be useful as a SOC can become a loud environment in certain circumstances. This is often easily managed by a supervisor kindly asking the noisy party to leave the room.

The Carrapateena SOC had its first and only window installed in 2024. This window serves two purposes.

1. The window allows for natural light to come into the room. This is important as natural light assist in improved productivity and focus by enhancing alertness and cognitive function. It is also known to enhance mood and well-being.

2. The window is not only to allow SOC technicians to see out, but it also importantly allows others at the mine site to see in. The SOC should be a centre piece of a mine site's non-processing infrastructure and office layout and should be visible and welcoming to others at the mine. Before the window went in at Carrapateena, many employees or contractors didn't realise there was a SOC hidden down the back corridors of the main office area.

The SOC should be easily accessible to the teams it services, specifically, mining operations, maintenance and reliability (fixed plant) and processing. The Carrapateena SOC is under the same roof as the underground mining muster room, shift supervisors and mining superintendents. This serves a valuable purpose as the SOC needs to work closely with these parties to achieve results. Making it easy for a shift supervisor or mining superintendent to pop their head in or talk face to face with a SOC technician and get direct access to live data and updates is extremely powerful. It also facilitates and fosters relationship building. As the mining team and supervisors spend time in the room, they build professional relationships with the technicians which facilitates enhanced communication and a level of trust between technician and supervisor. This trust can lead to the technician taking on more responsibility in the coordination of shift activities, on the supervisor's behalf creating a powerful working relationship that will ultimately lead to increased safety, productivity, and efficiency of the mining activities. This is true for all SOC functions but is generally a different case for the processing technician who often has worked their way up through the ranks and is already at or near a processing supervisor level.

THE SOC STRUCTURE

The Carrapateena SOC was run very lean on resourcing for a long time. This created some not so unexpected issues, such as a disgruntled workforce. Some of the primary issues stemmed from not having time for breaks due to radio calls coming in non-stop. Technicians needed to take their radios to the bathroom or be at risk of missing a critical call or instruction. Lunch breaks were difficult to manage, and managing personal or annual leave created a time-in-lieu balance across the department that was outside the business policies.

Fortunately, appropriately resourcing a SOC that operates 24/7, 365 is an exercise in logic. In other, non-operations centre staffing arrangements, if a person is absent, tasks can often wait until they return. This is not the case for the SOC technicians that staff a desk function. If the processing technician is absent, the Mill doesn't run. If the Material Handling System (MHS) technician is absent, the crusher and MHS cannot run. For the mining side it is not quite as dire, however critical operational data will be missed and functions like controlled area access and ventilation on demand cannot be facilitated. Essentially, the operation can't afford for a technician to be absent. As such, additional staff must be carried to account for leave, breaks and training/upskilling. A simple waterfall chart in Figure 9 can help demonstrate this.

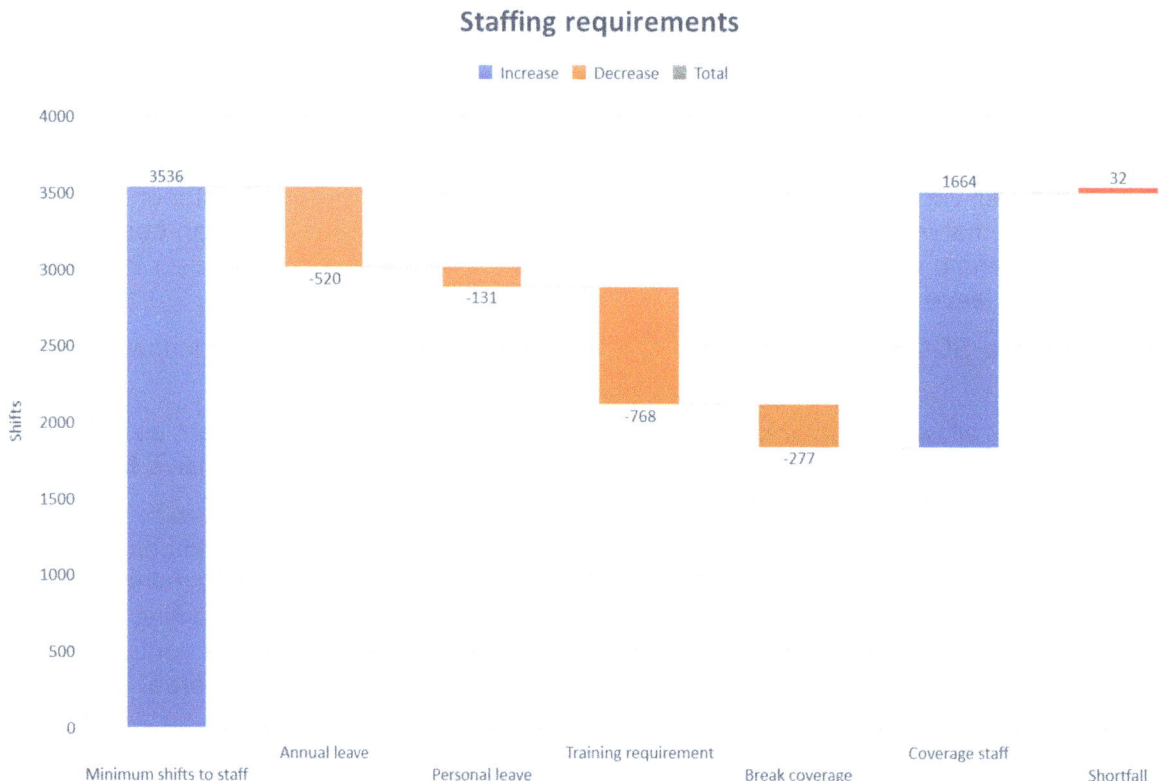

FIG 9 – SOC staffing requirements.

An alternative to carrying additional staff could be having trained personnel in other roles of the business to step in if required. In practice this isn't feasible. The SOC is ever changing and maintaining competencies in a 'reserve' workforce would be extremely difficult to manage. There are also contractual issues, as most staff roles are not night shift and as such requesting a day shift role to cover night shift could be create issues.

The graph in Figure 9 is based on calculating the minimum number of annualised shifts, in this case one supervisor and four technician functions, then deducting non-working time for leave, training and breaks.

In addition to 'frontline' technicians and just as critical to a successful SOC are the supervisors and support roles. The Senior Technician – Systems role supports all the underlying technology systems, databases and data quality that underpin a reliable operations centre. The quality of the data and reporting that comes out of the SOC can make or ruin the departments credibility and hamper the value harnessed in the operational data insights. This is why having a dedicated in-house data team reporting to the SOC Superintendent is immensely valuable. The data team not only facilitates reporting and data visualisation via dashboards for the SOC, but they are also available to the broader operational teams to assist with complex operations that often require external support or labour-intensive manual reporting and data collection.

The step change in resourcing from an overly lean, under pressure team to an adequately resourced team, coincided with the commissioning of the second underground crusher at Carrapateena. This step change for the operation, also required a step change in the SOC. An increase in production rate from 4.25 Mtpa to 6.0 Mtpa, the operation of three times as much material handling infrastructure, an extensive orepass system, and some additional LOM mine services infrastructure required a restructure and additional resourcing. Utilising the waterfall chart to justify the headcount changed the team structure from 25 (Figure 10) to 33 (Figure 11).

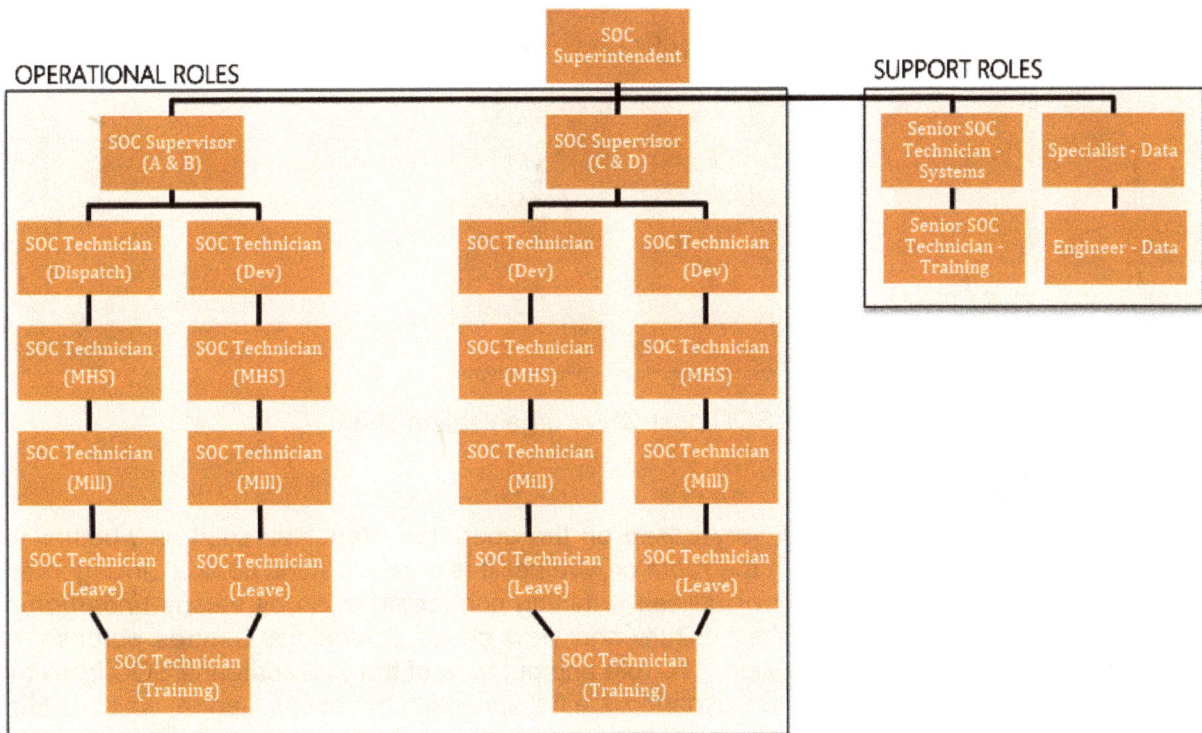

FIG 10 – SOC pre-CS02 organisation chart.

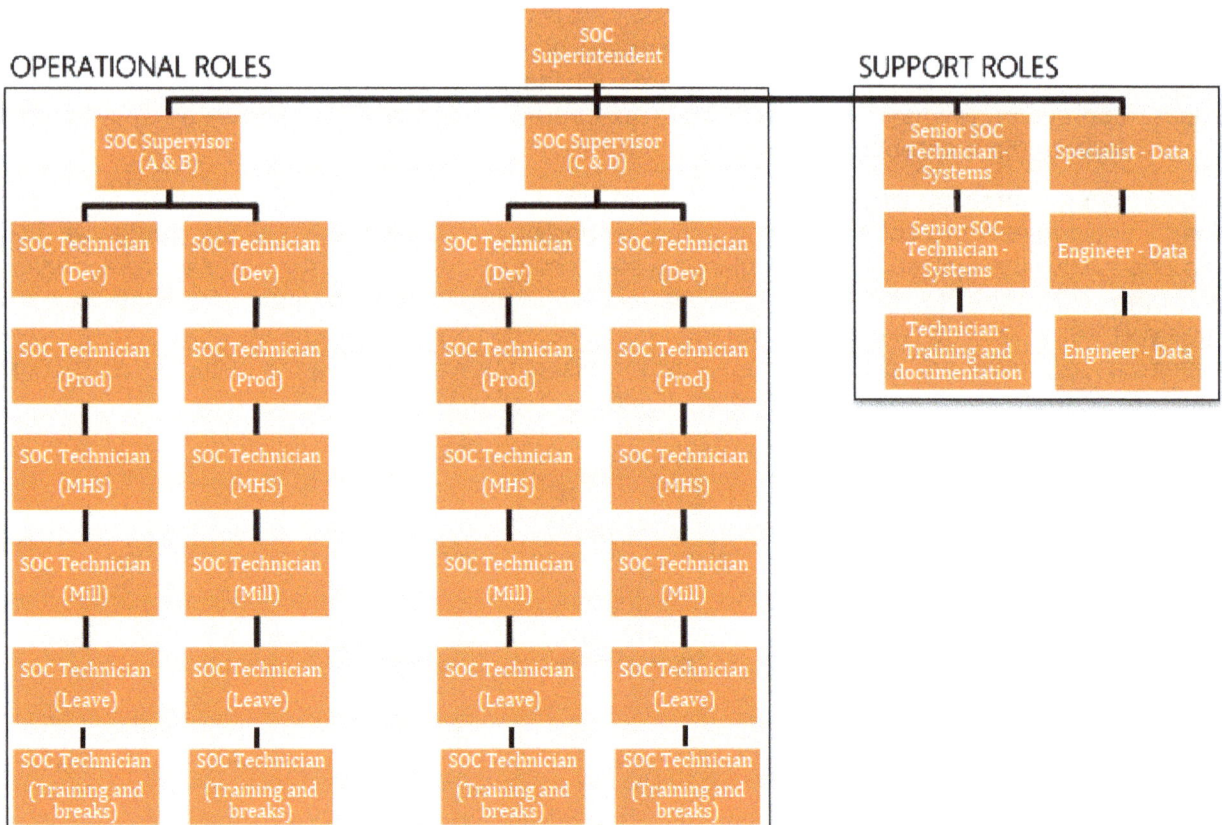

FIG 11 – SOC post-CS02 organisation chart.

THE SOC'S ROLE

The SOC's role in an operation will vary based on the operation. Variables such as whether the processing team has its technician in the room or whether it is purely a 'mine control' arrangement, whether the mine is owner operator or utilises a mining contractor, open pit versus underground even just specific companies and their culture and view on an Operations Centres purpose will impact the role it plays for your operation. The role and purpose of the Carrapateena SOC is always evolving and adjusting. Carrapateena's dynamic is a full services Operations Centre, which is Mine to Mill, servicing a mine utilising a mining contractor and a fixed plant and processing team that is company based. How the SOC interacts with each area will differ. The processing side of the world, as previously mentioned, generally employs a technician that is highly skilled and experienced in processing. Their role is truly 'control'. They are operating the Mill along with monitoring of all assets in their field. The underground roles vary from assistance and data collection to short interval control of assets in particular areas of the mine.

It's worth considering the detail when designing a SOC – even what you call each role. Often SOC's are referred to as 'Mine Control', this can sometimes be a difficult concept for a hardened shift supervisor or superintendent that believes that they control the mine. At Carrapateena, we are a site operations centre and the people that work it are technicians. In 2023 as an effort to improve the working relationships between the SOC and the underground supervisors and superintendents, Carrapateena focused on its use of language when working with the mining team, removing words like 'controllers' and instead focusing on their correct title of technician. It played a minor role as part of a broader strategy but was surprisingly effective. Ultimately, it was about refocusing on what Carrapateena's SOC's role is, and that is one of a service provider. The SOC is the central nervous system of the operation. If you want to know something, call SOC, if you have a problem or an emergency, call SOC. The SOC is there to help and facilitate the works happening in the operation ultimately making it safer, more efficient and more productive.

THE SOC'S FUTURE – APPLICATIONS OF ARTIFICIAL INTELLIGENCE

The recent adaptation of Large Language Models (LLM) such as ChatGPT and Claude for commercial use has captured the imagination of many. This opens the discussion – will Artificial Intelligence (AI) one day replace the role of control room technicians in the mining industry and how can we utilise AI in environments such as the SOC. In this context, it is important to define what counts as AI and what defines the role of a control room technician.

In the most advanced context, the total replacement of staff in a Site Operations Centre environment would require an Artificial General Intelligence (AGI) with highly specialised knowledge of mining and milling processes that can cooperate with various operations teams on-site. In addition, for a total replacement of our current Human Machine Interfaces (HMI), the AGI would also need to be able to overcome the uncanny valley effect (Rosenthal-von der Pütten et al, 2019) to allow such a solution to interact with other teams – likely still staffed by humans to make safe and efficient decisions on behalf of the operation.

Since the beginning of operations, the types of AI utilised to date at Carrapateena SOC has been limited to Reactive Machines and Limited Memory Machines based on the classification proposed by (Hintze, 2016). Table 1 outlines the list of applications currently used at Carrapateena.

TABLE 1

AI applications and classification in SOC.

Application name	Vendor	Purpose	General classification
Avigilon Control Centre	Avigilon	Security camera with limited facial and activity recognition with tracking	Reactive machines
Driver Safety System	Caterpillar	Driver fatigue monitoring and notification process	Limited memory
KAESM	Predict	Predictive reliability of processing infrastructure	Limited memory
ORB	Polymathian	Short Interval Control Optimisation	Limited memory
RockSense	Metso	On stream particle size analyser	Reactive machines

Hintze (2016) proposed that the key improvement of Limited Memory machines from Reactive Machines is the ability to learn from the past in some form of machine learning. Many off the shelf commercial AI solutions used in mining today are edge solutions that do not have automatic processes that updates or self-improves with historical data. For those that do, their purposes are limited into a very narrow field rather than providing general command, control and dispatch capabilities.

Researchers have been mapping the raising costs of developing and training AI models over the years. They have found that only a few large organisations can keep up with the significant costs of energy and hardware challenges in developing AGI or even Artificial Narrow Intelligence (ANI) (Cottier et al, 2024). It is safe to say that the mining industry is unlikely to develop its own (ANI) without collaboration from established organisations.

Another challenge of replacing technician roles with some form of ANI is the need for continuous adaptation of changing operational procedures and environments. Many mining operations including Carrapateena have changing operational environments, procedures and work instructions on a weekly to monthly basis. Some changes, such as the planned transition of Sub-Level Caving to Block Caving operations at Carrapateena would present significant challenges for an ANI to adapt to. Technicians can understand these changes and adapt to them in the context of other changes whilst ANI would require significant time and cost input from the developer organisation to adapt. It is likely that in the event a SOC operating environment is run by ANI agents, it would require a team of artificial intelligence specialists who would be dedicated to modifying and retraining ANI models to adapt. The economic costs of ANI deployment over technician staff in a SOC environment might be a topic of study for future operations to consider.

Another significant consideration of ANI deployment in SOC is the time required to train and develop ANI and where the machine learning process can source its training data set from. Unlike the numerous pretrained models today based on computer vision or natural language processing, an ANI model for SOC operations would need to rely on a combination of existing models and also train from a specific set of training data. This set of training data would likely be created by technicians in a simulated operational environment. The data must be meaningful enough such that it can be used to adjust the weights of whatever neural network models are applied for training.

The most significant barrier to the total replacement of technicians with ANI in a SOC environment would be for concerns of safety. Implementation of ANI with direct control over mining processes from fixed plant, to dispatch and coordination of mining operations as well as emergency response would likely require significant guardrails on its boundary outputs. This in-turn may require another type of technician training that constantly monitors the decision output of the ANI for approval and execution. Typical thought experiments for a SOC AI could include, how would an ANI agent react to a stench gas release? How would the agent react if part of the ventilation system did not respond, how would it react again if it received a radio call that the gas release was by mistake? Questions of AI safety in a SOC AI context could also involve conflicting input parameters such as a Surface Lightning Trigger Action Response Plan (TARP) that directs vehicles to take refuge underground and an Underground Fire TARP that directs some vehicles and people to the surface. How will the ANI react if both TARPS were triggered at the same time and gave conflicting objectives? Would or should a mining operation utilising a SOC ANI agent be insurable without a technician monitoring its outputs?

In summary the advances in AI have significantly reduced the cost of developing Limited Memory Machines and its off the shelf applications are critical to modern SOC operations. However, there are still many difficulties to overcome before a SOC ANI agent can be adopted and accepted by the industry, regulations, and society. Further research in the current applications of AI in SOC environments is needed to direct future applications of Limited Memory Machines or ANI in the industry.

ACKNOWLEDGEMENTS

Jonathan Rossiter, Mat Poyzer and Mitchell Leathem, for their significant contribution to the Carrapateena SOC story and for their photography.

REFERENCES

Cottier, B, Rahman, R, Fattorini, L, Maslej, N and Owen, D, 2024. How Much Does It Cost to Train Frontier AI Models?, *Epoch AI* [online]. Available from: <https://epochai.org/blog/how-much-does-it-cost-to-train-frontier-ai-models> [Accessed: 20 September 2024].

Hintze, A, 2016. Understanding the Four Types of Artificial Intelligence, *Government Technology* [online]. Available from: <https://www.govtech.com/computing/understanding-the-four-types-of-artificial-intelligence.html> [Accessed: 20 September 2024].

Rosenthal-von der Pütten, A M, Krämer, N C, Maderwald, S, Brand, M and Grabenhorst, F, 2019. Neural Mechanisms for Accepting and Rejecting Artificial Social Partners in the Uncanny Valley, *Journal of Neuroscience*, 39(33):6555–6570.

Establishing a strategic mine plan for a multi-centuries-old mine – challenges, implementations and optimisation process

B Fresia[1] and G Calzada[2]

1. Regional Lead of Technical Services, Kazzinc LLP., Ust-Kamenogorsk, Kazakhstan.
   Email: bastien.fresia@glencore.ca
2. Senior Mine Planning Engineer, Kazzinc LLP., Ust-Kamenogorsk, Kazakhstan.
   Email: giancarlo.calzada@glencore.ca

ABSTRACT

The Ridder-Sokolny Mine, located in Eastern Kazakhstan, is a historic site exploiting a Volcanogenic Massive Sulfide (VMS)-type deposit. The mine features a mostly exhausted polymetallic orebody underlain by a gold-copper (Au-Cu) stockwork of discontinuous narrow-veins. Over the past decade, the mine transitioned from manual operations to full mechanisation and advanced geological modelling, producing 1.8 million tons per annum (Mt/a) across 18 geological zones using Sublevel Stoping method in narrow veins.

Developing an integrated Strategic Mine Plan at Ridder-Sokolny has proven difficult due to its complex geological environment, extensive historical legacy areas, and operational constraints. This paper explores how the group addressed these challenges, focusing on the integration of short-, mid-, and long-term geological and mine planning information and working toward mine plan optimisation through scenario generation and risk rating.

The scenario generation allows the mine to simulate key production factors by varying equipment allocation and cut-off grades, balancing fleet requirements and production rates. Risk rating aims to de-risk the mine plan and enhance planning compliance by prioritising production areas based on uncertainties and operational constraints. These methods aim to stabilise the mine plan, balancing production rates, development metres, and operational costs.

The study reveals that production could increase to 2 Mt/a. within four years while managing geological and operational uncertainties. Strategic exploration and infrastructure development offer potential for further production increases. The investigation highlights the importance of addressing geological modelling stability and operational constraints through interdepartmental collaboration and iterative mine plan simulations.

INTRODUCTION

Background of Ridder-Sokolny Mine

The Ridder-Sokolny Mine, located in Eastern Kazakhstan as illustrated in the Figure 1, has been a cornerstone of the region's industrial and economic development since its establishment in 1786 (Kazzinc LLP, 2020). It quickly became a major hub for extracting lead, zinc, copper, and precious metals, significantly contributing to Kazakhstan's mining sector. Under Soviet control, the mine expanded rapidly to support war efforts and post-war reconstruction, while spurring infrastructure growth and transforming the town of Ridder into an industrial centre.

Today, the mine remains a key asset in Kazakhstan's mining industry, driving economic growth and providing employment, while fostering regional development and technological innovation. Despite its rich legacy, the mine now faces significant operational challenges.

For decades, extraction focused on the upper massive sulfide layers rich in lead, zinc, copper, and silver. However, as these orebodies were depleted over 200+ years of mining, the focus shifted to deeper horizons, targeting the stringer zone beneath the main ore lens. This zone consists of structurally controlled, discontinuous copper- and gold-bearing veins, primarily mineralised with chalcopyrite, with gold enrichment. The stratigraphic layering at Ridder-Sokolny is well-preserved, and the transition from mining the polymetallic massive sulfides to focusing on the Cu-Au stringer zone reflects typical VMS zonation, with copper and gold deposits dominating at greater depths due to high-temperature hydrothermal processes (Glencore Xstrata, 2013).

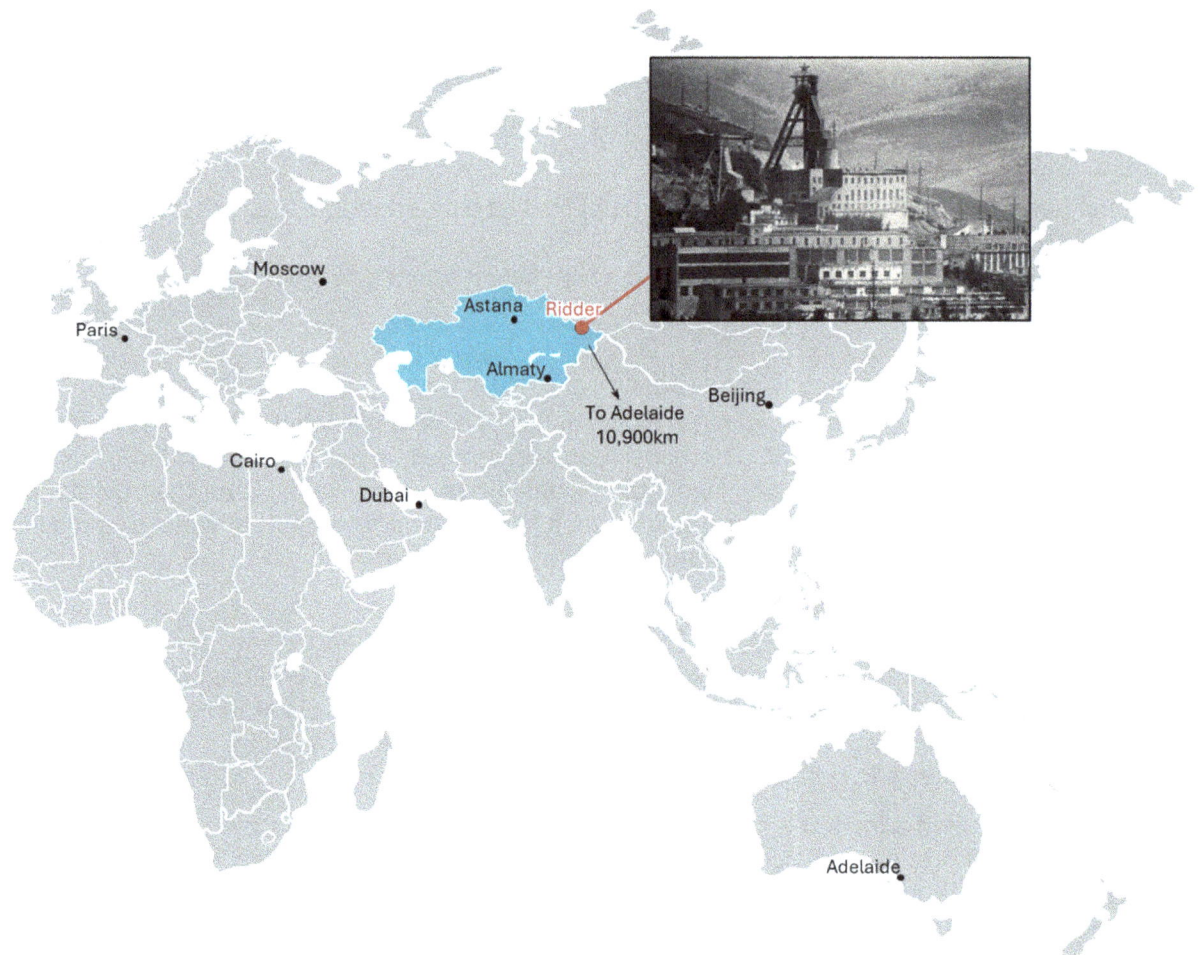

FIG 1 – Kazakhstan and Ridder's town location in Eurasian continent.

The introduction of mechanisation significantly transformed the operation, with large-scale equipment such as jumbo drills, load-haul-dump (LHD) machines, and mechanised haulage systems improving safety and reducing human exposure to hazardous environments. By the early 2020s, handheld equipment was entirely phased out. Ventilation systems were upgraded to support diesel-powered machinery, reducing dust and gas exposure, according to the Mining Plan for the Ridder-Sokolny Deposit (KazMinTech Engineering, 2021).

However, this shift brought new challenges, including the need for higher capital investments, maintenance of complex machinery, and a requirement of a skilled workforce to operate and repair the equipment. It also necessitated more robust mine infrastructure, with larger drifts and ramps to accommodate the machinery, which led to increased development costs and ore dilution.

The mine's extensive layout and complex geology complicate efficient production planning. The difficulty in identifying economically viable mining fronts far ahead of development results in a short-term approach, with veins often mined opportunistically after detailed geological mapping. Full mechanisation has added complexity, limiting the ability to extract the scarce high-grade ore distributed across the vast footprint of the mine. Additionally, inflationary pressures in the early 2020s have further squeezed profitability by driving up mining costs, pressing the need to develop a strategy that ensures the long-term sustainability of the operation.

This paper aims to highlight the challenges in implementing strategic mine planning and the strategies adopted to overcome them, as well as outline the process of constructing a strategic plan that enhances planning compliance and maximises profitability, using scenario generation and risk scoring.

GEOLOGICAL AND OPERATIONAL CONTEXT

Historic and geological overview of Ridder-Sokolny Mine

The Leninogorsk mining district (host of the Ridder-Sokolny Mine), located in the Altay Mountains near Ridder town, is a prolific sector in eastern Kazakhstan. It hosts several past and present mines, primarily VMS deposits rich in barium, zinc, copper, lead, gold and silver.

Ridder-Sokolny is a cluster of over 18 ore zones, predominantly veinlet-type deposits. Following the discovery of Ridder Mine in 1786, other nearby deposits, including Kryukovsky (1811), Filipovsky (1817), and Sokolny (1820), were soon identified. After the Russian Revolution, systematic exploration led to the first geological map of the Ridder region in 1925. During Soviet times, major underground lodes such as the South-Western, Bystrushinskaya, and Innokentievskaya were discovered, with deep-level mining starting in the mid-1950s. This era saw extensive research into ore composition, stratigraphy, and mineralogy. By 1964, the Ridder-Sokolny deposit was officially consolidated, and a comprehensive geological report estimated mineral resources for the unified lodes (Kazzinc, nd).

A typical cross-section of the deposit geology is shown in Figure 2. The Ridder-Sokolny deposit lies in the centre of a 20 km-long graben, oriented perpendicular to the regional north-west structural trend. The Silurian to Middle Devonian volcano-sedimentary sequence remains remarkably well-preserved. The deposit is multistage and polygenetic, featuring polymetallic sulfides and gold-rich sulfide-quartz veins in the upper levels. As these upper levels became depleted, mining shifted to deeper Cu-Zn- and Cu-bearing stringer-type veins (Glencore Xstrata, 2013).

To cover for the complexity of accurately modelling narrow veins systems, dense drilling, mapping and litho-structural information are compiled in the resource, exploratory and operation block models. Both manual and implicit modelling are employed for resource estimation, while unconstrained, purely statistical modelling is used for grades and volumes estimation where data is limited. Balancing these methods allows Ridder-Sokolny to incorporate both high-certainty areas and less-explored zones into its strategic planning, although care is taken to mitigate the risks associated with unconstrained modelling in metal estimation.

The mine has operated various extraction methods over time, including first open pit on surface then open-stoping, sub-level stoping, sub-level caving, and conventional shrinkage and breasting in underground. To date, it has produced approximately 207 Mt of ore with grades of 2.9 g/t Au, 27 g/t Ag, 0.3 per cent Cu, 0.8 per cent Pb, and 1.6 per cent Zn (Glencore Zinc Group, 2023). Infrastructure is well-developed, with power supplied by the near-by hydro-power station and direct transportation links to Ust-Kamenogorsk, regional administrative centre and metallurgical complex for the company.

I. Geological

A. Stratigraphic

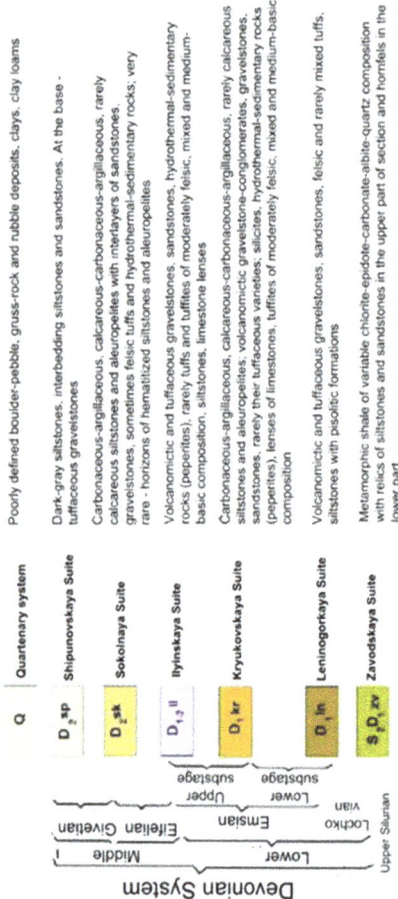

Q	Quarternary system	Poorly defined boulder-pebble, gruss-rock and rubble deposits, clays, clay loams
$D_2{}^{sp}$	Shipunovskaya Suite	Dark-gray siltstones, interbedding siltstones and sandstones. At the base - tuffaceous gravelstones
D_2 sk	Sokolnaya Suite	Carbonaceous-argillaceous, calcareous-carbonaceous-argillaceous, rarely calcareous siltstones and aleuropelites with interlayers of sandstones, gravelstones, sometimes felsic tuffs and hydrothermal-sedimentary rocks; very rare - horizons of hematitized siltstones and aleuropelites
D_{1-2}ll	Ilyinskaya Suite	Volcanomictic and tuffaceous gravelstones, sandstones, hydrothermal-sedimentary rocks (peperites), rarely tuffs and tuffites of moderately felsic, mixed and medium-basic composition, siltstones, limestone lenses
D_1 kr	Kryukovskaya Suite	Carbonaceous-argillaceous, calcareous-carbonaceous-argillaceous, rarely calcareous siltstones and aleuropelites; volcanomictic gravelstone-conglomerates, gravelstones, sandstones, rarely their tuffaceous varieties; silicites, hydrothermal-sedimentary rocks (peperites), lenses of limestones, tuffites of moderately felsic, mixed and medium-basic composition
D_1 ln	Leninogorkaya Suite	Volcanomictic and tuffaceous gravelstones, sandstones, felsic and rarely mixed tuffs, siltstones with pisolitic formations
S_2-D_1 zv	Zavodskaya Suite	Metamorphic shale of variable chlorite-epidote-carbonate-albite-quartz composition with relics of siltstones and sandstones in the upper part of section and hornfels in the lower part

Devonian System — Middle: Eifelian, Givetian; Lower: Emsian (Upper substage, Lower substage), Lochko-vian; Upper Silurian

B. Intrusive and extrusive-subvolcanic formations

C_2-P_1 z	Zmeinogorsky intrusive complex. Fine and medium-grained granites, granite-porphyres, granophyres.

$\beta\alpha$ D_3-C_1 — Late Devonian - Early Carboniferous gabbroid intrusive complex. Diabase porpyrites, gabbro-diabases

$\lambda\alpha$ D_2 — Middle Devonian extrusive-subvolcanic liparite, liparite-dacite porphyres (so called quartz keratophyres, quartz albitophyres)

$\alpha\beta\sigma\epsilon$ D_{1-2} — Early - Middle Devonian extrusive-subvolcanic andesite, andesite-basalt porphyrites

D_{1-2} S — Sinyushinsky intrusive complex. Biotite granites, plagiogranites, granodiorites, vein series: felsitod porphyres, diabases, granite-porphyres

$\lambda\pi$ D_1 — Early Devonian extrusive-subvolcanic liparite, liparite-dacite porphyres

C. Secondary alterations and metamorphic rocks

q — Microquartzites, quartzites, sericite microquartzites, sericite-quartz and quartz-sericite rocks, brecciated varieties, sericiolites, quartz-barite rocks.

D. Ores

🟥 — Veinlet-disseminated polymetallic

🟩 — Veinlet and disseminated copper

Barite

II. Actual

▨ — Mined-out areas of ore bodies and lodes

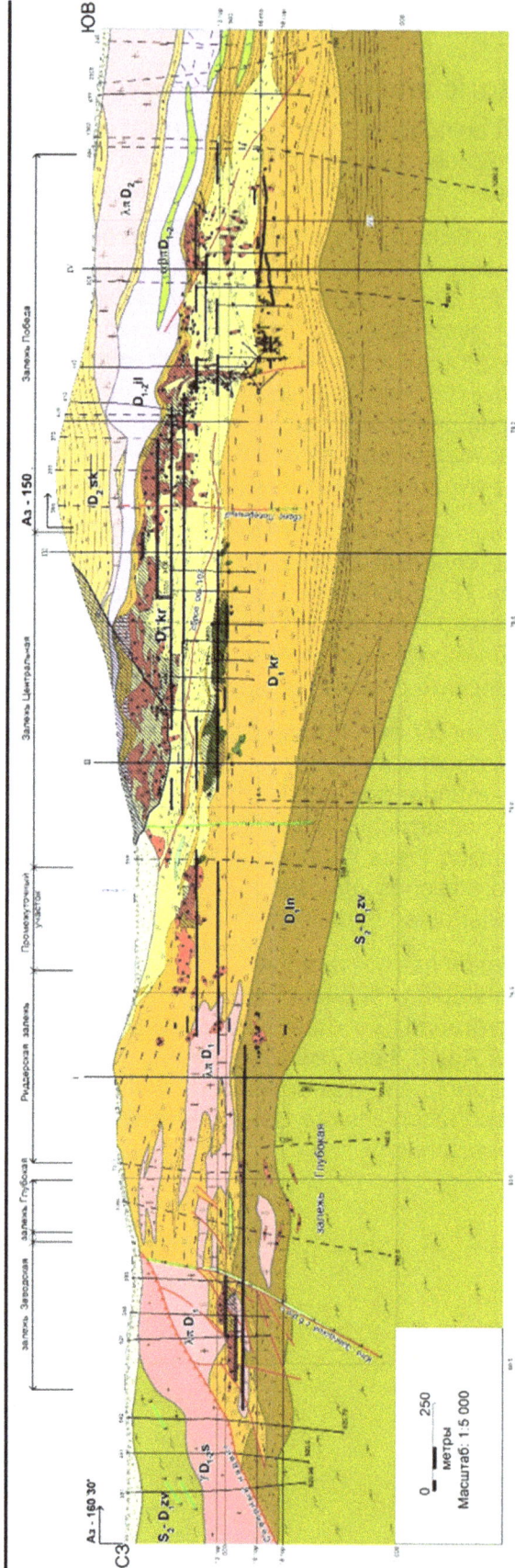

Масштаб: 1:5 000
0 — 250 метры

FIG 2 – Geological cross-section, looking NE of the Leninogorsky basin illustrating the position of the Ridder-Sokolny Deposit in the lithographic sequence. The historical Open pit and actual Underground levels of exploitation are overprinted in black. Red orebodies are of polymetallic nature, green orebodies are dominated by copper.

Operational overview of Ridder-Sokolny Mine

The current and historical underground operation at Ridder-Sokolny Mine spans an area of 4 km by 4 km, extending to depths of 0.5 km. Originally starting as an open pit, the mine evolved into underground bulk mining before transitioning to selective techniques. Historical underground voids remain unfilled, contributing to surface subsidence issues in some locations. A large part of the mine is situated beneath the township of Ridder, which adds managing subsidence and vibrations that could affect local infrastructure. Subsidence monitoring is permanent and a comprehensive modelling in 2021–2022 identified areas of 'non-entry' and sterilised extensive parts of the Mineral Resource.

The mine is accessed via a ramp in the south-west and via a shaft, central to the deposit. Two production shafts are directly linked to the concentrator plant above. Although many historical shafts have been decommissioned, ventilation shafts, the ramp, and production raises maintain adequate airflow across the extensive underground network. Material handling relies on ore and waste passes near active mining fronts, connecting production areas to levels equipped with trains that efficiently transport materials to the haulage shafts. This system supports cost-effective operations as long as production remains concentrated near the existing infrastructure and above the lower level equipped with trains (KazMinTech Engineering, 2021).

Production has fluctuated between 1.6 and 1.8 Mt/a over the past decade, with no noticeable improvement from mechanisation. Gold remains the primary product, though grades have steadily declined since 2020 due to challenges in accessing high-grade zones on time and due to increasing dilution (Glencore, 2023a). Current production aims for 1.8 Mt/a at 1.8 to 2.0 g/t Au, although the mine's full potential is limited by operational inefficiencies, scattered development, and infrastructure constraints. The current production rate leaves spare capacity at both the shafts and concentrator level (Glencore, 2024).

Operationally, the mine is divided into 18 geological lodes (Glencore, 2022); and substantially more working zones, with teams working independently under the central mine planning and production departments. In addition, historical development in small sections limits mechanised equipment movement, creating logistical challenges for both personnel and equipment, and limiting planning options.

The geological complexity of the deposit introduces variability in mining results, with material losses, dilution, and the need for opportunistic mining of newly mapped veins. This is one of the main causes for increased development demands since mechanisation, as establishing multiple scattered mining fronts and confirming vein geometry ahead of production is critical. Consequently, development now accounts for the majority of the mining costs, without a corresponding increase in output. Reducing reliance on development is key to lowering costs, simplifying operations, and improving overall efficiency.

BUILDING A STRATEGIC MODEL FOR A LARGE SCALE AND MULTI-CENTURY OLD OPERATION

Geological modelling approaches and challenges

At Ridder-Sokolny, the traditional approach focuses heavily on reducing dilution and relies in manual interpretation and modelling methods for short-term. Geology was drawn on paper sections and plans and digitalised only to provide a frame for the drill and blast design. Geologists are closely updating maps and diamond drilling data to ensure short-term planning aligns closely with the geological wireframes.

In 2018, the first block model was produced for the most 'active' sections of the mine, initially divided into 40 sub-lodes for computational efficiency. Over time, the model was extended to cover the entire known deposit through careful digitisation and the integration of historical data. By 2022, the first fully unified block model and a large-scale digital Life-of-mine (LOM) plan were completed (Glencore, 2022). Despite these achievements, fully transitioning to a computer-supported process has faced resistance due to concerns about the reliability of 'bulk-interpretation' models when compared to the perceived precision of manual mapping.

There are numerous technical challenges in modelling such a complex deposit, as illustrated in Figure 3. These include interpreting narrow, structurally controlled veins—sometimes only centimetres thick—and dealing with the heterogeneous nature of the mineralisation, which is further complicated by the heterogeneity of the gold concentrations. Additionally, historical data has variable reliability, as robust QA/QC protocols and accurate diamond drill hole deviation measurements were only recently implemented.

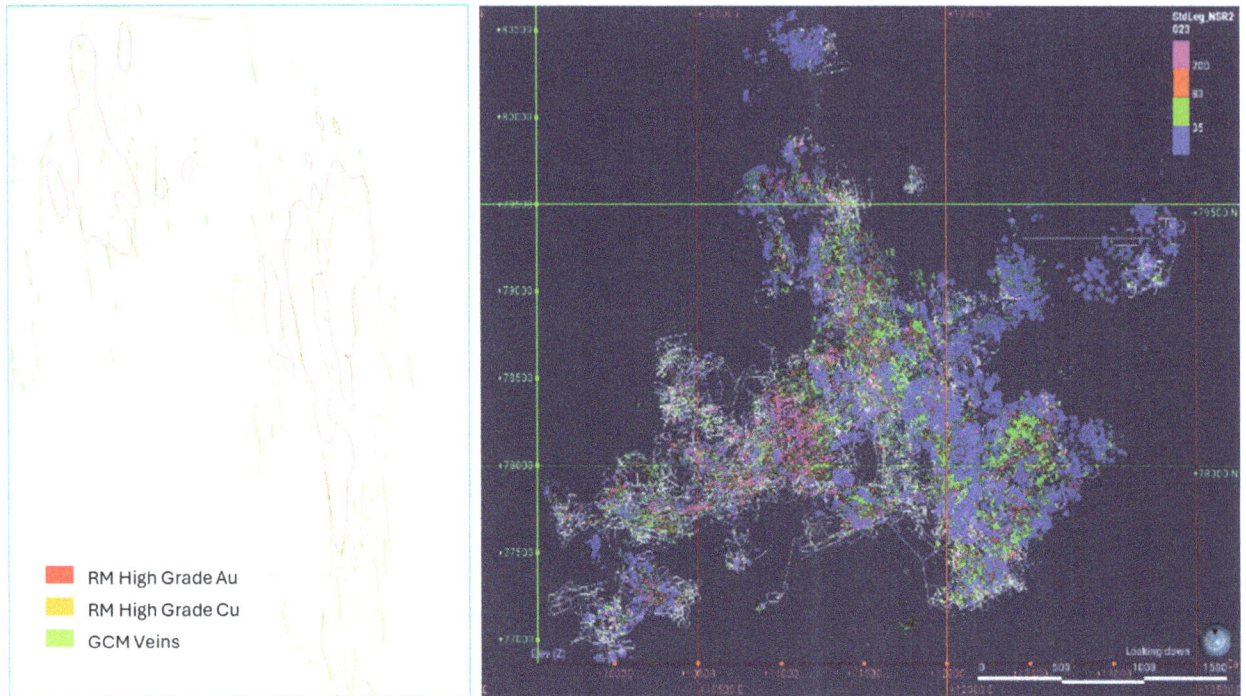

RM High Grade Au
RM High Grade Cu
GCM Veins

FIG 3 – Illustration of multiple interpretation of the same metric veins, using different economic factors for implicit modelling in Red and Orange, and manual interpretation in green (left). Plan view of the whole implicit model use as base for Mineral Resource Estimation at Year-End 2023. Blocks are coloured by NSR value, the historical development is displayed in grey (right).

In late 2022, to build a more reliable and integrated block model for Ridder-Sokolny, the team launched a comprehensive internal and external audit of the modelling approach. The goal was to look for opportunities to streamline block modelling, enhance connections between different modelling and planning levels, and create robust mid-term volume estimates that reflect the current operation, while incorporating scarcer and less reliable data for long-term resource development opportunities (Glencore, 2023b).

This led to the adoption of two approaches:

1. Implicit modelling, building on the existing model but with stronger controls via a detailed litho-structural model and feedback from manual modelling in operational areas.

2. Unconstrained modelling, applied to supplementary information, without predefined wireframes and only limited by geostatistical and structural criteria.

The dual approach—combining implicit and manual modelling—allows a balance between efficiency and detail, providing a robust foundation for Mineral Resource estimation despite the complexity and heterogeneity of the deposit.

Integration of operation constraints and infrastructure

The integration of information into a reliable 3D planning model of the Ridder-Sokolny Mine has been an iterative process, advancing through several stages over time:

- 2017–2020: Collaborations with consulting firms facilitated the digitalisation of geotechnical and void models, converting historical paper sections of surface and underground mining footprints into digital format. A key focus was the detailed digitalisation of pillars that protect vital underground infrastructure and the surface concentrator. 'Sterilisation shapes' were also created to exclude areas with uncertain void data or remnant pillars from Mineral Resource and Ore Reserve estimates. After the first block model was constructed, stope optimisation tools were utilised to estimate the mine's operational inventory. However, challenges such as the incomplete digitalisation of infrastructure and preliminary geotechnical data limited the development of a comprehensive mine plan.

- 2020–2021: Further consulting work analysed the correlation between underground historical voids and surface deformations, enabling refined modelling of subsidence-affected zones (Beck Engineering, 2021). This was particularly crucial due to historical voids left unfilled or mined using caving methods, practices which have since been prohibited. The mine closely monitors operational vibrations beneath the city to safeguard both the population and infrastructure, with certain mine areas excluded from production for protection.

- Life-of-mine (LOM) Design (2020–2021): During this period, the first LOM centrelines design was completed for the most active parts of the mine. This provided a framework to estimate the development and infrastructure necessary to access the dispersed ore volumes identified through stope optimisation.

- Full Ground Support (2020–2021): With the major implementation of higher safety standards, the transition to full ground support across the operation required larger development sections and led to a decrease in development rates. This period also marked the complete phasing out of manual development equipment and crews, further shifting towards mechanised operations (KazMinTech Engineering, 2021).

- 2022–2023: The void models were enhanced with additional data and measurements, additional work on development centreline design led to an expanded underground LOM model. For the first time, a fully sequenced LOM plan covering the known extent of the deposit was generated, providing a clearer picture of the mine's long-term potential (Glencore, 2023a).

- 2023 Optimisation Program: A strategic initiative was launched to identify long-term opportunities in light of increasing mining costs and stagnant metal throughput. All design criteria were reviewed and compared against current geotechnical and operational data. As presented in the RSM Optimization Study (Glencore, 2023b), this review aimed to refine design parameters, test the feasibility of both bulk and selective mining approaches, and identify potential improvements in planning and operational efficiency.

- Void Model and Rehabilitation (2023): A thorough review of the void model was conducted to pinpoint areas requiring rehabilitation or re-access. These rehabilitation headings were incorporated into the LOM design, ensuring they were integrated with the overall infrastructure (Glencore, 2023c).

- Material Handling System: The mine's material handling network was fully digitalised, allowing for the inclusion of its constraints into the mine plan. This digitalisation enabled the identification of required modifications to support current operations and future production targets.

These steps collectively enabled the development of a more reliable and integrated mine plan, with the aim of enhancing operational efficiency, maintaining safety, and ensuring the strategic long-term viability of the Ridder-Sokolny operation.

Next step

Figure 4 illustrates the strategy underpinning the ongoing optimisation program, highlighting the importance of developing a strategic model to navigate the complex network of constraints affecting production efficiency and the long-term outlook of a deposit with such scale and historical significance.

The next step leverages this model to identify opportunities for overcoming the geological and operational challenges faced by the mine. This is achieved by addressing the root causes of deviations from planned production as well as using mine planning process to propose long-term strategy and enhance operational efficiency.

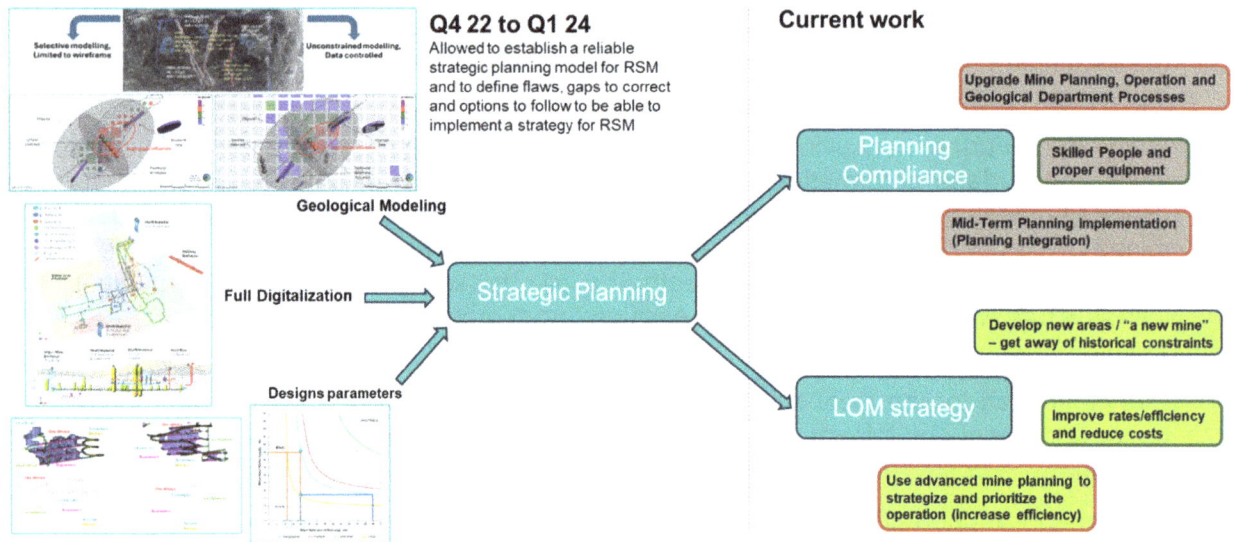

FIG 4 – Summary of the optimisation program strategy.

To maintain focus, the following sections of this paper centre on developing a comprehensive long-term strategy for the operation, employing scenario generation and risk rating to optimise the mine plan for the years ahead.

MINE PLAN OPTIMISATION PROCESS – SCENARIO GENERATION AND RISK RATING

Strategic mine planning process

The strategic mine planning approach at Ridder-Sokolny Mine focuses on reducing development metres per tonne of extracted metal, enhancing operational efficiency, and de-risking the mine plan. The strategy is centred around four guiding principles:

- Prioritising Geological Certainty: Target high-confidence zones for operational decision-making.

- Maximising High-Grade Zones Early: Boost economic returns and cash flow by focusing on high-grade ore.

- Reducing Mine Dispersion and Enhancing Efficiency: Centralise operations to optimise logistics and reduce costs.

- Risk Quantification and Scenario Development: Adapt to market and operational changes through scenario generation and risk evaluation.

The strategic process was organised into stages, each designed to optimise planning and align with current mine conditions. Initial stages, including comprehensive data collection and mine zoning, form the backbone for accurate and adaptable mine planning. The block model is refined to include key geological classifications, such as Measured, Indicated, Inferred, and mineral potential categories, as well as the grades of primary and secondary metals. The design layout integrates the interaction between mining methods and infrastructure, balancing detailed stope geometry with operational constraints. Equipment performance is assessed for its influence on mine dispersion, ensuring utilisation and availability are aligned with the overall strategic plan. Finally, economic assumptions such as the Net Smelter Return (NSR) are modelled using both historical values and forward-looking projections, providing a robust economic framework for cut-off grade scenario generation.

Mine zoning as a strategic tool

Mine Zoning is particularly pivotal to the strategic plan, organising the mine around shared infrastructure, mining methods, geometallurgical domains, and operational conditions. For Ridder-Sokolny, the predominant zoning criterion is shared infrastructure, such as ramps, orepasses, and ventilation systems, enhancing operational efficiency and reducing unnecessary dispersion.

Understanding how to adequately model the constraints related to people, material, and equipment movement is crucial not only for unlocking operational opportunities but also for securing a practical and executable mine plan. This approach also enables the evaluation of the economic viability of developing specific new areas within the mine and is primary to identify and prioritise the most prospective zones for future production, and preparation activities such as exploration and geotechnical investigations.

The Figure 5 illustrates how the mine is divided in lodes to facilitate geological modelling in one hand and in areas of productions, searching for operational efficiency, on the other hand.

ZONING BY GEOLOGICAL LODE

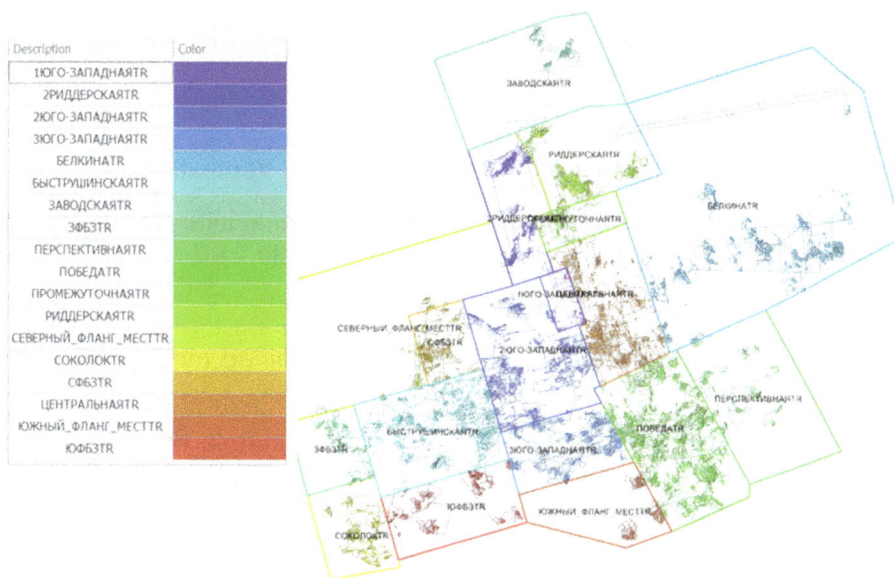

STRATEGIC ZONING

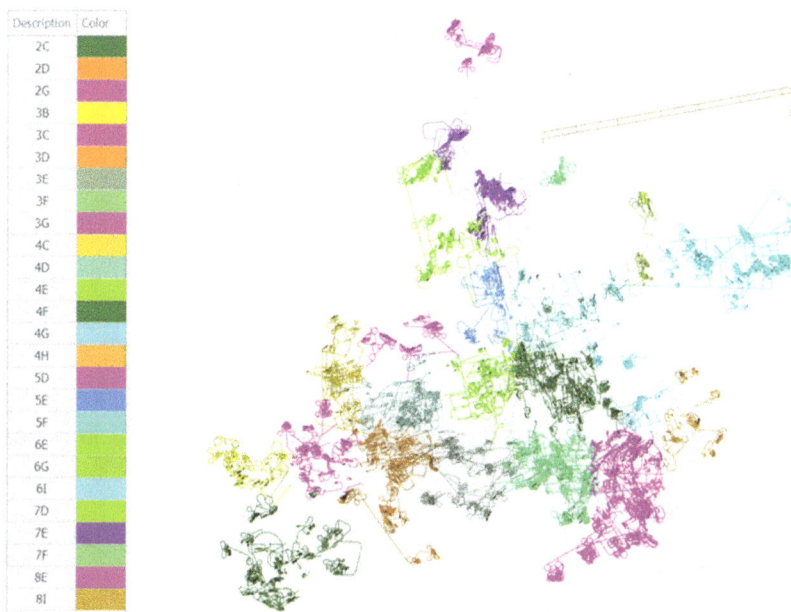

FIG 5 – Illustration of a geological and mine planning zonation of the mine.

By prioritising these principles and ensuring the planning model is adaptable, Ridder-Sokolny is aiming building a strategic mine plan that secure the current operations profitability while remaining flexible for future uncertainties. The processes applied here — though tailored to the mine's unique context — are broad enough to be of benefit to other companies considering mining legacy areas or restarting historical operations.

Risk scoring

Rooted in the strategic planning process and scenario development, risk scoring plays a critical role in multiple stages of the mine planning process. It serves two key purposes: classifying mining inventory volumes based on risk mitigation factors and quantifying the potential risks associated with different planning scenarios.

Risk classification matrix

A Risk Classification Matrix is developed to categorise mining volumes according to a range of 'risk mitigation' factors, which include:

- Classification of Mineral Resources.
- Availability of additional geological and operational information (mapping, grab samples, litho-structural and geotechnical data etc…).
- Proximity to actual active production fronts.
- Position relative to major infrastructure (Material handling, Ventilation etc…).
- Geotechnical risk or need for specific investigation (potential interaction with voids, proximity to surface and to the town protective pillar, water inflows etc…).
- Metal content.
- Potential for Mineral Resources growth (Is the Unconstrained model prediction for the area showing potential?).

This classification enables the prioritisation of mining zones across different planning horizons, aligning technical services and production activities with a clearly defined strategy. Figure 6 illustrates this integrated approach.

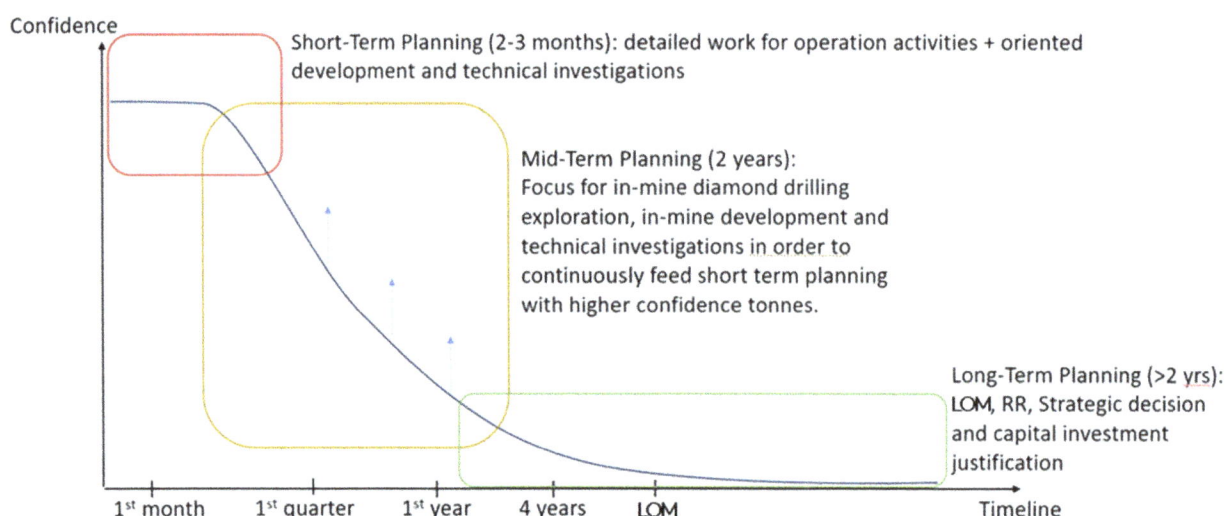

FIG 6 – Integrated mine planning strategy, using inventory risk-based classification and variable horizons of planning.

- Priority 1: Zones designated for short-term planning, focusing on detailed drill-and-blast designs and monthly/quarterly production plans.
- Priority 2: Target areas for the first two quarters, requiring immediate geological and technical investigation to upgrade them for short-term operational readiness.

- Priority 3: Zones earmarked for mid-term horizons (6 to 18 months), with plans to upgrade them according to the sequenced mid-term strategy.

- Priority 4: Areas reserved for Life-of-mine (LOM) planning, typically excluded from mid-term plans but identified as opportunities for future growth and long-term production.

The strength of this approach lies in its ability to orientate mine planning and exploration activities to unlock future growth while efficiently managing current production. However, the challenge is in quantifying the various risk factors and integrating activities and time horizons into a cohesive model. Calibration of weights and scores is critical to the risk scoring process and requires ongoing collaboration between strategic planning, operational planning, and on-the-ground mining teams.

Risk scoring for scenario evaluation

Risk scoring is also applied to quantify the volatility associated with different production scenarios, accounting for how mining zones react to economic fluctuations (eg metal prices, operating cost variations) and operational changes (eg equipment availability). This is achieved through a multi-step process.

The production plan is separated into short to mid-term and long-term horizons to assess stability and sustainability over different timeline. This allows for specific analysis across both time frames.

A sensitivity factor measures the responsiveness of each zone to key parameters, such as cut-off grade adjustments and equipment changes. In simpler word, the production throughput of different scenarios is compared to highlight how sensitive to parameters each production fronts of the mine is. Higher sensitivity zones are considered more volatile, as their production tonnage is more affected by these variables.

The final risk score is derived by normalising the sensitivity score, as illustrated in the Figure 7, to the amount of ore available for each areas. Higher scores indicate zones that are more volatile and, thus, presenting higher economical or operational risk.

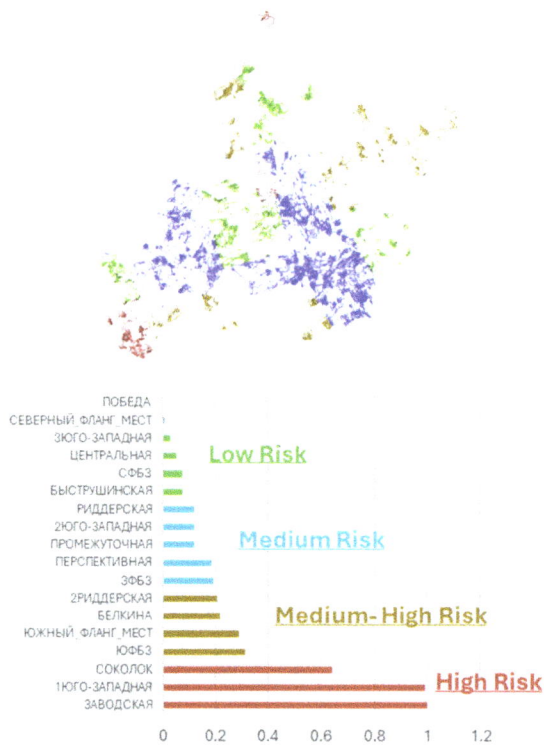

FIG 7 – Example of risk scoring for a series of scenario, focusing on operating costs variation.

A different risk scoring version is also use on a very short-term basis, to highlight the stopes and headings at risk to not produce as planned on a monthly and quarterly basis. Mostly based on the geological confidence on a very short time scale, it aims to inform management of the risk on production as well as to prepare mitigative actions and alternative plans in case of planning compliance failure during the period.

For the Ridder-Sokolny Mine, this use of risk scoring — both for inventory classification and scenario evaluation — aims to provide a structured way to prioritise stable, low-risk zones for production, ensuring efficient use of resources and enabling proactive planning. It also allows to evaluate the economic viability of new development areas and to identify key opportunities for future production, preparation, and exploration work. By integrating risk assessment into both the strategic and operational phases, the mine is able to align its short-, mid-, and long-term planning efforts, enabling data-driven decision-making toward operation efficiency.

Developing production scenarios

The development of production scenarios is a critical aspect of strategic mine planning, designed to understand how changes in operational and economic variables impact production, equipment needs, and overall mine performance. Scenarios are generated by analysing the interactions between key operational factors (such as equipment availability) and economic parameters (such as cut-off grades). These scenarios provide insights into production patterns within mine zones and help prioritise areas based on their contribution to overall output.

A type of Hill-of-Value™ analysis is employed to support the creation of optimal scenarios and establish a balanced approach. In the context of Ridder-Sokolny, the economic variable is primarily manipulated by adjusting cut-off grades. Lower cut-off scenarios expand the mining inventory to include lower-grade ore, while higher cut-offs reduce the inventory but enhance average grade quality and value.

Operationally, equipment availability plays a pivotal role in defining production scenarios. Specifically, the number of development jumbos—crucial for influencing mine dispersion—is varied to evaluate how changes in equipment affect production. Various equipment levels (eg 22, 24, and 26 jumbos) are tested to identify the optimal balance between development requirements and metal production, aiming to maximise efficiency and identify the 'sweet spot' for the operation.

Each scenario aligns with the strategic objectives set earlier, focusing on early access to high-grade zones, minimising initial development costs, smoothing development requirements over time, and maximising equipment utilisation. Operational and base constraints include prioritising high-confidence Mineral Resources categories, maintaining stable production within plant capacity, adhering to material handling limits, and balancing capital requirements against production targets.

Scenario generation also aids in assessing the contribution of each area to the production plan, determining its criticality for operations. Zones with low productivity are evaluated to confirm the economic viability of developing new mine areas that may only provide marginal metal supply or economic returns, while weighing all constraints and associated risks. By generating various scenarios, mine planning can focus on maximising value, balancing risks, and enhancing operational efficiency. Figure 8 provides an example of scenario generation and analysis.

This approach to scenario development and zone prioritisation establishes a structured framework for long-term planning, allowing for responsiveness to both market conditions and operational realities.

Average distribution of Mineral Resource sources over the different iterations.
Potential mineral is included as a strategic vision

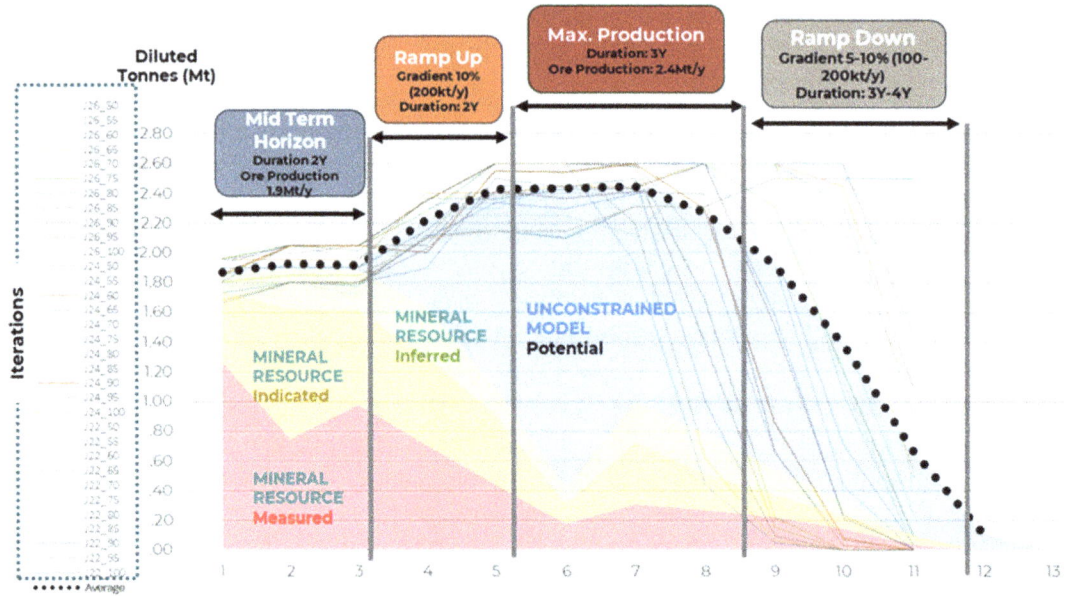

Proportion of areas classified by production efficiency in the "preferred scenario

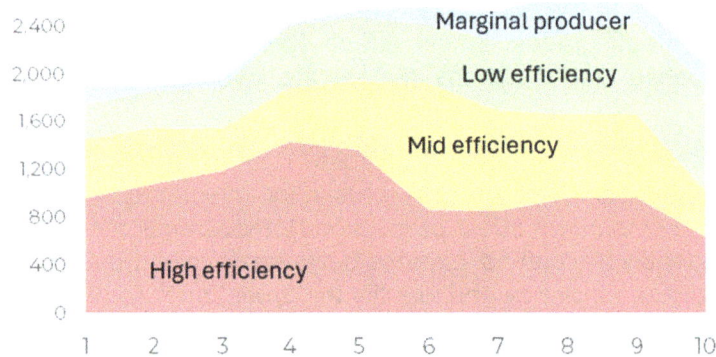

FIG 8 – Ridder-Sokolny production profiles analysis of scenarios depending on a set of 11 cut-offs and three numbers of jumbos.

RESULTS AND DISCUSSION

The strategic planning framework in implementation at the Ridder-Sokolny Mine addresses key challenges including the geological complexity of narrow vein mining, operational inefficiencies due to scattered development, and the need for more reliable and integrated data for decision-making.

Challenges addressed

Geological Uncertainty and Data Integration: The adoption of a dual modelling approach and continuous auditing processes provided solutions to managing geological uncertainty and diverse data quality, balancing precision with scalability:

- Operational Constraints and Legacy Infrastructure: Digitalising material handling systems and infrastructure constraints enabled a more realistic and practical mine plan that aligns with the physical limitations and historical complexities of the mine layout.

- Adaptive and Forward-Looking Planning: The strategic framework incorporates risk assessment into both short-term and long-term planning horizons, allowing the mine to respond proactively to economic and operational variations and to refine planning parameters as new data and operational insights are acquired.

- Potential Applications for Similar Operations: While tailored to Ridder-Sokolny's unique context, the strategic framework offers valuable insights for other mining operations facing similar challenges of complex geology, legacy infrastructure, and multi-century exploitation. The emphasis on integrated geological modelling, risk-based scenario generation, and flexible strategic planning can be adapted to improve operational efficiency and long-term sustainability in other underground mines. As the framework is further developed and refined, it provides a blueprint for enhancing planning compliance, reducing risks, and ensuring consistent alignment with strategic business objectives.

Ultimately, the Ridder-Sokolny case demonstrates the potential of a holistic strategic approach to transforming long-standing operational challenges into opportunities for growth, operational efficiency, and sustainable resource development.

Lessons learned

The strategic planning framework at Ridder-Sokolny Mine has provided valuable lessons for refining adaptable and effective planning in complex mining operations:

- Mechanisation Requires a Strategic Shift: Introducing mechanisation goes beyond purchasing equipment; it requires a fundamental change in planning and operations. Success comes from redesigning workflows, adjusting scheduling, and ensuring workforce training aligns with new mechanised methods to improve efficiency.

- Strategic Plan Input Quality: A reliable strategic plan starts with accurate, well-controlled data. Geological block models and void/sterilisation models must be built on high-quality data with robust QA/QC standards. Older or uncertain data should be modelled conservatively, and potential opportunities should be aligned with infrastructure plans while applying risk controls.

- Balancing Detail Across Time Horizons: The strategic plan should balance detailed short-term objectives with broader long-term goals, preventing minor deviations from impacting the entire mine plan. Focusing on key activities minimises inefficiencies, and plans must remain adaptable without becoming overly complex.

- Scenario-Based Planning for Flexibility: Developing multiple scenarios with varying cut-off grades prepares the operation to respond to market fluctuations and cost changes. This flexibility helps optimise value by exploiting high-grade zones in favourable markets and protecting against risks during downturns.

- Risk Rating and Compliance for Informed Decision-Making: The implementation of a risk classification matrix ensures strategic alignment by prioritising zones based on risk factors. Regular calibration and monitoring of risk scores help address uncertainties proactively, allowing for better operational planning and execution.

While infrastructure complexities, exploration alignment, and equipment management are critical, the main takeaway is the importance of proactive planning and collaboration across teams. Accepting uncertainties and continuously refining the plan fosters adaptability, enhancing operational efficiency and supporting long-term sustainability. The collective focus should be on simplifying strategies to ensure clear, effective decision-making and risk mitigation.

CONCLUSIONS

The strategic planning process at Ridder-Sokolny Mine has demonstrated the benefits of a holistic approach that incorporates robust geological modelling, risk assessment, and operational planning. By aligning the strategic plan with real-world operational constraints and economic variables, the mine is now able to enhance its planning precision, reduce development inefficiencies, and position itself for future expansion.

While some aspects are unique to Ridder-Sokolny, the processes are broad enough to benefit other companies considering the mining of legacy areas or the re-starting of historical mining operations.

The importance of scenario-based planning, flexibility in operations, and continuous collaboration between teams cannot be overstated. This methodology ensures that strategic planning is not just a long-term vision but a dynamic process adaptable to changing conditions and opportunities.

ACKNOWLEDGEMENTS

The authors wish to express their gratitude to Kazzinc and to Glencore Zinc Technical Services for their support and provision of resources that made this work possible. We are particularly grateful to Adrian Penney, Jody Todd and Vyacheslav Kalinichev for their collaboration and invaluable insights throughout the development of these frameworks. Their expertise and constructive feedback contributed significantly to the refinement of the results.

We also extend our appreciation to the operational teams at Ridder-Sokolny, led by Vladimir Skuratov, Vyacheslav Kalinichev, and Azamat Mukhamedkaliyev, as well as to Asturmine group, including Manuel Varela and Juan Fernandez, whose assistance in data collection and analysis, as well as the numerous discussions we had, were essential to this study.

We sincerely thank the management of the Eastern Kazakhstan Division of Kazzinc, under the leadership of Igor Anisimov, for their trust and support.

Lastly, we extend our appreciation to the organisers of the AusIMM Underground Operators 2025 for providing a platform to share and discuss this work.

REFERENCES

Beck Engineering, 2021. Crown Pillar Assessment and Study for the Ridder-Sokolny Mine: Final Report, internal report, 17 December 2021.

Glencore Xstrata, 2013. Annual update of mineral resources and ore reserves for Ridder-Sokolny as of 31 December 2013, internal document, Wardell Armstrong.

Glencore Zinc Group, 2023. Ridder-Sokolny 2023 Ore Reserve Statement – Section 4 of Table 1, internal technical document, prepared by Bastien Fresia.

Glencore, 2022. YE 2022 Resources – Ridder-Sokolny, internal presentation, slide 26, December 2022, prepared by Juan Fernández and Manuel Valera.

Glencore, 2023a. Asset Summary Ridder-Sokolny Mine – LoA 2024: Pre-read Presentation, internal budget document, Slide 11: Context and Constraints.

Glencore, 2023b. RSM Optimization Study – Steering Committee Presentation: Slide Block Model Review and Resource Modelling Approaches, internal presentation, 13 October 2023.

Glencore, 2023c. YE 2023 Ore Reserves – Ridder-Sokolny, internal presentation, December 2023, prepared by Bastien Fresia and Giancarlo Calzada, Slide 5: Void Model Update.

Glencore, 2024. LoA 2025 – Ridder-Sokolny Mine, internal presentation, June 2024.

KazMinTech Engineering, 2021. Mining plan for the Ridder-Sokolny deposit, Technical Memorandum, internal document, RGOK.

Kazzinc, 2020. RGOK-HISTORY-2020, internal document, Planning Team, Kazzinc.

Kazzinc, nd. Brief geological and minerogenic characteristics of the work area, internal document.

Twin strand cables replaced with Falcon Bolts at Tomingley Gold Operations

R Galluzzi[1], C Rademeyer[2], A Brown[3] and C Kenny[4]

1. R&D Engineer, Jennmar Australia, Smeaton Grange, NSW 2567.
   Email: rgalluzzi@jennmar.com.au
2. Geotechnical Engineer, Alkane Resources, Tomingley NSW 2869.
   Email: cornelius.rademeyer@alkane.com.au
3. Mine Manager, Alkane Resources, Tomingley NSW 2869.
   Email: andrew.brown@alkane.com.au
4. Mine Services Representative, Jennmar Australia, Smeaton Grange, NSW 2567.
   Email: ckenny@jennmar.com.au

ABSTRACT

This case study, conducted between Tomingley Gold Operations and Jennmar Australia, explores the full-scale adoption of pre-tensioned, post-grouted self-drilling bolts in lieu of manually installed cable anchors for intersection support, brow support, and other applications where cable bolts are typically implemented. The data required to justify and drive the change was collected during early trials and is also presented in this case study. The removal of cable cutting, pushing, plating and tensioning stages, as well as eliminating reliance on elevated work platforms for cable installations reduced the risk of associated injury to workers. Improvements in installation efficiency, the adoption of fast setting resins in place of cement grouts and the development of a high-capacity steel provide further opportunity for schedule optimisation and rapid development.

INTRODUCTION

Rockfall hazards present safety and production risks in underground mining, particularly near production areas where ground vibrations and changing stress conditions can destabilise structures. The larger spans of wide drives and intersections have increased risk of rockfall as they expose more structures. Primary ground support does not always provide sufficient capacity or depth of embedment to support these wedges. A typical intersection wedge failure is shown in Figure 1 along with a kinematic analysis used to perform ground support assessment.

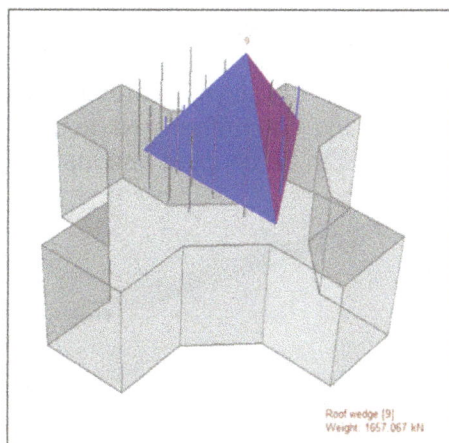

FIG 1 – (Left) intersection wedge failure leading to abandonment of development and bypass development. (Right) Typical kinematic wedge assessment to aid ground support design or failure back analysis.

Underground mines typically use 6.0 m cable bolts to achieve the required deeper embedment length and higher capacity to support these structures. Cables provide the length and strength to anchor a potential wedge into stable ground beyond wedge limits. While cable bolts are versatile and cost-effective, their installation can be challenging. A range of mechanised cable bolting machines are available that can circumvent these challenges, however if the planned amount of cable bolts to be installed does not justify the cost of purchasing and maintaining a dedicated cable bolting machine, or there are physical restrictions in mine design, mining operations may pursue manual installations. In this case, cable bolts are manually pushed into holes that are pre-drilled by Jumbos or production drills. This method of installation comes with its drawbacks, risks and challenges. Jumbos are not well suited to extension drilling 6.0 m holes vertically, and a variety of methods are employed to achieve this. Utilising production drills for development introduces additional tramming and they drill oversized holes, requiring additional grout. Hand installing cable bolts is a time-consuming process with increased potential for manual handling, chemical and eye injuries. These drawbacks may compel mining operations to pursue alternative options.

Small trials investigating replacement of cables with extension drilled SDAs (self-drilling anchors) have been conducted. In 2019, a limited number of extension drilled 15 m SDAs were installed at Malmberget in Sweden. These SDAs were drilled and grouted from an Epiroc Boltec M using a resin modified with an increased hardening time (Bray, 2019). It was noted that mechanising the installation process greatly improved operator safety and productivity. The trial concluded that the success of the extension drilled SDAs coupled with injected resin could provide an alternative to typical cable bolting.

In 2023, a small trial was conducted exploring replacement of cables with standard extension drilled SDAs at Kanmantoo Mine (Jardine, 2023). SDAs were installed using a typical Jumbo drill rig then post injected with resin from an IT Basket. It was found that installing SDAs in development intersections with a fast-setting resin could improve cycle time since the schedule did not need to account for grout cure time and SDA installations reduce manual work. A case study at St Barbara Gwalia Mine investigating how SDAs that are installed and resin injected from an Epiroc Boltec M compare with typical ground support products (Safari, 2023). During this trial, extension drilled 2.4 m SDAs were substituted for cable bolts with 24 7.2 m installations and nine 9.6 m installations. The case study found the extension drilled SDA installation process to be more efficient than the cable installation process.

Falcon Bolt

Falcon Bolts are differentiated from typical SDAs. The Falcon Bolt features a mechanical anchor at its toe enabling it to achieve pre-tension and provides some capacity before the bolt has been grouted. This innovation allows tensioning to be completed with drill rotation prior to resin injection, removing the plating and tensioning process. Further, the Falcon Bolt is driven via a hex as opposed to a threaded coupling, simplifying connection and disconnection from drilling dolly This feature assists extension drilling, the drifter is retracted while the anchor also holds the bolt in position and a second or third bar can be coupled and installed. The Falcon Bolt is depicted in Figure 2.

FIG 2 – Falcon Bolt.

Tomingley Gold Operations and project background

Tomingley Gold Operations (TGO) currently comprises several underground hard rock mines over a 3.8 km (north–south) × 1.0 km (east–west) footprint in the greater western plains in New South

Wales, Australia. Underground production depth ranges from 100 m to 450 m. Ultimate compressive strength of intact core ranges from 50 MPa to 300 MPa; and typically lie between 150 MPa to 200 MPa. Stress ranges from low, to medium confining stress at these depths. Given these conditions, site observations and numerical modelling, ground support and reinforcement are designed for static conditions.

At TGO, cable bolts were manually installed and predominately employed for development intersections and drawpoints. The number of cable bolt installations is growing in response to increased production and changing ground conditions as summarised in Table 1. The quantity of cable bolts required per month can be unpredictable, which presents operational challenges; particularly with less time available to service crew to attend to other tasks essential to supporting the development cycle. Constant feedback and requests were made to management to review hand-installation of cable bolts due to the nature of the work.

TABLE 1

Overview of Tomingley Gold Operations production and schedule overview.

	FY22	FY23	FY24	FY25 est.
Ore tonnes ('000s)	806	835	1047	1104
Development metres	5500	6500	7250	7000
Cable metres	13 900	14 700	16 000	18 450

In 2023 it was identified that additional resources would be required to meet the increasing number of cable bolt installations. TGO considered three options to meet this demand:purchasing a second-hand cable bolter to take over the task of installing cable bolts with contractor operation, or employing additional service crew and equipment to continue hand-installation. A summary of the cost comparison is shown in Table 2. Estimates for options 2 and 3 assumed that there would be no fleet changes.

- Option 1 – Purchasing a refurbished cable bolter to take over the task of installing cable bolts with contractor operation.
- Option 2 – Employing additional service crew and equipment to continue hand-installation.
- Option 3 – Using Falcon Bolts to replace cable bolts.

TABLE 2

Cost analysis carried out by TGO to assess how to meet requirements for additional cable bolts in FY24 and forward.

	Purchase cable bolter	Additional service crew	Falcon Bolts
Capital estimate	$ 800 000 (Refurbished cable bolter)	$ -	$ 100 000.00 (Resin basket + other)
Additional operational cost per annum	$ 940 000 (Contractor operation and maintenance)	$ 1 150,000 (Service crew + IT hire per annum)	$ 50 000.00 (Resin basket + other maintenance)
Additional cost of Falcon Bolts over cable bolts	$ -	$ -	$ 510 000
Year 1	$ 1 740,000	$ 1 150,000	$ 660 000
Year 2	$ 940 000	$ 1 150,000	$ 560 000

Through this period, Jennmar and TGO conducted a series of trials focusing on single length Falcon Bolt installations. As Falcon Bolt development progressed, trials shifted to focus on extension-drilled variants, with an extensive campaign of over 100 6.0 m and 9.0 m lengths installed successfully, proving the bolt's capacity to provide deep embedment ground support. Based on the overall costs and the required capital outlay, TGO began looking closer at the Falcon Bolt as a replacement for secondary support, rather than a rapid development resin bolt.

As cable bolt use increased, three back strain injuries occurred over a six-month period. These injuries, and a string of successful trials and initial business case comparison, accelerated the decision to change secondary support install methods. In early 2024 TGO changed methods and commenced process to change from twin-strand cables Falcon Bolts. The flexibility of using production rigs to install cable bolts for stope brows meant cables were still hand installed in these applications. Subsequently In June 2024, another back strain occurred when pushing cables for an interim brow. As an outcome, the mine moved away from cable bolts entirely and transitioned to Falcon Bolts from July 2024.

During initial implementation, 50 per cent more Falcon Bolts were required to reach equivalent cable support capacity for a given intersection. Through a collaboration process to enable greater viability, Jennmar investigated and produced the R32X, a stronger variant that would allow a one to one replacement rate for cables.

In May 2023 health monitoring showed that a worker was exposed to respirable crystalline silica (RCS) above acceptable limits. The review identified the potential cause of the exceedance was the volume and manner of grouting. This has driven the development of a fit-for-purpose dedicated IT basket retrofitted with a polyester resin injection system to replace cementitious grouting as part of the second phase of Falcon Bolt implementation. The system also provides additional productivity and quality control benefits. Jennmar has designed and developed this system, at time of writing the Injection System has been successfully trialled, and is due to be commissioned in December 2024.

Technical comparison of twin strand cables and Falcon Bolts

Several ground control management plans for Australian mines use the parabolic arch or dome method for ground support design in wide accesses and intersections. The design aims to support the dead weight within a volume defined by an arch above the excavation. This dead weight is anchored to the competent rock mass beyond this theoretical arch by deep embedment rock bolts. This method provides a conservative estimate for ground support design. Based on extensive application in Australian underground mines for several decades this system appears to mitigate the majority of gravity/structurally driven roof failures (Potvin and Hadjigeorgiou, 2020). With sufficient site data, kinematic analysis on probable formed wedges is also used to verify and design bespoke intersection support systems allowing for variations in installation requirements. Figure 3 shows a visualisation of these respective methods.

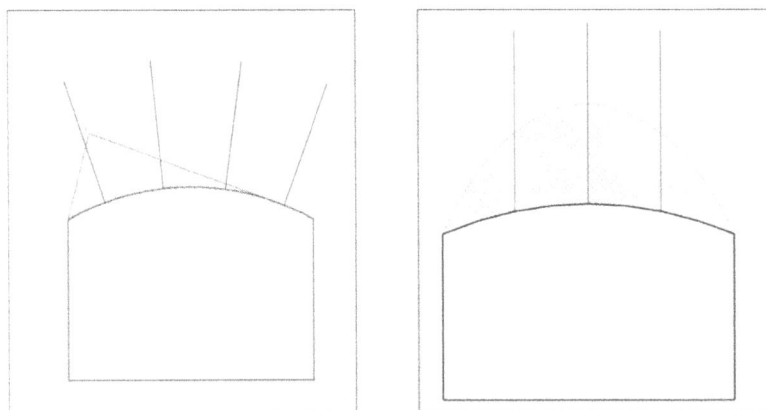

FIG 3 – (Left) Kinematic wedge analysis, (right) parabolic arch method.

Intersection and wide span ground support at Tomingley Gold is predominantly based on the parabolic arch or dome method to determine the embedment length and system capacity to control rockfall posed by increased wedge potential. As further data has been collected with progressing underground development, kinematic wedge analysis is being used to verify intersection and wide span ground support design. Intersection deep support at TGO consist of a two-pass system, with the majority of support installed prior to taking a drag and the creation of a wide-span (1st pass) and the remaining support installed as 'infills' in this drag cut (2nd pass).

Cable bolts and extension drilled Falcon Bolts are both capable of achieving the required deep embedment lengths. However, the behaviour of the flexible cable compared with the stiff, hollow bar in a ground support role can be contrasting, and the corresponding advantages and limitations must be explored. Testing showed no discernible changes in mechanical properties of Falcon Bolts that had been drilled to a depth 6 m with a typical jumbo.

Working from the yield strength, the R32S bolts have 43 per cent less tensile capacity than a twin-strand cable installation. This was initially managed by installing 1.5× more Falcon Bolts per supported area than 15.2 mm twin strand cables. As a permanent solution, Jennmar has developed a high strength 'X' grade R32 threaded steel, capable of supporting at least 480 kN in yield, reducing the number of Falcon Bolts needed to offer equivalent support as twin strand cables. These performance differences are presented in Table 3. The R32X has been successfully trialled with 24 bolts installed. TGO have switched from R32S to R32X in December 2024.

TABLE 3

Bolt properties.

	15.2 m twin strand	Falcon Bolt E	Falcon Bolt S	Falcon Bolt X
Min ultimate strength	530 kN	280 kN	360 kN	570 kN
Min yield strength	490 kN	230 kN	280 kN	480 kN
Elongation	5%	15%	6%	5%
Number of Falcon Bolts to replace 1 twin strand cable		2×	1.5×	1×
Critical embedment length in grout	2 m	0.3 m	0.3 m	untested
Status	Not in use	Not in use	In use	In trial

Shear strength

When a ground support element is subjected to shearing rock planes, bolts do not tend to fail in shear, but respond with a bending moment. Since the softer supporting material (the rock mass) cannot support the displacing steel, the bolt will bend and ultimately fail in a combination of bending, shearing and tensile loading (Knox, 2022). This is supported by a review of shear test results. Aziz (2015) conducted double shear tests on a range of cables and observed that shear failure occurred between 95–100 per cent of tensile load. Stjern (1995) found that the shear strength of a rock bolt is between 80 per cent and 100 per cent of its tensile strength. Bending stiffness is thus a critical parameter when comparing the response of cables and self-drilling bolts to radial loading induced by shearing rock planes. Cables are not stiff in bending. They are formed with a weave of smaller, flexible strands (Figure 4), and each strand has a degree of freedom relative to its counterparts, allowing the network to flex and move as it is cantilevered. Once the system is loaded, these properties allow the cable to bend until the individual strands fail in tensile loading. Hollow bolts used in Falcon Bolts on the other hand are stiff in bending. They are formed with rigid steel and a consistent cross-section. While this stiffness enables the self-drilling property, the structure resists bending, resulting in bending stress.

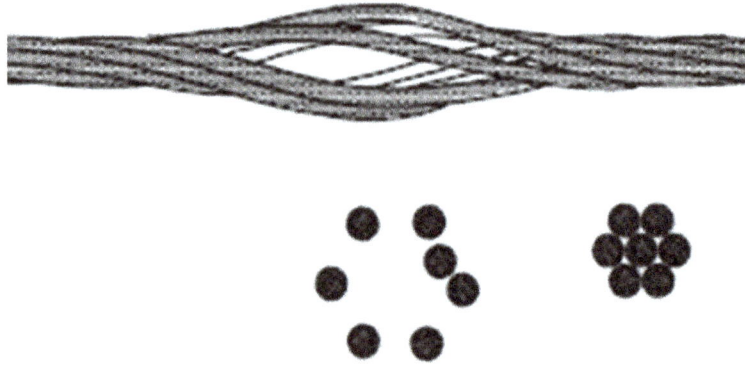

FIG 4 – Typical cable bolt (Hutchinson and Diederichs, 1996).

To evaluate the different responses of each bolt in shear loading, double shear tests were conducted in collaboration with University of Wollongong (UOW). UOW developed a double shear system fitted with a Lateral Truss (LTS), which enables the tested tendon to respond as if it were supporting shearing rock planes. Each bolt sample is fit inside an assembly of three concrete cylinders with a UCS of 60 MPa. The concrete cylinders are cast with an internal reinforcing cylinder of 200 mm diameter 5 mm wall steel tube inside it. To replicate the Falcon Bolt, concrete blocks are cast with a 51 mm hole to replicate the cable bolt, samples are cast with a 65 mm hole. There are two blocks 300 mm in length, and a third block, which is 450 mm thick, is placed between the 300 mm blocks. There is a gap between these blocks to prevent frictional forces interfering with results. The bolt tendon is manually pushed into the cast hole. Hollow bar samples are pre-tensioned to 50 kN, then grouted with Jennchem TD80 grout to represent the Falcon Bolt installation process. Cables are grouted, then tensioned after the grout has cured. The cables samples are prepared so that each cable has a bulb grouted within each block, thus a twin strand cable has two bulbs cast within each block. The central block is then loaded radially relative to the external blocks, and the applied radial load is entirely reacted by the tested tendon. Load cells record applied radial load and axial load at each end of the tested tendon, and displacement of central block is measured. This test set-up is depicted in Figure 5.

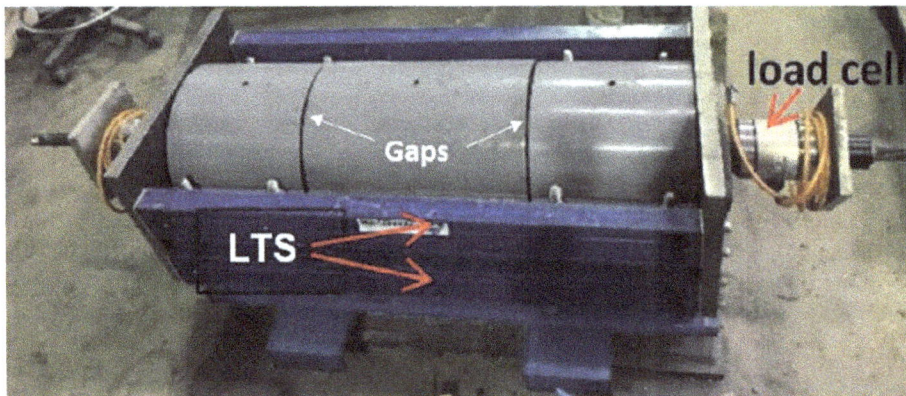

(a)

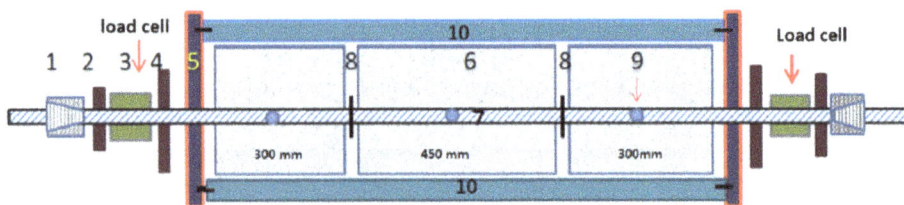

1: Barrel & Wedge. 2: Outside plate. 3: Load cell. 4: Inside plate. 5: LTS Side steel plate.
6: Concrete block. 7: Bolt (cable bolt). 8: Ring Packers (10 mm). 9: Grouting hole. 10: LTS

FIG 5 – Double shear test set-up.

Results are presented in Table 4. The cable reacted more radial load and permitted greater deformation than the tested hollow bars. The lack of bending stiffness allowed the cable to transfer a large portion of the radial load to axial load. The applied shearing force was reacted axially back to the plate securing it to each side. The 2 m critical embedment length means axial load reacts back to barrel and wedge assembly, allowing a large bend radius. The cable can respond to radial load by bending until it ultimately fails in tension. This suggests that if a cable is sheared *in situ*, it would convert a large percentage of the radial load into axial load within the cable, thus its shear capacity is far more dependent on its critical embedment length and the quality of its encapsulation.

TABLE 4

Falcon Bolt double sheer test results.

Bolt	Axial pre-tension (kN)	Max axial load at failure (kN)	Radial load at failure, (each side)	Corresponding displacement (mm)	UTS (kN)	SF/UTS (%)
R32X	50	0	271.87	37.82	570	48%
R32X	50	0	258.61	34.92	570	45%
R32S	50	0	198.21	28.95	360	55%
R32S	50	0	203.21	31.43	360	56%
15.2 twin strand cable bolt	0	360 kN	466.9	95.57	2 × 250	93.4%

The hollow bolts have a much stiffer interaction between the bolt and the grout, requiring less than 300 mm to react the bolt's full strength. This stiffer interaction factor allows a much smaller bend radius, concentrating bending force and ultimately increasing stress in the steel. A weaker encapsulation medium would correspond to longer critical embedment length, which could allow a greater bend radius, corresponding to greater resistance to shearing rock planes. The testing should be repeated with J Lok P 1:1 to confirm this theory. These results indicate a SF/UTS ratio of 48 per cent and 45 per cent for R32X steel, with failure patterns suggesting the bolts may not be failing in shear but failing in bending. Note this test does not account for friction between sliding planes. Geotechnical calculations must consider the relative shear capacities of the R32 hollow bar used in the Falcon Bolt and an equivalent cable bolt. Additional testing will be conducted to assess the impact of the encapsulation medium on the bolt's bend radius and how this impacts bending capacity and therefore response to a shearing rock mass, and effort can be invested in developing a steel that allows greater bending capacity. Further testing could explore how the Falcon Bolt's pre-tensioning property may enhance friction between sliding rock faces to improve shear resistance *in situ*, and ensuring that the shear test accurately models the response of a rock bolt to a shearing rock mass.

Site trials

Initial trials conducted early in the development of the Falcon Bolt focused on single length installations up to 3 m, aiming to direct the design path and verify engineering decisions. Early trials investigated and resolved issues such as drill bit design, grouting issues, and overall installation consistency. As Falcon Bolt development progressed and TGO expressed interest in cable replacement, trials aimed to assess specific performance parameters defined by TGO to ensure the Falcon Bolt can effectively replace cables.

Critical embedment length

Once the bolt has been installed and the threaded profile is encapsulated in grout, the grout moulds a physical inverse of the bolts thread. If the bolt is loaded axially, the loading force shears the grouted thread profile. Assuming the bolt geometry remains consistent, the resulting stress is a function of tensile force on the bolt, bolt geometry, and length of bolt in contact with grout. The critical embedment length defines the length where the bond shear strength of a given grout or resin is greater than the tensile strength of the bar, at which point the bar can be fully supported by the grout.

The critical embedment length is a crucial parameter for the TGO ground support design methodology.

A campaign of short encapsulation pull tests is conducted aiming to determine the critical embedment length of an R32 self-drilling bolt in J Lok P pumpable resin and TD80 grout. This campaign aims to verify the critical embedment length of the bar itself, and the capacity of the toe anchor inclusive of drill bit. It was anticipated that the 51 mm drill bit would greatly increase the capacity of the system. Note the shell anchor is not tested, as the mechanical interface could skew results. This approach focuses solely on assessing the encapsulation strength, with any additional strength from the mechanical anchor considered a supplementary benefit. All tests are performed at the same location.

A summary of results to date are presented in Table 5. Due to the large number of tests and the logistics involved in allowing pre-defined cure durations, tests are conducted in batches.

TABLE 5

Critical embedment results.

J Lok P 1:1 R32X Critical Embedment to reach 520 kN (90% UTS)

	Cure time		
	1 hr	**3 hr**	**24 hr**
R32X with drill bit	600 mm*	600 mm*	600 mm*
R32X	1200 mm – 1800 mm**	1200 mm – 1800 mm**	1200 mm – 1800 mm**

J Lok P 1:1 R32S Critical Embedment to reach 320 kN (90% of UTS)

	Cure time		
	1 hr	**3 hr**	**24 hr**
R32S with drill bit	600 mm*	600 mm*	300 mm
R32S	*Not yet tested*	900 mm	450 mm

TD80 Grout R32S Critical Embedment to reach 320 kN (90% of UTS)

R32S with drill bit	300 mm	-	-
R32S	300 mm	-	-

* – Shorter lengths not tested. ** – Length within range.

Test process as follows:

- Boreholes are predrilled to 51 mm and bars are manually inserted with centralisers to maintain position in the borehole, without compromising testing (Figure 6).

- Bars are fully grouted with either TD80 or J Lok P 1:1 resin.

- Each sample is pull tested to 90 per cent of UTS.

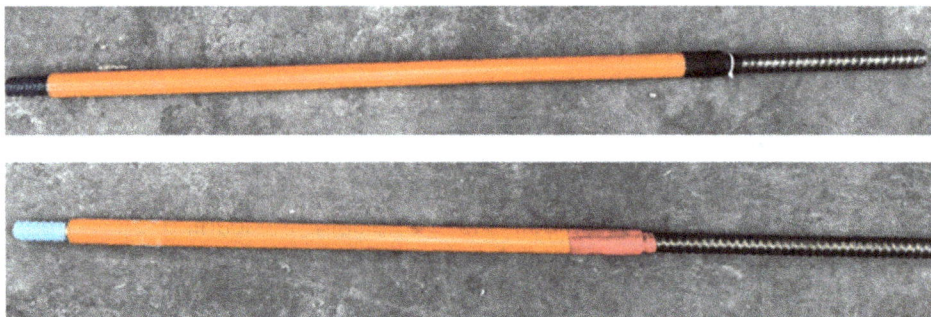

FIG 6 – Bolt configuration for embedment length testing.

Testing needs to be completed to a greater resolution to identify the critical embedment length for the R32X bar, however it is known that 1800 mm can easily support 90 per cent of UTS after 1, 3 and 24 hrs. R32S with drill bit also needs to be tested with shorter lengths, to identify the minimum encapsulation required, however it is known that the critical embedment length for R32S bar with a drill bit fitted encapsulated with J Lok P 1:1 is 300 mm after 3 hrs. 600 mm will support 90 per cent of UTS after 1 hr, however further testing is needed to determine the minimum length required. An R32X bar with a drill bit fitted can be supported to 90 per cent of UTS with 600 mm of embedment after 1 hr. To achieve an equivalent anchor, a plain strand cable bolt requires a critical embedment length of 2.0 m with cement grout (Bawden, 1994). Since 2 m of the 6 m cable bolt is needed to support the other 4 m, a 6 m cable can conveivably be replaced with a 4.3 m Falcon Bolt injected with J Lok P in a 1:1 ratio after 1 hr cure time. Pull tests of installed 6 m Falcon Bolts encapsulated in J Lok P 1:1 consistently pull test to 28 t (typical ram maximum) after 1 hr cure with no sign of displacement or failure.

In-situ anchor testing

The key advantage of the Falcon Bolt is the mechanical anchor at the bolt toe. This allows the bolt to pre-tension, facilitates extension drilling and provides confidence the bolt will not move while waiting for the grouting process. The initial Falcon Bolt design brief required that the mechanical anchor alone without grout or resin should hold at least 100 kN capacity in typical conditions. As part of the trial process, 6 m installations were pull tested prior to grouting to test this criteria, providing a large quantity of data in a range of rock conditions.

Early designs of the mechanical anchor typically exceeded the required 100 kN load, with some anchors exceeding the capacity of the ram, and others only achieving 20 kN before pulling out of the borehole. In lab conditions, all anchors consistently achieved 200 kN or more, consequently a campaign of trials was conducted to identify the cause for the inconsistent anchor strengths. As part of the campaign, five 1 m Falcon Bolts were installed in hard and soft ground and the anchors were pull tested. Each anchor achieved more than 100 kN, with three anchors exceeding 250 kN. This particular trial indicated an issue that was occurring during the 6 m installation process. Following a thorough investigation, a new anchor design was developed to address the identified issue. Since implementing the revised anchor, 100 per cent of pull tests have achieved 100 kN or more prior to grout injection. Testing is ongoing to monitor the performance of the improved mechanical anchor.

Pre-tension

The advantages of rock bolt pre-tension are well researched. For example, a study found that pre-tensioned bolts improve the distribution and transfer path of compressive stress in a fractured rock mass, ultimately enhancing the shear strength of joints (Yongshui Kang, 2023). A laboratory shear test (Roberts, 2013) found that the application of 50 kN pre-tension applied sufficient friction along the sliding shear plane to increase the shear capacity of the system by 42 per cent. Simulations conducted by (Fu-Qiang Gao, 2008) also showed that when pre-tensioned rock bolts are installed in the backs (roof), the preload applied by the bolt subjects the supported rock mass along the bolt's axis to compression. This stiffens the structure, allowing load to transfer and distribute to a greater area and thus reduce stress. When pre-tensioned bolts are installed in the walls, Fu-Qiang found that the increased horizontal stress stiffens the wall creating a strengthened boundary that assists in supporting vertical roof stress, reducing load on bolts in the backs. The simulation found that the benefits of pre-tension were proportional to the amount of pre-tension applied.

The Falcon Bolt pre-tensions the supported rock mass, prior to resin or grout injection. The amount of pre-tension the Falcon Bolt applies is dependent on torque output of the drill rig. Typical jumbos will achieve around 50 kN, however newer systems can theoretically achieve up to 100 kN (Galluzzi, 2023). The 15.2 mm twin strand cables cannot apply pre-tension. Instead, the cables are tensioned after the grout has cured. This post-tension only loads the plate against the rock face and cannot apply compression along the entire length of cable, instead only compressing a shallow depth about the collar.

The stiffer ground support system afforded by the decreased critical embedment length and pre-tensioning during installation is considered favourable for ground support design condition at TGO. In particular, frictional forces between potential failed wedges and intact rock bridges or asperities that prevent kinematic sliding can allow structures to self-support, and this can be a critical component to maintaining stability. However, such wedges may be rendered unstable later with changing conditions and stresses from adjacent mining or production voids, weathering of surfaces or the ongoing effects of water. These low stress regions are susceptible to wedge failures as the normal 'clamping' forces acting on the wedge faces are insufficient. The compressive load applied by the bolts pre-tension can assist this 'clamping' force.

Pre-tension can also work to provide resistance to shear and tension across pre-existing discontinuities by preserving the confinement and shear resistance across these faces. Allowing the faces of a wedge to maintain this self-supporting element by limiting any initial movement of a wedge can be critical. If a bolt system with increased dynamic capacity is required, Falcon Bolt variations optimised for dynamic loading can be implemented. These variations feature a smooth debonded section, and offer more than 50 kJ dynamic capacity (Galluzzi, 2023).

Installations process and time comparison

Hand installed cable bolts at TGO

The hand installtion of cable bolts consists of three key phases:

1. Drilling cable holes with either a Jumbo or production drill depending on location, purpose and timing of the ground support cycle.

   o Drilling cable holes with a production drill allows for more precise installation around production holes but also results in a larger diameter hole which requires more grout and grouting time. A greater grout anulus between the cable and hole also places the cable at risk of reduced strength due to increased sensitivity to grout strength.

2. Pushing cable bolts into these pre-dilled holes whilst working from an integrated tool carrier (IT) basket and grouting them.

   o This step includes first configuring cable bolts: unfurling them from a coil (stored energy with potential for injury), cutting grout tubes, and joining these to the cable lengths using tape.

   o These cables are then pushed vertically into the pre-drilled holes from an IT basket, which is a manually intensive operation (back injuries occur) with the added risk of disloddged rock framents falling from the pre dilled holes (eye injuries occur).

   o The hole collar must then be blocked before grouting can commence, another time consuming proccess usually carried out using cotton wadding. This wadding also results in a section of ungrouted hole at the cable collar, allowing cables to debond. (Plates are installed later in the process to prevent this.)

   o Manually mixing and pumping cementitious grout using pnuematic equipment and a breather tube means the quality is highly dependent on several work processes. Often, grout is mixed thinner than specified as it is faster and easier for operators. This results in weaker grout. Stringent QA/QC is required to constantly verify mix and grout strengths, with weak cable bolts requiring replacement.

3. After sufficient time for the grout to cure, the cables are plated and tensioned, again using an IT and basket.

   o Although the plating and tensioning is not time consuming, the IT must again be mobilised. This can mean significant tramming time for a small task, meaning ITs are not always effectively utilised. Additionally, the portable hydraulic jacking units are susceptible to high wear and tear from the underground operators.

Falcon bolt installation at TGO

Falcon Bolt installation for deep intersection support consists of two steps:

1. Drilling, installing, and tensioning the Falcon Bolt.

- o Falcon Bolts are drilled and tensioned, and are considered effective to 10 t capacity.
- o Without the flexibility and precision of using a production rig to drill and install Falcon Bolts, additional planning and coordination is required when planning brow support to stop interaction with production drilling and holes.
- o As Jumbo operators learn to work with a new bolt type and installation methodology, it is observed that initially Falcon Bolts are wasted at an average rate of 15 per cent. This is accounted for in all proceeding calculations. As more time was spent installing Falcon Bolts, this improved to 5 per cent.

2. Service crew grouts Falcon Bolts using a thixotropic grout.

- o Thixotropic grout pumped in a 'top down' injection principle removes the need for wadding and breather tubes. A quick release grout lance also increases efficiency of process.
- o Thixotropic properties prevent grout from dripping from borehole and can be mixed thicker than cementitious grout resulting in increased grout strength

Process improvement times are demonstrated through a reduction in time required by service crew to install Falcon Bolts compared to cable bolts (Figure 7 and Table 6). The additional required 'installation' and 'plate/tension' time for cable bolts adds an additional 4 to 5 hrs to the cable installation time to support to a three-way intersection at TGO. When the curing time is included in the comparison, (12 hrs for grout versus 1 hr for resin) major process improvements can be achieved.

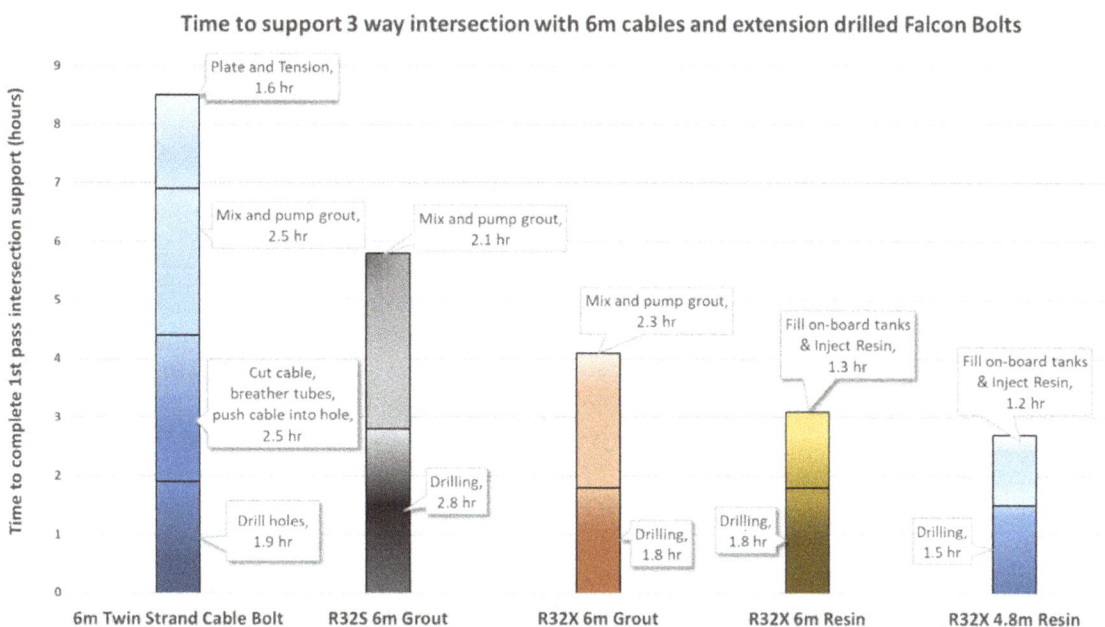

Time to support 3 way intersection with 6m cables and extension drilled Falcon Bolts

FIG 7 – 'Time on Bolt', Installation steps and times for 6.0 m twin strand cable bolt and various Falcon Bolt configurations.

TABLE 6

Deep support installation time comparison for a three-way intersection at TGO.

			Falcon Bolt			
		6.0 m Twin Strand (hrs)	R32S 6.0 m Grout (hrs)	R32X 6.0 m Grout (hrs)	R32X 6.0 m Resin (hrs)	R32X 4.8 m Resin (hrs)
1st pass (hours)	Drill holes	1.5	2.2	1.3	1.5	1.2
	Install	4				
	Grout/resin		2.1	1.6	0.8	0.7
	Cure	12	12	12	1	1
	Plate and tension	1	(Completed during installation)			
2nd pass (hours)	Drill holes	0.4	0.6	0.4	0.4	0.3
	Install	1				
	Grout/resin		0.9	0.8	0.6	0.6
	Cure	12	12	12	1	1
	Plate and tension	0.6	(Completed during installation)			
Total time (hours)	Jumbo time	1.9	2.8	1.8	1.8	1..5
	Service crew time	6.6	3.0	2.3	1.3	1.3
	Total install incl. Curing	32.5	29.7	28.1	5.1	4.7

Resin injection

The implementation of injected polyester resin provides further improvements to overall development efficiency, cost and safety. Using injected J Lok P, Jennmar's pumpable polyester resin, the total calendar time for installation and curing time for an intersection or heading was reduced from 33 hrs to 4 hrs. Jennmar and TGO collaborated closely in the design of the J Lok P resin injection system, culturing an environment that considered both the practical requirements of the mine and the engineering constraints of the manufacturer. The synthesis of these perspectives facilitated the development of an effective and usable system that integrated well into the existing mining infrastructure. This success underlines the importance of collaboration between mining operations and equipment manufacturers.The J Lok P Resin Injection System is retrofitted to an Integrated Tool (IT) Basket (Figure 8) and aims to replace conventional cement grouting with a rapid resin injection process.

The Injection System is primarily designed to streamline the installation of rock bolts by reducing the time, labour, cost and risk associated with cement-based grouting. The J Lok P Resin Injection System is powered with compressed air and delivers resin injection via two piston pumps. J Lok P polyester resin is delivered in a 1:1 ratio via an injection lance equipped with a mixing system, ensuring that the resin components are thoroughly combined before injection. Once mixed, the resin solidifies within 10 mins, achieving usable strength in 1 hr and full structural strength within 24 hrs. Through testing, it was found that the high-pressure injection enables the encapsulation of a 6 m Falcon Bolt in approximately 1 min with a single operator. The design removes the labour intensive mixing and wash-down procedures, and the risk of silica exposure inherent to cement grouting. Resin is transferred to the basket from IBCs using a specially developed transfer system.

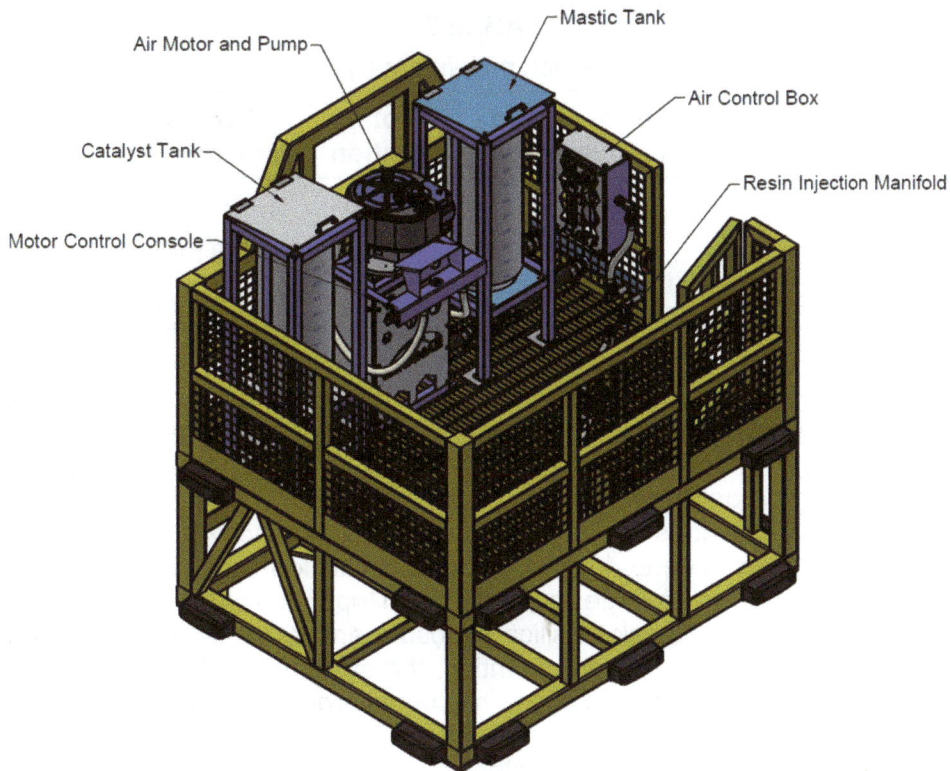

FIG 8 – Resin injection IT basket.

Prior to comissioning the Resin Injection System, three trials were undertaken to test the baskets functions and identify and rectify potential issues. During the first trial 11 6 m Falcon Bolts were injected. Some improvements were identified to prevent minor resin spills from occurring with the trial considered as successful. It was determined that injection time could be reduced from over 4 mins to 1 min if operators ensured that mixed resin is flushed from the injection lance if the system remained idle for more than 2 mins. This is due to the exponential increase of material viscosity over time once the resin components are mixed. Following the incorporation of this adjustment into the injection procedure, trials 2 and 3 demonstrated a clear improvement in operational efficiency. In trial 1, 11 bolts were injected in 1 hr, corresponding to an average duration of 6 mins per bolt. In trial 3, eight bolts were injected in 17 mins. This corresponds to an average duration of 2.1 mins per bolt to capture the entire process, with approximately 1 min dedicated to injection alone. This data is presented in Figure 9 and Table 7.

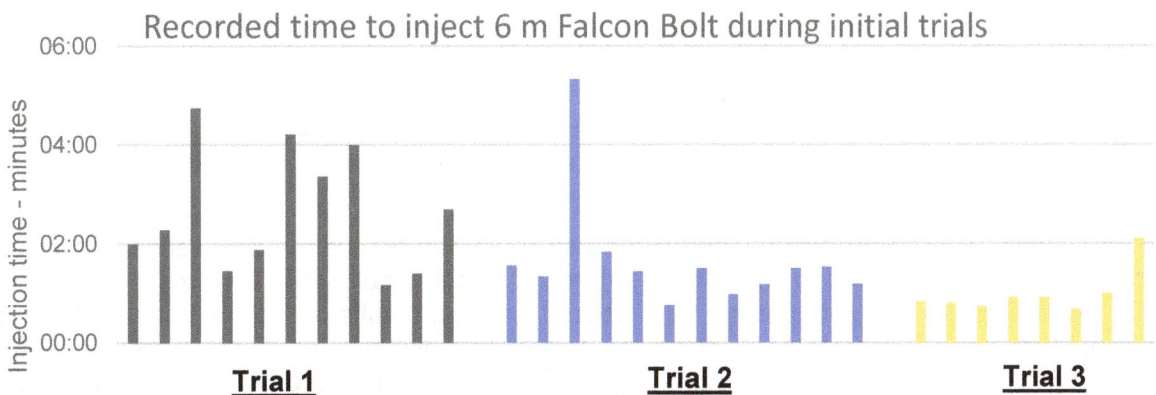

FIG 9 – Injection time per bolt.

TABLE 7

Injection trial summary.

Trial	Number of Falcon Bolts injected	Total duration	Average time per bolt
1	10	60 min	6 min
2	13	34 min	2.6 min
3	8	17 min	2.1 min

In trial 3, operators had become profficient in the operation of the resin injection system. The task consisted of tramming to worksite, connecting air and water services, then completing injection of eight bolts in 17 mins with a crew of one IT operator and one service crew operating the injection system. Note the resin injection process is very clean, the system does not require wash-down and mixing is automated, thus operators do not need to come into contact with raw resin. An equivalent grouting process can require three to 4 hrs to complete multiple mixes, injection, wash-down and plating/tensioning processes with a larger crew of one IT operator and two service crew workers. Further, the use of resin removes risk of silica exposure and dependence on cement and water ratios, which can be inconsistent. At time of writing, the system is in trial stages, however given success of J Lok P injection the system is due to be commissioned at Tomingley Gold Operatons in December of 2024. After a period of use the impacts on schedule optimisation, service crew utilisation, safety and cost will be further evaluated.

Large reductions in critical embedment length and consequently reductions in cable bolt performance can occur if there are deficiencies in grouting. In particular, cables encapsulated using the breather tube methodology can be grouted with a low water:cement ratio, greatly reducing bond strength, and therefore shear strength. The manual installation and grouting of ground support systems is becoming less common, presenting challenges in the effective transfer of skills and adherence to proper procedures, particularly during periods of staff turnover. Strict quality assurance testing and control protocols are essential to uphold the integrity of ground support performance. It is not uncommon for grout testing results to fall below established standards, necessitating the reinstallation of ground support in affected areas. The transition to Falcon Bolts and the implementation of an automated resin dosing and injection system to replace the traditional method of manually mixing and injecting cement grout is designed to enhance encapsulation consistency. This shift aims to ensure consistent installation quality and reinforces the assurance that ground support is being installed in accordance with design specifications.

Installation costs

A comparison of the operating costs of supporting a three-way intersection at TGO using 15.2 mm twin strand cables and Falcon Bolts with grout and resin is presented in Table 8. Installation costs build on the recorded process durations presented in Table 6, and data recorded during resin injection trials presented in Table 7. The study found that the direct costs to purchase ground support equipment does not represent the total cost of intersection reinforcement; machine time and labour are the most substantial cost contributors.

A thorough cost analysis shows that the Falcon Bolt R32X can reduce the direct costs of supporting intersections when compared to typical twin-strand 6.0 m cable bolts. The final costs in Table 8 are presented as a range calculated using TGO labour rates and results of Falcon Bolt installation time and motion studies. Contractor labour rates can be higher, which has a major influence on total cost per intersection, in this case the cost advantages of the Falcon Bolt become more pronounced. The analysis shows while Falcon Bolts have a higher unit cost, particularly when resin is included, their enhanced installation efficiency corresponds to an overall improvement in the operating cost of intersection support. Overtime, operators are gaining proficiency in Falcon Bolt installations, therefore installation efficiency and cost can be expected to improve. Further, as Falcon Bolt production numbers increase, the unit cost is also anticipated to decrease.

TABLE 8
OPEX cost comparison of different support options, total cost for a three-way intersection.

	6.0 m Twin strand cable bolt	Falcon Bolt			
		R32S 6.0 m grout	R32X 6.0 m grout	R32X 6.0 m resin	R32X 4.8 m resin
	No longer used by TGO	Currently used by TGO	-	In Trial	Not yet trialled
Element Supply, Grout/Resin and Consumables	Reference price (100%)	281%	238%	307%	276%
Drilling cost		122%	97%	97%	78%[+]
Service crew and IT cost		30%	27%	10%	10%[+]
Cables: Cut, attach breather, push, grout, plate and tension					
Falcon Bolt: Inject grout or resin					
Total cost per intersection (% of reference)		88–100%*	74–83%	80–89%	70–78%[+]

* – Drilling and installing 1.5× more in the case of R32S 6.0 m Falcon due to lower steel strength. [+] – Not yet trialled, projected based on available measurements.

The introduction of the R32X steel, which offers equivalent strength to twin strand cables, allows a bolt pattern similar to the previously used twin strand cables. During trials, the time needed to drill and tension a pattern of R32X Falcon Bolts was found to be nearly equivalent to the time required to pre-drill holes for cable bolts. The additional time spent tensioning the Falcon Bolt is offset by the reduced hole size and the elimination of drill steel extraction and decoupling processes. The greatly decreased critical embedment length provides further opportunities for optimisation through the use of a 4.8 m Falcon Bolt, with projected installation times detailed in Figure 6. Eliminating the plating and tensioning process negates the need for a return visit, further alleviating pressure on the development cycle. As summarised in Figure 6, the time required from the service crew is substantially reduced. The direct savings are presented in Table 8.

The broader benefits to schedule optimisation and the additional time now available to service crew are more complicated and harder to quantify. A brief comparison of TGO's scheduled versus actual cable installations and scheduled versus actual development metres before and after implementation of Falcon Bolts attempts to investigate these benefits. Since the adoption of Falcon Bolts as secondary intersection support in April 2024. Although Jumbos spend more time installing Falcon Bolts (1.5× drill metres due to lower strength R32S steel and production rigs no longer pre-drilling brow support) the development metres have not decreased, and the variance to schedule has actually improved. Similarly, improvements are evident in the variance between scheduled versus installed cable/Falcon Bolts. Further practical benefits are expected with the adoption of the stronger R32X Falcon Bolt and resin injection. Development and intersection support schedules are influenced by a broad array of factors, and a more comprehensive analysis will be conducted to explore the impacts of this ground support strategy on the overall mine development process.

CONCLUSIONS

This case study presents key learnings from the development and implementation of a self-drilling bolt to replace cables at Tomingley Gold Mine. The project was driven by the impracticality, inefficiency and high-risk exposure of manual cable installations. After investigations, it was found that the geotechnical design methods and conditions were favourable for the change, however Jennmar was requested to improve the strength of the R32S bolt, and developed the R32X. Upon the full-scale implementation of the 6 m Falcon Bolt R32S as a replacement for 6 m twin-strand cables, it was determined that the overall cost per installed intersection remained comparable despite requiring more bolts, and the implementation of the R32X Falcon Bolt in 6 m format is expected to reduce costs of ground support installation. The efficiency of the installation process provided opportunity to optimise priorities for service crews and jumbos, correlating with an improvement to scheduling. For the first time, actual development and number of intersections installed exceeded schedule. Testing is currently underway to justify a change to a 4.8 m Falcon Bolt, which is anticipated to reduce costs further still. The commissioning of a purpose developed resin injection system is expected to offer further benefits to scheduling, installation efficiency, risk reduction, and QA/QC improvements.

Feedback from operators across all four crews involved in the development process has been positive, and the active pursuit of input from workers across these crews at TGO was crucial to the success of the change management and the product development process. The safety improvements offered by the removal of the cable installation process are significant. In six months, three injuries occurred as a direct result of the cable installation method, these hazards are now entirely removed. This opportunity emerged from the collaboration between Jennmar and TGO management and geotechnical team, and the project's success highlights the importance of such collaborative efforts.

ACKNOWLEDGEMENTS

The Authors thank Naj Aziz, Adel Mottahedi and technical staff at University of Wollongong whose expertise in managing laboratory shear testing was a crucial to the completion of this project.

REFERENCES

Aziz, N, Craig, P, Mirza, A, Rasekh, H, Nemcik, J and Li, X, 2015. Behaviour of Cable Bolts in Shear; Experimental Study and Mathematical Modelling, in *Proceedings of the 15th Coal Operators' Conference*, pp 146–159 (University of Wollongong, The Australasian Institute of Mining and Metallurgy, and Mine Managers Association of Australia).

Bawden, W, 1994. Integrated Seismic-Stress-Geomechanical Analysis of a Cable Bolted Back Failure, Mines Gaspe, Canada, 1st North American Rock Mechanics Symposium.

Bray, P, Johnsson, A and Schunnesson, H, 2019. Rock reinforcement solutions case study: Malmberget iron ore mine, Sweden, in *Deep Mining 2019: Proceedings of the Ninth International Conference on Deep and High Stress Mining*, pp 191–204 (The Southern African Institute of Mining and Metallurgy: Johannesburg).

Galluzzi, R, Holden, M, Dodds, A, Matthews, J and Bennett, A, 2023. Design, development and testing of the Falcon Bolt, in *Ground Support 2023: Proceedings of the 10th International Conference on Ground Support in Mining*, pp 271–284 (Australian Centre for Geomechanics: Perth).

Gao, F-Q and Kang, H-P, 2008. Effect of pre-tensioned rock bolts on stress redistribution around a roadway—insight from numerical modeling, *Journal of China University of Mining and Technology*, 18(4):509–515.

Hutchinson, D J and Diederichs, M S, 1996. *Cablebolting in Underground Mines* (BiTech Publishers: Canada).

Jardine, J, Roache, B, Thomas, S and Jere, P, 2023. Intersection reinforcement with self-drilling anchors for improved productivity at Kanmantoo, in *Proceedings of the Underground Operators Conference 2023*, pp 369–379 (The Australasian Institute of Mining and Metallurgy: Melbourne).

Kang, Y, Hou, C, Xu, C, Liu, B and Xiao, J, 2023. Investigation on mechanical behavior of pre-tensioned bolt in fractured rock mass using Continuum Discontinuum Element Method (CDEM), *Engineering Analysis with Boundary Elements*, 151:30–40.

Knox, G and Hadjigeorgiou, J, 2022. Shear performance of yielding self-drilling anchors under controlled conditions, in *Caving 2022: Proceedings of the Fifth International Conference on Block and Sublevel Caving*, pp 201–212 (Australian Centre for Geomechanics: Perth).

Potvin, Y and Hadjigeorgiou, J (eds), 2020. *Ground Support For Underground Mines* (Australian Centre for Geomechanics: Perth).

Roberts, T and Dodds, A, 2013. Design, development and testing of the JTech bolt for use in static, quasi-static and dynamic domains, in *Ground Support 2013: Proceedings of the Seventh International Symposium on Ground Support in Mining and Underground Construction* (eds: Y Potvin and B Brady), pp 305–321 (Australian Centre for Geomechanics: Perth).

Safari, A and Jere, P, 2023. Improving underground development cycle time using performance mine grouts, in *Ground Support 2023: Proceedings of the 10th International Conference on Ground Support in Mining*, pp 593–606 (Australian Centre for Geomechanics: Perth).

Stjern, G, 1995. Practical performance of rock bolts, PhD thesis, University of Trondheim, Trondheim, Norway.

Ventilation challenges and opportunities at tunnelling construction operation – Snowy 2.0 Powerhouse Complex Ventilation

M Jamieson[1], C Giordano[2] and M Hooman[3]

1. Senior Engineer, BBE Group, Perth WA 6027. Email: mjamieson@bbegroup.com.au
2. Senior Engineer, Future Generation Joint Venture, Cooma NSW 2630.
   Email: c.giordano@futuregenerationjv.com.au
3. Principal Engineer, BBE Group, Perth WA 6027. Email: mhooman@bbegroup.com.au

ABSTRACT

The Snowy 2.0 project involves linking two existing dams, Tantangara and Talbingo, through 27 km of tunnels and building a new underground power station. Water will be pumped to the upper dam when there will be surplus renewable energy production, and the energy demand will be low and then released back to the lower dam to generate energy when the electricity demand is high. It will provide power while reusing the water in a closed loop and maximise the efficiency of other energy sources to pump water to the higher dam, which will be stored for later use.

This paper focuses on the ventilation challenges, implementation, legislative requirements and planning on one of the sites, taking upfront designs into the execution phase with development and construction taking place before handover for long-term use. Lobs Hole consists of the two main tunnels, the Main Access Tunnel (MAT) and the Emergency Cable Ventilation Tunnel (ECVT), both developed by an 11 mØ Tunnel Boring Machine (TBM) that form the accesses to the two main powerhouse chambers developed by drill and blast techniques. These chambers, or 'Halls', will house the electric power station and turbines and will be two of the largest underground chambers in the world (60 mH × 30 mW × 250 mL).

INTRODUCTION

Snowy Hydro Ltd is constructing Snowy 2.0 as the next stage in the Snowy Hydro Scheme's Project. It is a renewable energy Project that will provide on-demand energy and large-scale power storage, the largest in Australia. Snowy 2.0 will provide an additional 2200 megawatts (mW) of generating capacity and approximately 350 000 megawatt hours (mWh) of large-scale storage to the National Electricity Market. This is enough energy to power three million homes over the course of a week (Snowy Hydro Ltd, 2024b).

Future Generation is a joint venture created specifically to establish Snowy 2.0 on behalf of Snowy Hydro Limited (Snowy 2.0), bringing the combined engineering expertise of three companies consisting of Italy's Webuild (formerly Salini Impregilo), Australian-based Clough, and US-based Lane Construction. Combined, they form the contractor to build the project as Future Generation Joint Venture (FGJV) completed the planning and engineering phases and is to undertake the construction phase that aims to deliver the Snowy 2.0 Project on behalf of Snowy Hydro Limited (Snowy Hydro Ltd, 2024a).

The Project involves linking two existing dams, Tantangara and Talbingo, through 27 km of tunnels and building a new underground power station 800 m below surface. Water will be pumped to the upper dam in a closed circuit when there is surplus renewable energy production utilising excess solar and wind energy when the demand for energy is low. Water will be released back to the lower dam utilising the hydraulic head between the dams to generate energy when electricity demand is high (Figure 1; Snowy Hydro Ltd, 2024b).

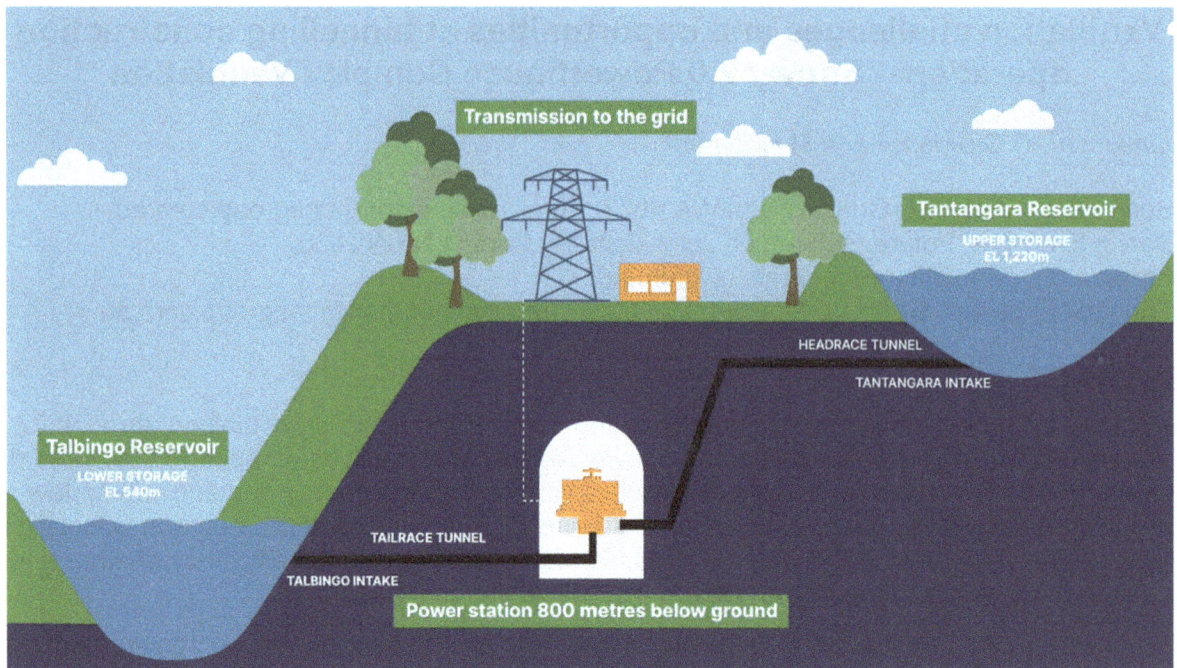

FIG 1 – Snowy 2.0 pumped-hydro system overview (Snowy Hydro Ltd, 2024b).

At the Snowy 2.0, there will be four development sites (Figure 2):

1. Talbingo – Tailrace tunnel inlet/outlet.
2. Lobs Hole – Main Access Tunnel (MAT) and Emergency/Cable/Ventilation Tunnel (ECVT), Tailrace Pressure Shaft and Powerhouse complex.
3. Marica – Headrace pressure shaft.
4. Tantangara – Headrace tunnel inlet/outlet.

FGJV contracted BBE Consulting Australasia to work with Systra™, an engineering company based in Italy, to provide on-site specialist ventilation consulting work with the planning, implementation, and industry compliance of the ventilation scope. FGJV started the development of the ECVT access declines to the subsurface caverns (Transformer Hall and Machine Hall) in 2021. Cavern excavation started in Jun-2023 and tunnel boring machines (TBMs) started development in 2022 at the tailrace and headrace tunnels.

This paper aims to discuss the challenges and critical solutions employed to change the primary and secondary ventilation circuits at the Lobs Hole site (Figure 2) to adequately ventilate the Transformer and Machine Halls, which will be two of the largest subsurface chambers in the world. The paper will also discuss value-add engineering practices used to conceptualise, design and install ventilation network controls to maintain the legislated level of safety and health for personnel during development and construction.

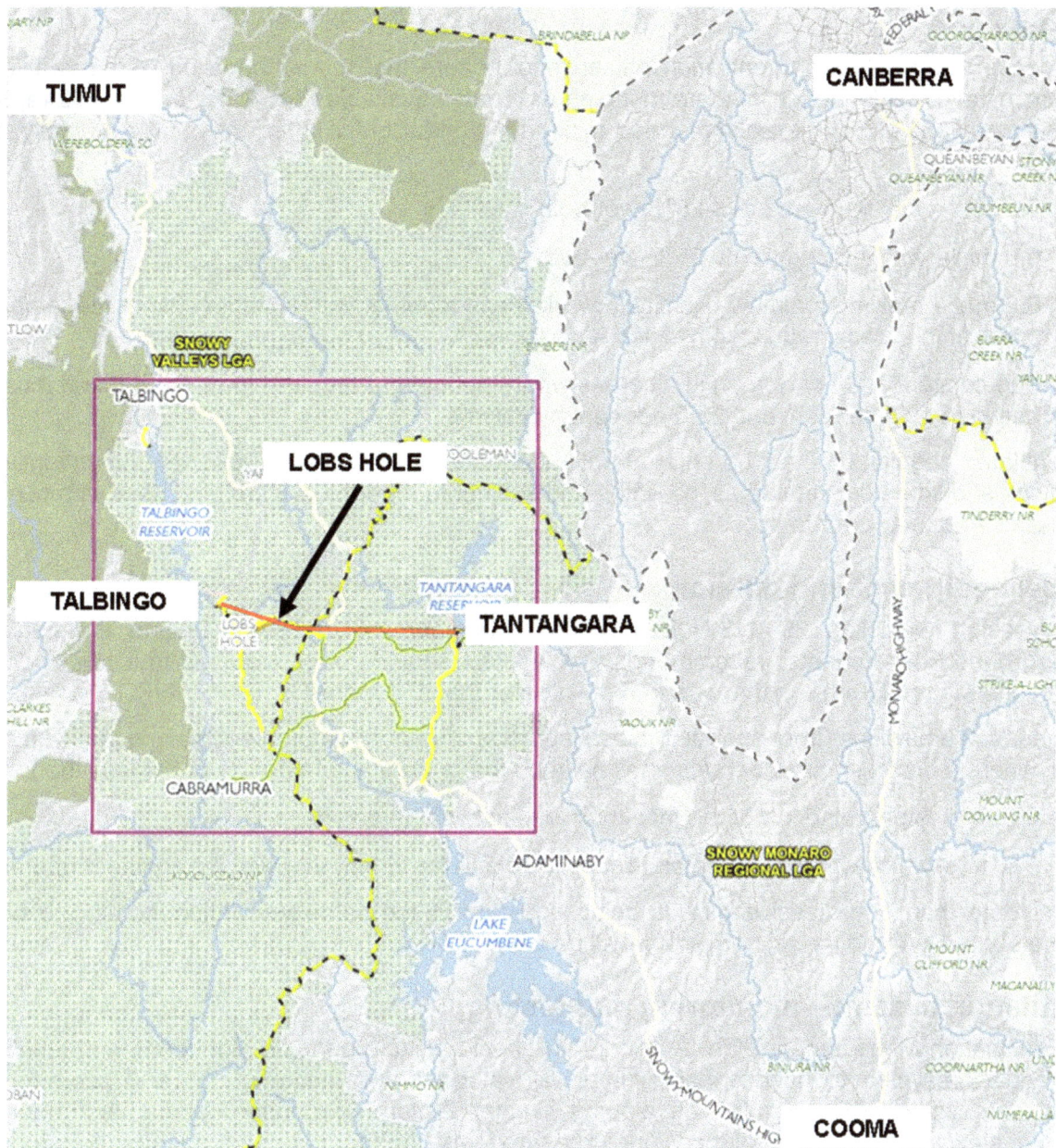

FIG 2 – Snowy 2.0 and Lobs Hole location (Snowy Hydro Ltd, 2024b).

VENTILATION DESIGN CONSIDERATIONS

Tunnelling guidelines and industry best practice standards for air quality, contaminant dilution and monitoring are discussed in this section. A summary of the ventilation criteria has been provided.

Standards

The Project complies with the SafeWork Australia – Tunnelling Guide and industry-regulated exposure limits, and it is based on the following reference documents:

- SafeWork Australia – Guide for Tunnelling Work (Safe Work Australia, 2013)

- SafeWork Australia – Workplace Exposure Standards (Safe Work Australia, 2024)

- MDG 29: Guideline for the management of diesel engine pollutants in underground environments British standard, BS 6164 code of practice for safety in tunnelling in the construction industry (New South Wales Department of Primary Industries (NSW DPI), 2008).

Air quality

The construction team and ventilation officer must ensure that the air supplied to the ventilation system in the tunnel(s) are obtained from the purest source available (Snowy Hydro Ltd, 2024a; Safe Work Australia, 2013). The recommended minimum standards for sufficient volume, velocity, and quality are:

- 0.3 m/s – with temperature ≤ 25°C wet-bulb

- 0.5 m/s – with temperature ≥ 25°C wet-bulb.

- 6.0 m/s – maximum airflow velocity in work areas together with suitable dust management in place (if in excess, regulated by Risk Assessment).

- The concentration of oxygen in all areas where people are required to work, or travel must be between 19.5 per cent and 23.5 per cent by volume.

Unventilated headings will not be entered until adequate ventilation is established. All unventilated headings should be designated as 'no entry' and subject to a permit system of access with controls to assess air quality prior to access.

Diesel equipment and dilution

Mine Design Guideline 29 (NSW DPI, 2008), 'Management of Diesel Engine Pollutants in Underground Environments', recommends that workers should not be exposed to levels of Diesel Particulate Matter (DPM)—in the form of EC—greater than 0.1 mg/m^3.

For gaseous emissions (from operating diesel equipment), the minimum ventilation quantity in each place where a diesel engine operates shall be such that a ventilation quantity of not less than:

- 0.067 m^3/s per rated kW of the maximum capacity of the engine, or

- 3.5 m^3/s, whichever is the greater, is supplied along the airway in which the engine is operating.

When more than one diesel engine is being operated in the same ventilating circuit, the diesel engine's rated kW and the related airflow will be cumulative.

Ventilation system – monitoring and testing

The Ventilation Management Plan (Anon, 2024) should include periodic monitoring and testing of the ventilation system to validate that minimum air velocities and dilution of dust and gases remain within safe limits. This includes developing Trigger Action Response Plans (TARPs) where emissions exceed 50 per cent of the Work Exposure Limit (WEL).

The ventilation management plan should include the following:

- Diesel gas and DPM management plan.

- An emergency preparedness management plan, together with current, up-to-date ventilation plans showing refuge chambers, critical ventilation infrastructure fans, and control devices with directions and magnitudes of airflows throughout the workings.

- Specifies the maximum rated diesel engine power that may operate in each tunnel section and how this will be managed.

- Stench gas system, leaky feeder and radio communication system to alert workers in an emergency.

- Welding standard work practices to minimise workers' exposure to welding fumes.

The summary of ventilation criteria is continuously monitored to ensure compliance as each of the sites that will be discussed further.

THE MAIN FOCUS – VENTILATION REQUIREMENTS AT LOBS HOLE (MAIN ACCESS TUNNELS)

The Lobs Hole site will consist of two 9.9 mØ concrete lined TBM excavated tunnels that will be up to 3.5 kms each when complete. These tunnels (yellow) provide connections to the Powerhouse Complex (Transformer and Machine Halls) as shown in Figure 3. Both Halls (pink) will be 250 m long, 30–40 m wide and 80 m high, and will connect to the penstock (olive) and draft tube tunnels (cyan) to power the turbines. The tunnels established perpendicular off the main TBM tunnels will be an average size of 9 m wide × 9 m high and will be established by drill and blast techniques.

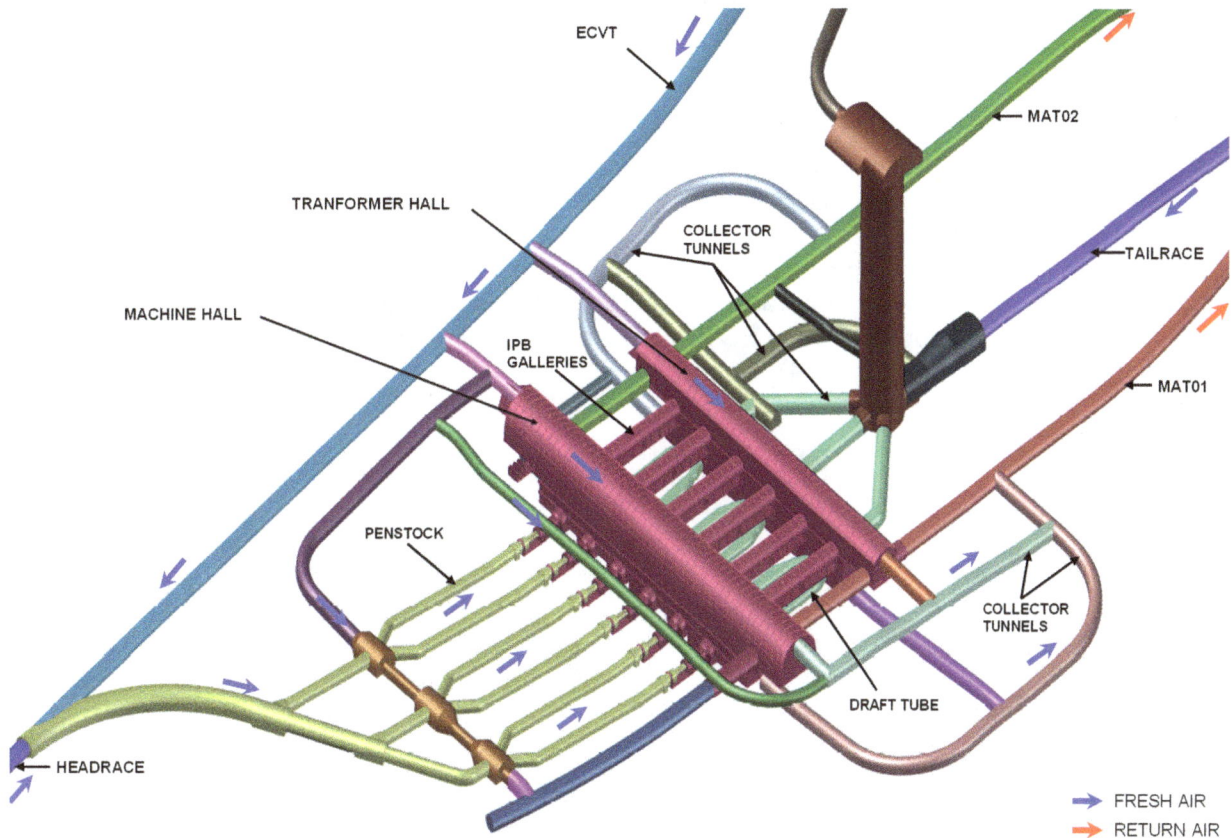

FIG 3 – Lobs Hole MAT and ECVT – Powerhouse Complex (isometric view; Fratello and D'Ulisse, 2024).

Challenge – establishing a primary ventilation network during development

Each of the MAT and ECVT tunnels will be developed from surface. Force ventilation ducts from surface in up to 3.5 km will have to supply adequate ventilation to the back of each TBM train which will be challenging to achieve. Connecting cross-cuts between the intake airway, ECVT, and return airway, MAT, will be established while the TBMs progress.

During the development phase, the ECVT and MAT tunnels will be force-ventilated via two twin-stage 315 kW fans (four fans installed) and one single-stage 500 kW fan from the portals utilising 2.4 mØ and 3.0 mØ ducting. As soon as through-ventilation is established at the Machine and/or Transformer Halls, the ECVT fans will be used as Primary Fans, and the MAT fans will be repurposed as Secondary Development Fans. Air will naturally ventilate to the surface via the MAT tunnel as all ventilation control devices will be commissioned. The ventilation requirement during the development phase is approximately 480 m³/s and will be increased to ~900 m³/s to serve the construction phase.

The total airflow of 480 m³/s can serve a total diesel power of 7.2 MW (at 0.067 m³/s per kW) to achieve an average velocity of 1.5 m/s throughout underground. The total rated power and number of equipment required during the development phase will be 3.26 MW. The ventilation layout included the following as indicated in Table 1.

TABLE 1

Development phase infrastructure requirements.

Ventilation control/fan	# of units
Primary fans 315 kW	8
Primary fans 500 kW	1
Auxiliary fans 55–160 kW	3
Regulated bulkhead	1
Brattice wall	1

Establishing a primary ventilation network during construction

The main challenge with establishing minimum ventilation requirements in large, excavated tunnels is that there are no shafts to use as intake or return airways to provide 200 m³/s of fresh air into each of the Machine and Transformer Halls to satisfy the minimum airflow quantity and velocity criteria of 0.3 m/s would be managed on the working floor via strategically placed air-movers. Therefore, it was recommended to use the TBM headrace and tailrace tunnels as intake and return in lieu of shafts. Fresh intake air will be supplied via the ECVT tunnel and will be returned via the MAT tunnel creating a primary ventilation circuit (Figure 4).

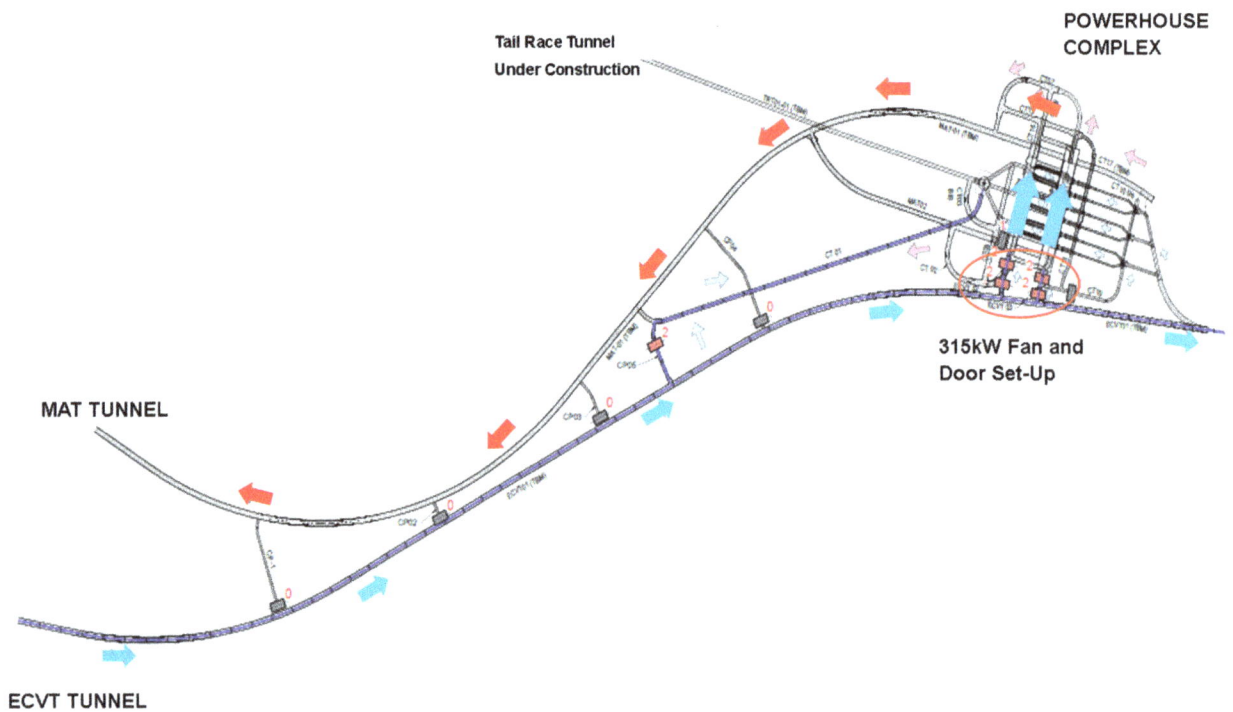

FIG 4 – Primary ventilation circuit (plan view; Fratello and D'Ulisse, 2024).

This is achieved by installing airlock doors to the access of each hall supported by two single-stage 315 kW fans each (Figure 5). These fans act as the Primary Fans as they are the driving force behind the primary circuit. To minimise contamination, personnel and light vehicles access the complex only via the fresh air intake ECVT. All other heavy vehicles access via the MAT as velocities are expected to reach close to 9.0 m/s towards the surface sections of the MAT.

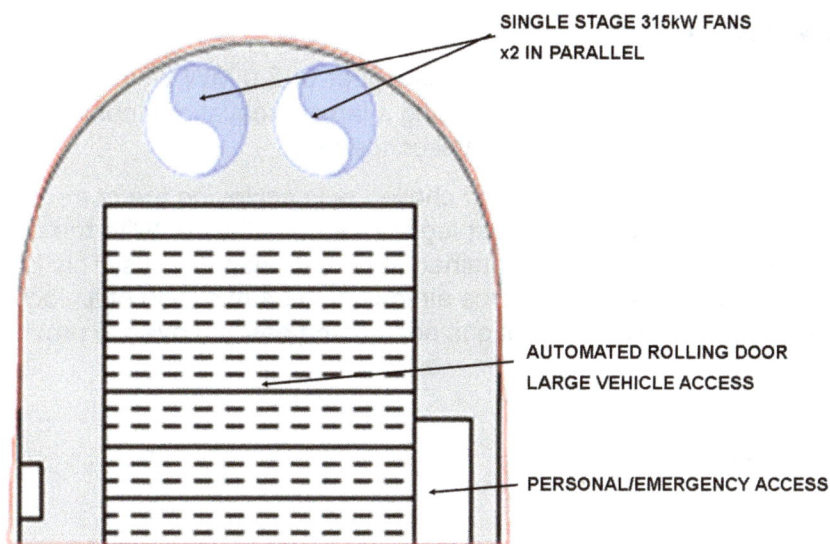

FIG 5 – Door and fan arrangement (section view; Snowy Hydro Ltd, 2024b).

Ventilation of halls during development and construction

Once the primary ventilation circuit is established, the challenge will be to establish a secondary system to adequately ventilate the large halls. Secondary rigid steel ducting teed off from the primary ventilation ducts from the ECVT tunnel will supply air to each of the Halls. Reduced sections of ducts will be installed in each Hall and will be extended vertically down until it reaches each active development level (Figure 6).

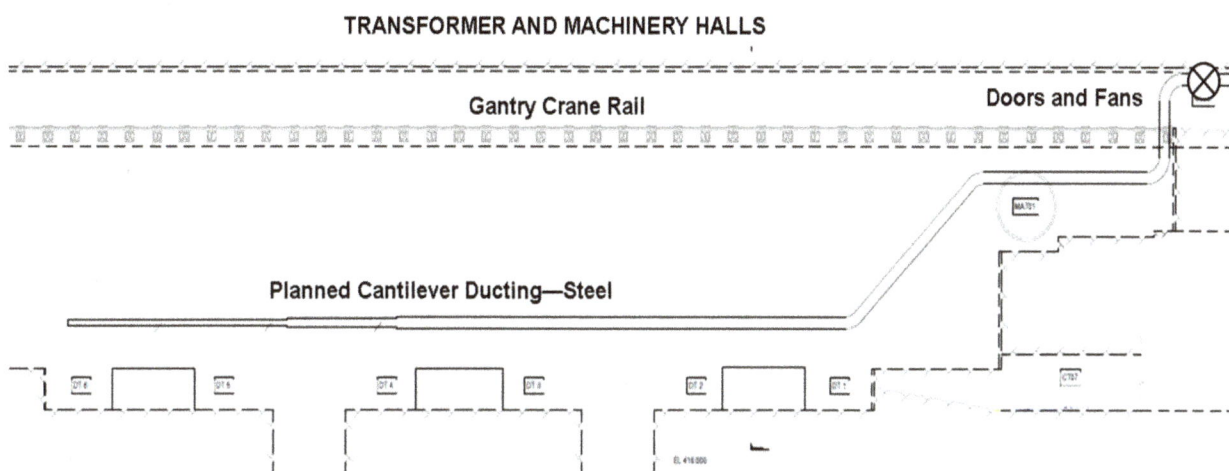

FIG 6 – Hall cross-section (Snowy Hydro Ltd, 2024b).

Ventilation distribution to each Hall will be achieved by using step-down duct sections at strategic locations along the Hall. Modelling and duct calculations were completed to ensure the set-down cantilever duct system is established at the correct locations as indicated in Figure 6. The step-down cantilever duct will comprise flexible ducting to 1.6 mØ, 1.4 mØ and finally 1.2 mØ.

Each section will have two smaller openings ventilating the development floor as the duct extends. The pressure in the duct is planned be at ~1000 Pa to 2000 Pa to ensure equal airflow quantities out of each opening along the length of the steel duct.

With both ducts set-up in this configuration in each Hall, the 200 m³/s can be evenly distributed along the Hall's development floors. The schedule and distribution of works were calculated using diesel fleet loading to create a base to justify predicted contamination dilution rates.

Meeting airflow requirements

The quantity at 200 m³/s gives an allowable diesel fleet of 3.0 MW which is ~200 per cent of the planned diesel fleet in a Hall at any one time. Being a large excavation double the amount of airflow was designed to mitigate dust and heat potential concerns.

Once the ventilation system is set-up, the next challenge is achieving the correct velocity to support personnel on the construction floor and meet legislative compliance. With the required quantity in each Hall, the velocity is required to be maintained at no less than 0.3 m/s. This is then achieved via the use of large-scale air movers that will force airflow up to 34 m ahead of its location. Strategically placed fans ensure compliance with regulations and can be easily moved to provide adequate air to active work areas (Figure 7).

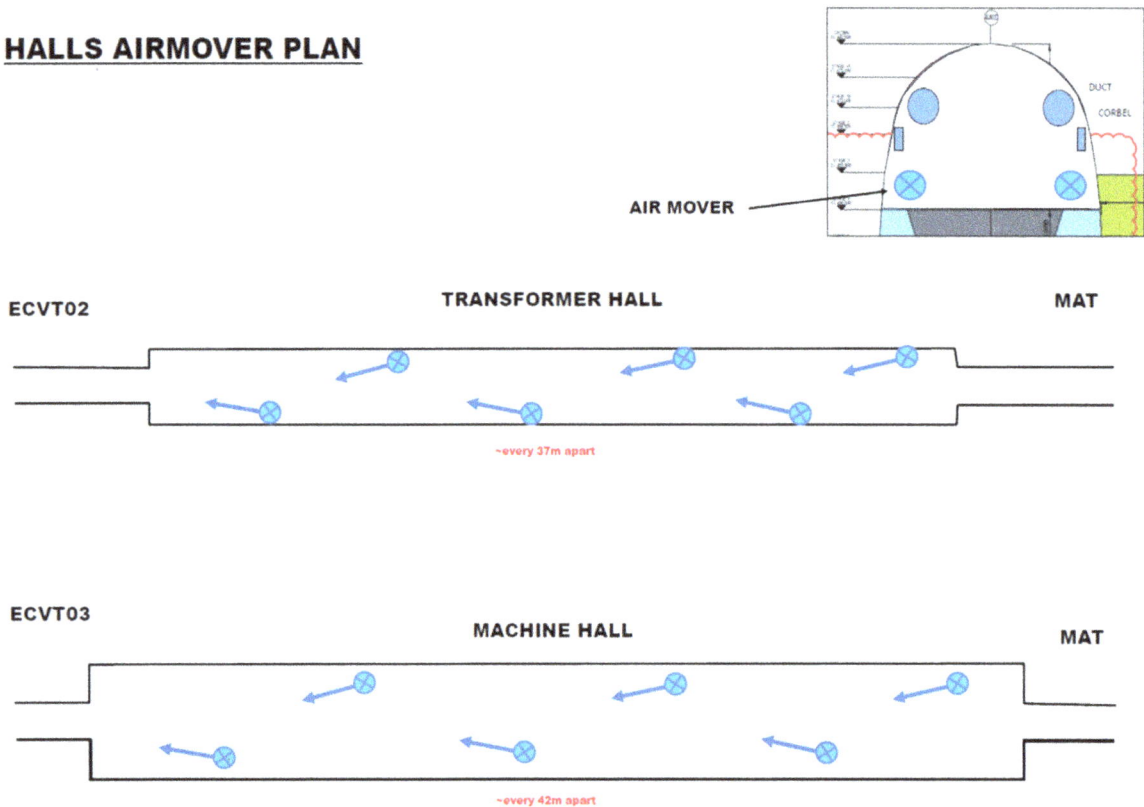

FIG 7 – Air mover placement (top view).

Ventilation modelling software was used to evaluate airflow movement, velocity, and heat mapping and to minimise any potential recirculation/reuse of airflow in the Halls. In addition, software is used to ensure air moves from south (ECVT) to north (MAT) through the Halls. This was necessary to identify compliant airflow movement and to ensure the proposed system set-up could meet the optimal working conditions.

Modelling is also used to identified if ventilation environmental conditions would be acceptable for any upcoming change in the development and construction works when mobile fleet would also increase.

Ventilation controls and system optimisation

To achieve the circuit, the main driving force behind the network is the 315 kW force fans from the ECVT portal and the airlock doors at the ECVT side access to each Hall. These are essentially the system's primary fans that deliver the 200 m³/s to the Halls. Figures 5 and 8 shows cross-sectional views of the Transformer and Machine Halls detailing the essential ventilation system requirements and controls to adequately ventilate the halls to acceptable guidelines.

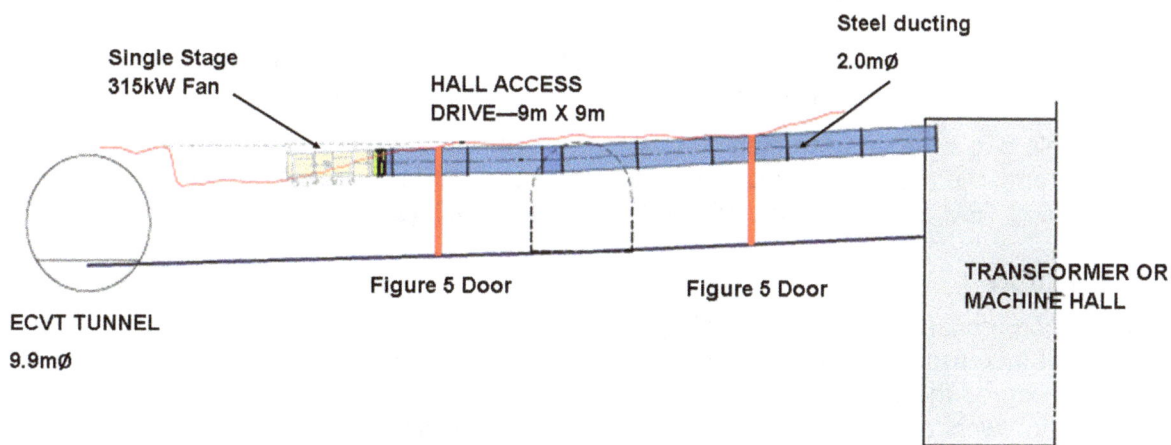

FIG 8 – Fan/door layout (Snowy Hydro Ltd, 2024b).

The pressure ratings for the doors/walls used in this application required an operational design load of ~500 Pa to suit locations with areas up to 100 m^2 in size. Little to no access will be through these doors as the development profile of the halls have a 'drop-off' 20–30 m past them as part of the hall design. All access for personnel and machinery will be via the MAT side.

Overall, the project was constrained in the extent ventilation controls could be implemented, as nearly all areas required full unrestricted access. Breakthroughs occurred in the halls from the penstock and draft tubes, which made ventilation planning more challenging.

During the ventilation set-up there was an opportunity to acquire certain fans/infrastructure readily available from other finished tunnelling projects (WestConnex, Rozelle Interchange etc). The airlock system for the halls was evaluated for infrastructure movement prior to fabrication and required a temporary works engineering design to manage the pressure load of <500 Pa. The doors also needed to be easily installed, maintained and removeable.

A set of doors was required at each breakthrough into the halls. They were designed with automation for frequent access and ease of access for large machinery movements (eg trucks). These doors were designed to be relocatable, robust in high demand shift changes, operate with high open/close cycles per hour, and withstand easy damage, etc. A robust control system was required to minimise ventilation disruption during operation and provide smooth operation.

A total of 32 ventilation monitoring stations were installed, which allows project personnel to monitor the airflow quantity and quality recorded. Velocity, temperatures, and respirable dust are also monitored at these stations.

SUMMARY

Whilst the Snowy 2.0 project is overall a basic ventilation system, the project around the halls faced several challenges, which were modelled, engineered, planned and executed with careful analysis, leaving space for on-site adjustments as required.

Changing the primary circuit to a dedicated intake and return system in lieu of shafts enabled clean fresh airflow to be distributed more efficiently at the required air quantity to reach the halls. Shafts were not an option this late in the project as the site is in a federal national park and surface environment disturbance approval take a long time to be granted.

Providing sufficient airflows to support hall completion and meet legislative compliance was planned by utilising ventilation modelling software via Ventsim™ and VUMA3D to evaluate airflow quantity and work area velocity (Fratello and D'Ulisse, 2024). Another challenge is that the project is restricted to utilising suitable ventilation controls, ie doors, regulators, brattice walls, parachutes etc, as unrestricted access is required by development crews and machinery/infrastructure (as per construction projects, Sydney tunnelling etc). Modelling to control and manage air distribution to each work area ensured airflow came directly from the surface, which maintained temperature and quality.

Due to the size of the halls and limited control devices utilised due to gantry crane infrastructure, the industrial cantilever ensured an adequate quantity of airflow was distributed in the halls, with on-the-ground air-movers ensuring a minimum velocity of 0.3 m/s for personnel.

The Lobs Hole site is required to consider its environmental footprint, resulting in minimal surface disturbance and the complex's requirement to have no raised bored shafts by utilising the predevelopment TBM tunnels as primary ventilation infrastructure.

CONCLUSION

Communication, consultation and engaging all stakeholders were crucial in overcoming the main challenges of increasing airflow and ventilation of two of the largest underground caverns in the world. This ensured the safety and success of the ventilation network and distribution of airflow to the construction work areas. By addressing each stage of ventilation change with development on-site and consultants in Torino, Italy, with a systemic and innovative approach, the Lobs Hole site was able to create and implement a primary and secondary ventilation system that supported the project development and construction life and contributed to the overall success of the Snowy 2.0 Project.

By utilising mining, tunnelling and industrial ventilation solutions, the permanent works team were able to successfully identify locations for the ventilation infrastructure with minimal changes as the project progressed, especially with the halls developing deeper.

Overall, the designed solutions implemented for the ventilation at Lobs Hole highlight the importance of thorough planning, analysis and innovation in such a large-scale development project.

ACKNOWLEDGEMENTS

The authors would like to thank the Snowy Hydro Ltd, Snowy 2.0 and Future Generation JV management teams, engineers from Systra, Torino, Italy, for their collaborative and continued support and permission to present this paper. They also acknowledge the assistance and support provided by the various drill and blast and construction managers and engineers.

REFERENCES

Anon, 2024. Ventilation Management Plan, Health and Safety Management Plan, Future Generation JV.

Fratello, L and D'Ulisse, G, 2024. VUMA3D modelling, Torino, Italy.

New South Wales Department of Primary Industries (NSW DPI), 2008. Guideline for the management of diesel engine pollutants in underground environments (MDG 29), Mine Safety Operations Division, New South Wales Department of Primary Industries, Available from: <www.dpi.nsw.gov.au/minerals/safety>

Safe Work Australia, 2013. Australian Guide for Tunnelling Work, Chapter 5, Safe Work Australia, Available from: <https://www.safeworkaustralia.gov.au/system/files/documents/1702/guide-tunnelling.pdf>

Safe Work Australia, 2024. Workplace exposure standards for airborne contaminants, Safe Work Australia, Avail from: <https://www.safeworkaustralia.gov.au/sites/default/files/2024–01/workplace_exposure_standards_for_airborne_contaminants_-_18_january_2024.pdf>

Snowy Hydro Ltd, 2024a. Snowy 2.0: Introduction and Lobs Hole Location, Snowy Hydro Ltd, Available from: <https://www.snowyhydro.com.au/snowy-20/about>

Snowy Hydro Ltd, 2024b. Snowy 2.0 documents and drawings, Snowy Hydro Ltd, Available from: <https://www.snowyhydro.com.au/snowy-20/documents>

Prominent Hill Wira Shaft sink

D Lagacé[1], D Kilkenny[2] and C Hill[3]

1. Package Manager – Shaft, BHP Prominent Hill Operations, Adelaide SA 5095.
   Email: daniel.lagace@bhp.com
2. General Manager, Raising Australia Pty Ltd, Perth WA 6150.
   Email: dave.kilkenny@raisingaustralia.com.au
3. Project Manager – Shafts, Byrnecut Australia Pty Ltd, Perth WA 6150.
   Email: christopher.hill@byrnecut.com.au

ABSTRACT

BHP's Prominent Hill Operations in South Australia is undergoing a major mine expansion project. The scope includes the construction of orepasses, a gyratory crusher, a loading station, a hoisting shaft, an overland conveyor system, and a ventilation upgrade. In Australia, the use of hoisting shafts for materials handling has been relatively limited compared to regions with developed underground hard rock mining industries. Shaft sinking projects have been carried out sporadically over recent decades, often drawing expertise from areas where shaft sinking is more common. Nevertheless, several large-capacity hoisting shafts have been successfully constructed in Australia and continue to operate today. Mine shafts remain a highly effective method of materials handling, particularly as an option for electrified material haulage.

This paper presents a case study on the application of the strip-and-line method for sinking a 7.55 m diameter, 1326 m deep, concrete-lined hoisting shaft. Constructed by Byrnecut Australia Pty Ltd, this shaft is pivotal in increasing underground production to 6.5 Mt/a by establishing a materials handling and hoisting system. The project leveraged prior shaft sinking experience to incorporate key improvements in shaft sinking arrangements, stage design, and shaft sink mining practices. Many of these enhancements have resulted in notable safety and productivity improvements, offering a credible path forward for similar projects.

The paper discusses the design, planning, and execution of the main shaft sinking, highlighting improvements to existing practices, the use of innovative stage designs, and the successes and challenges encountered during the project.

BACKGROUND

Located 650 km north-west of Adelaide, South Australia, Prominent Hill Operations form part of BHP's Copper South Australia portfolio, alongside Olympic Dam and Carrapateena. Initially established in 2009 as an open pit copper-gold mine, Prominent Hill transitioned to underground mining in 2012, starting with the Ankata deposit and later expanding to the Malu ore reserves beneath the open pit. With a processing capacity of 10 Mt/a, the mine currently produces approximately 4.0 Mt/a of ore from underground sources, supplemented by stockpiled material.

In 2020, OZ Minerals launched the Prominent Hill Expansion (PHOX) Project to increase the underground mining rate to 6.5 Mt/a, leveraging available process plant capacity and accelerating mine electrification. The project scope includes orepasses, an underground materials handling system (gyratory crusher, loading station), ventilation and refrigeration infrastructure, and the Wira Shaft. This shaft will transform Prominent Hill into a 6.5 Mt/a underground mining operation.

Detailed engineering began in 2021, and after a competitive tender, Byrnecut Australia Pty Ltd (BAPL) was awarded the shaft sinking contract. The pre-sink phase commenced in late 2022 and was completed in 2023, with the main sink starting at the end of the same year. Following OZ Minerals' acquisition by BHP in 2023, the project continued uninterrupted.

The Wira Shaft is a 1326 m deep, concrete-lined production shaft designed with two rope-guided skips (39.1 t payload each) and powered by a 9.5 MW ground-mounted friction winder. The main sink utilises the permanent headframe equipped with kibble and stage winders, with temporary sheaves to be replaced with permanent sheave clusters after completion. The Skyshaft structure, housing skip dumping, conveyance catch gear, and arrestor systems, will be installed using a Self-Propelled Modular Transporter (SPMT) upon completion of the sink.

Drawing on designs from similar Australian shafts, the Wira Shaft will achieve a nameplate capacity of 1100 t/h (6.5 Mt/a). With completion and hoisting plant commissioning expected by 2025, it represents the latest evolution in Australian shaft construction, setting new benchmarks in-depth, capacity, and engineering innovation.

DESIGN AND ENGINEERING

Geotechnical

Below the pre-sink, the Wira Shaft is entirely sunk through a complex sequence of volcanic basement rocks, primarily consisting of mafic volcanics, andesite, hematite, and hematite breccias. During the Early Contractor Involvement (ECI) phase, a review of the rock mass quality played a key role in the decision to adopt the strip and line sinking method.

A diamond drill borehole was drilled at the shaft location and core samples were logged to gather geotechnical data, which informed the assessment and design recommendations. Rock mass quality was evaluated using the Q system (Norwegian Geophysical Institute (NGI), 2022). The shaft walls were generally classified as fair to good, with occasional poor ground conditions. Very poor ground was encountered in some sections, particularly around the midshaft area.

TABLE 1
Rock properties (AMC Consultants, 2021).

Rock type	Uniaxial Compressive Strength (MPa)				Young's modulus	Poisson's ratio	Density (t/m^3)
	No. tests	Min	Mean	Max	Mean	Mean	Mean
Volcanics	6	50	122	187	71	0.23	3.06
Andesite	10	27	118	234	66	0.24	2.97
Dacite	1		131		70	0.14	2.77
Basalt	1		87		81	0.27	2.80

Stress conditions at shaft bottom are estimated to be as follows:

- σ_1 64 MPa @ 01/252.
- σ_2 39 MPa @ 12/343.
- σ_3 24 MPa @ 78/107.

The uppermost 94 m of the shaft were composed of very poorly consolidated, weak sedimentary material, favouring a non-entry sink methodology for the pre-sink, which is detailed in Kilkenny, Clark and Hill (2023). For the remainder of the shaft, conditions were favourable for raise boring and strip-and-line sinking. Unlike previous strip-and-line sinks in Australia, a larger raise bore disposal hole—up to 4.0 m in diameter—was feasible, which significantly improved drill-and-blast cycle times and provided better stress relief ahead of the final lining.

Overall shaft

The shaft location was selected within the footwall of the Malu orebody, outside the open pit crest. The site was initially covered by approximately 21 m of overburden, which was removed during site establishment. The shaft is positioned in footwall waste rock and passes near the horizontal centre of gravity of the known ore reserves in the Malu orebody, extending to a depth near the lower limit of these reserves.

Shaft depth is influenced by factors such as the desired loading elevation, rope length, skip cycle time, haulage level distance, and hoisting capacity. After evaluating these criteria, the final shaft depth was determined to be 1326 m, as this depth met all design and production requirements while remaining within hoisting design limits.

The Wira Shaft has an internal diameter of 7.55 m and is concrete lined. It is designed to accommodate two rope-guided skips, along with air, water, and electrical services. The shaft has four plats (shaft stations), which include:

1. Midshaft level (for head rope changes).

2. Workshop level (for ventilation breakthroughs).

3. Loading station level (for skip loading).

4. Shaft bottom (primarily for tail rope and guide rope changes).

A schematic of the Wira Shaft design is provided in Figure 1.

Pre-sink

A Herrenknecht Vertical Shaft Sinking Machine (VSM) was used to complete the pre-sink to a depth of 94 m, passing through the cover sequence rocks overlying the basement. The VSM is a non-entry shaft sinking method that employs mechanised rock cutting techniques in combination with pre-cast concrete segments to form the final shaft liner. Excavation and shaft lining occur concurrently, allowing for efficient progress.

The VSM is equipped with a cutting drum, similar to a road header, which is mounted on a telescopic boom. During operation, the shaft is filled with slurry, which helps to float the rock cuttings to the surface for separation and processing. For every 1250 mm of vertical advance, four pre-cast concrete panels are installed at the top of the pre-sink liner. Strand jacks then lower the entire column of concrete liners into position, after which cutting resumes to advance the shaft by another 1.25 m.

Details of the pre-sink operation are discussed in Kilkenny, Clark and Hill (2023).

Sinking plant

The major equipment associated with the sinking plant includes (see Figure 2):

- Four single-drum stage winders with integrated brake and control systems. Manufactured by Timberland Equipment Limited Canada, with maximum speed of 0.34 m/s.

- One dual clutched double-drum winder utilised as the Kibble Winder for people and materials. Maximum speed of 10 m/s. Mechanically configured to operate as a single-drum winder. Original equipment manufacturer (OEM) is E.A. Marr, driven by two 600 kW 1500 rev/min DC motors through two five-stage gearboxes.

- Other equipment including Hydraulic Power Units for braking systems, Winder Building structure, messenger winch to enable power supply to stage and temporary sinking sheaves.

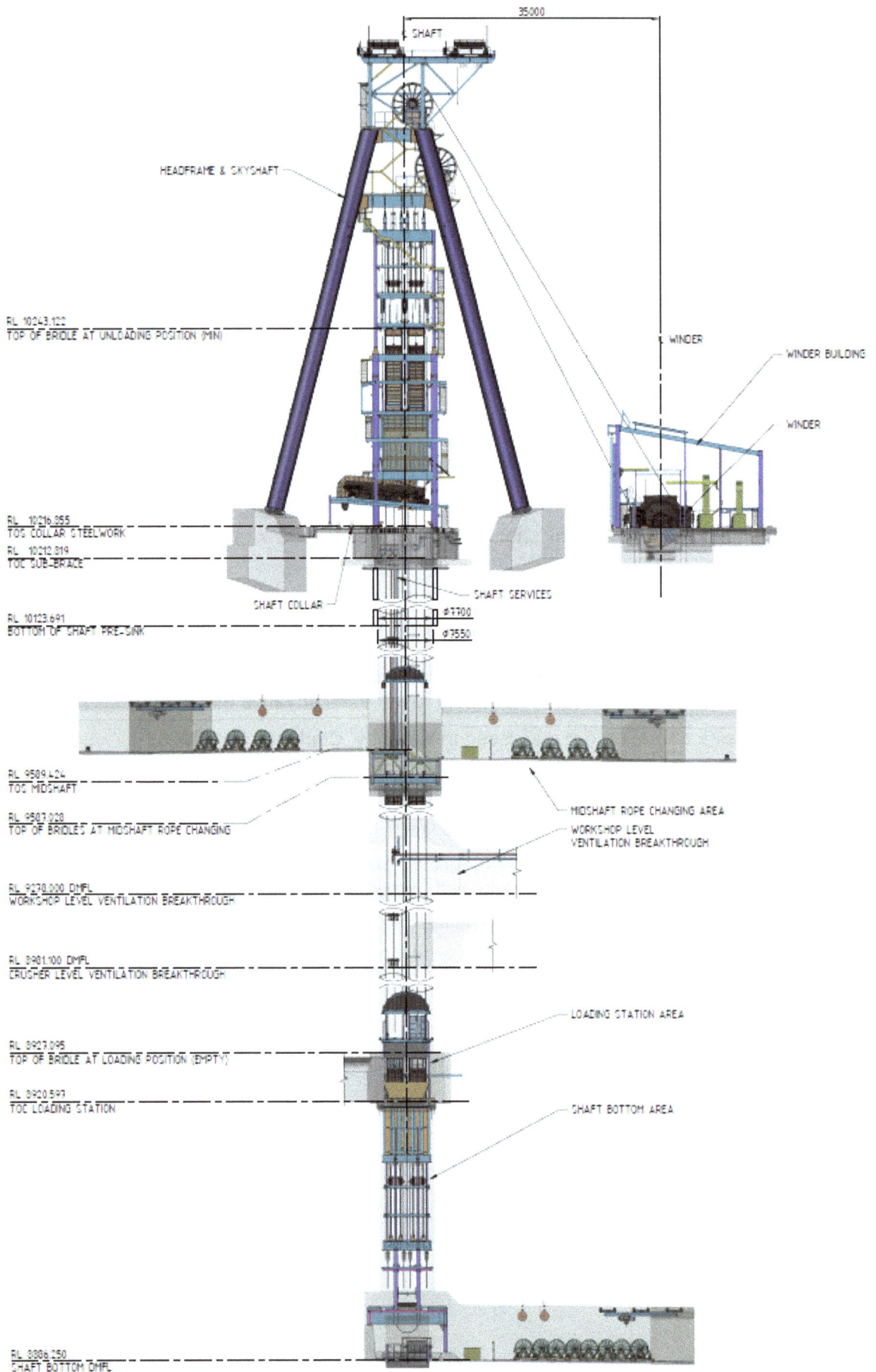

FIG 1 – Wira Shaft schematic – final configuration looking west.

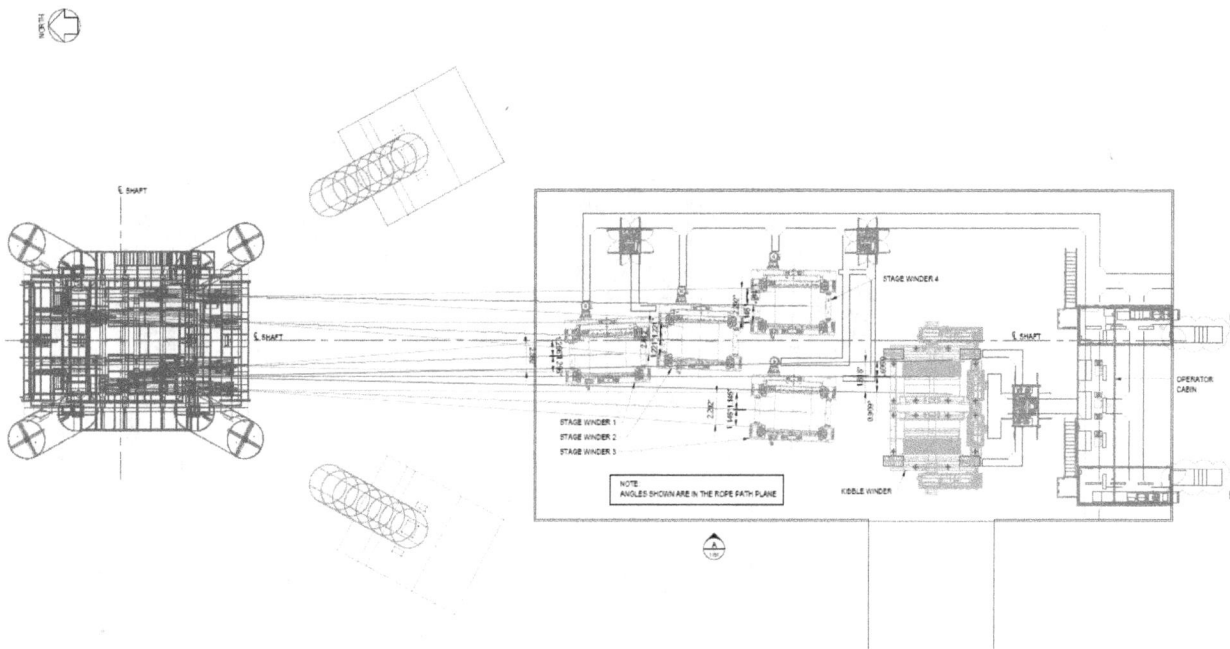

FIG 2 – Plan of sinking arrangement and fleet angles.

Sinking stage

The sinking stage was designed collaboratively by Byrnecut Australia (BAPL) and Thyssen Mining and Construction Canada (TMCC). It consists of five main decks, each with specific functions as outlined and shown in Figure 3:

- A-Deck: Provides overhead protection and houses the high-voltage (HV) electrical and hybrid fibre optic cable baskets. It serves as the upper access level for extending and installing in-shaft services.

- B-Deck: Houses the control system PLC, HV and electrical distribution equipment, receive and reconstitute concrete and fibre-reinforced shotcrete. It also serves as the lower deck for installing shaft services.

- C-Deck: Shotcrete pumping and housing the hydraulic power unit (HPU). Additionally, it supports the suspension and hydraulic slewing arrangement for E-Deck.

- Energy Chain Platform: Positioned below C-Deck, this platform facilitates the 330° rotation of hydraulic hoses and electrical cables that support the drilling and spraying operations on E-Deck. It also functions as a control station for the air chain hoists during the concreting cycle.

- D-Deck: Used for setting the curb and pouring concrete. It includes provisions for opening the shaft shutter (barrel) doors and aligning and levelling the shutters.

- E-Deck: Drilling and charging the bench, as well as applying temporary ground support. Secondary functions include controlling the shaft plug and providing access to the bench. The following equipment is on this deck:

  o Three stage-mounted custom Sandvik split-feed electric-hydraulic drills, capable of drilling vertically for blastholes and at up to 45° from vertical for angled or horizontal ground support.

  o Normet Minimec shotcrete spray boom, mounted below E-Deck when in use, which sprays a wet mix via remote control operation of the Normet Norstreamer pump on C-Deck.

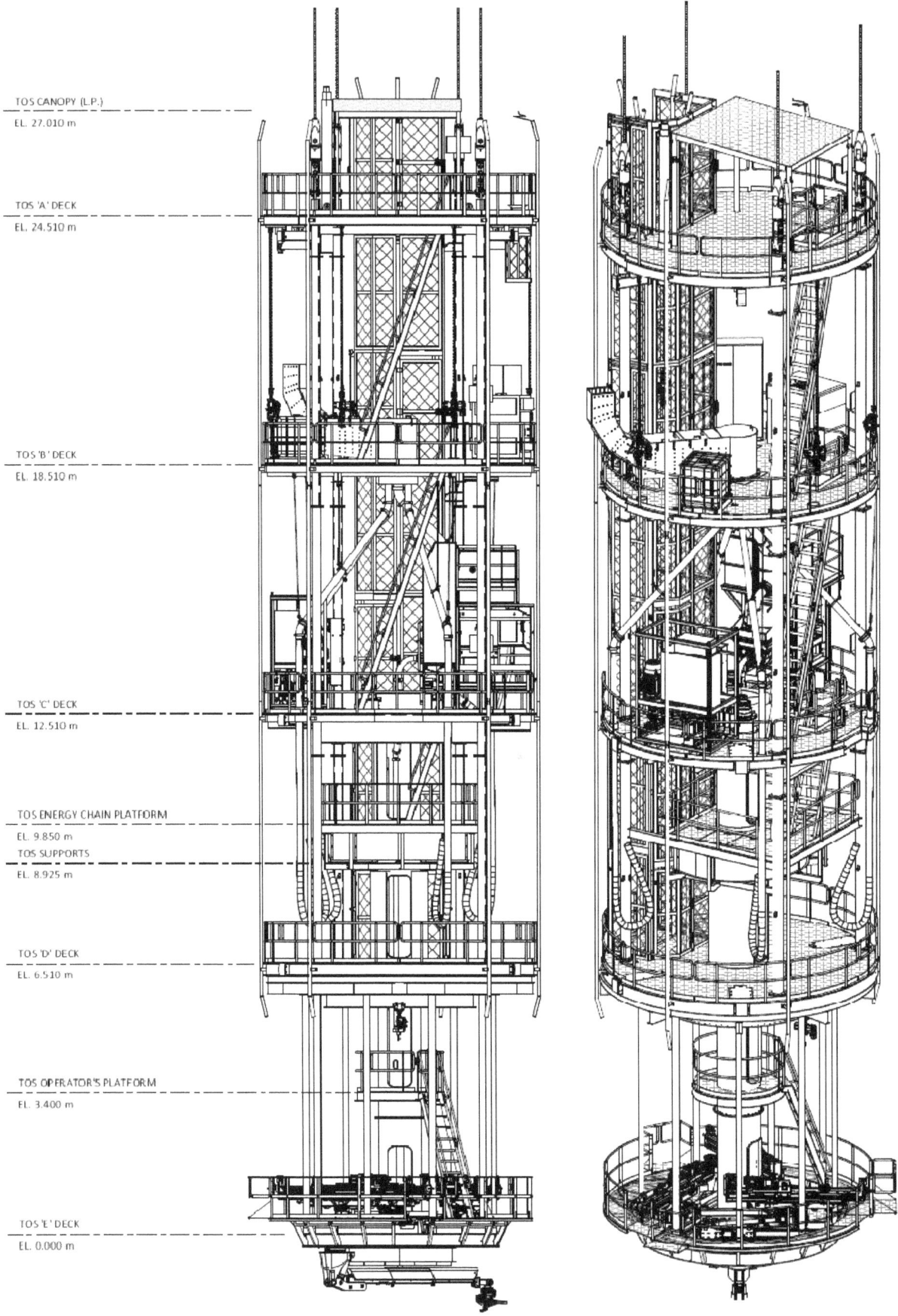

TOS CANOPY (L.P.)
EL. 27.010 m

TOS 'A' DECK
EL. 24.510 m

TOS 'B' DECK
EL. 18.510 m

TOS 'C' DECK
EL. 12.510 m

TOS ENERGY CHAIN PLATFORM
EL. 9.850 m
TOS SUPPORTS
EL. 8.925 m

TOS 'D' DECK
EL. 6.510 m

TOS OPERATOR'S PLATFORM
EL. 3.400 m

TOS 'E' DECK
EL. 0.000 m

ELEVATION - LOOKING WEST

ISOMETRIC

FIG 3 – Sinking stage schematic.

The improved stage design is based on an existing concept from TMCC, used successfully in two previous shaft sinks in North and South America. Several new design features contributed to a safer and more productive shaft sinking cycle:

- Maximising equipment kept on the stage, to enable rapid transitions between cycles and reduce logistical requirements.

- Rotating Lower Deck: The lower deck, containing the jumbo drills and shotcrete sprayer, can rotate up to 330° around the stage. This allows the drill booms to remain in a fixed position relative to the stage while moving from hole collar to hole collar. This set-up facilitated accurate perimeter drilling and operational ease. The shotcrete sprayer, which remains fixed to the underside of E-Deck, provides full 360° wall and floor coverage for spraying and blow-over. The use of a bottom slewing and drilling deck (which removes personnel from the shaft bench) enables larger diameter raise boring and blasting of an angled 'V' bench, improving muck disposal efficiency.

- Absence of Fixed, Stage-Mounted Mucking Equipment: Due to the open disposal hole, which allows blasted rock to clear the shaft bottom after firing, conventional mucking equipment (such as a clamshell mucker) was not installed on the stage. Instead, the remaining muck on the bench is cleared during the bench blow-over cycle using compressed air and water via the Minimec spray boom. Key benefits of this approach include:

  o Weight savings by eliminating the need for conventional mucking equipment.

  o Cost reduction by removing the need for conventional mucking equipment.

  o Space savings by eliminating the need for a second kibble well, which would otherwise occupy significant floor space.

  o Efficiency gains from incorporating a stage-mounted shotcrete pump and sprayer.

  o Simplified materials handling by removing the need for kibble dumping and materials handling in the headframe.

  o The design enables the use of a rotating drill deck with three single-boom drills, optimising operational flexibility.

- Temporary ground support: Typical initial ground support was 75 mm fibre-reinforced shotcrete. Where required, resin bolts or friction bolts were installed using the stage jumbos (up to 3.0 m long). Additionally, cable bolts up to 6 m in length could be installed manually when required. This shotcrete-only design philosophy significantly enhanced the speed of the ground support cycle, particularly in competent ground conditions, which was most of the shaft.

- Stage power supply via surface-mounted cable reeler: Deep shafts (eg more than 1000 m) have typically employed either fully pneumatic sinking systems, or the use of large diameter medium voltage electrical supply (up to 1 kV). Australian mining typically uses 1 kV supply cables, up to 120 mm², due to the prevalence and preference for electric/hydraulic shaft drills, stage hydraulic controls and electrical lighting and power requirements. As shafts get deeper, and voltage drops considerations become important, heavy electrical cables create hazards in the shaft associated with their installation; maintenance and fault rectification – particularly in wet conditions. They also naturally interact with the moving stage, which can generate loads that will damage wall mounted electrical cables unless safety interlocks or rigorous administration controls are implemented. The Prominent Hill sinking stage utilised:

  o A 1500 m long type 241.3 trailing cable at 3.3 kV which allowed a smaller diameter cable on a custom built Cavotech reel with a torque motor providing constant back tension.

  o A 20 t hoist with 28 mm Gr 1960 non-rotating wire rope.

  o Custom designed and tested connecting swivels.

  o 3.3 kV/415 V stage step down transformer.

Shaft sink plan

The shaft sinking method was determined to be strip-and-line. This decision was primarily driven by expected rock mass conditions, the proximity of existing underground development, and the significant cost and schedule advantages it offered over blind sinking. Initially, the plan involved two sinking stages or legs, but this was later revised to three. These stages were divided from the bottom of the pre-sink to the midshaft level, then to the workshop level, and finally to the shaft bottom. The strip-and-line methodology proved advantageous, reducing the overall project duration by approximately six months, or 15 per cent, compared to blind sinking.

In the strip-and-line process, the excavation begins by establishing a raise bored disposal hole at the shaft's centre. Once this hole is in place, excavation progresses by drilling and blasting (stripping) directly into the disposal hole from the sinking stage. The final concrete liner installation lags the active bench, depending on rock mass quality and stress conditions. The mining sequence involved taking two 3.0 m deep bench rounds, followed by a 6.0 m concrete liner pour. Services were extended after each liner pour to support ongoing sinking.

Ground support was installed in two stages. Temporary support was provided through the application of 75 mm fibre-reinforced shotcrete (fibrecrete) and spot bolting where necessary. These measures were guided by detailed geotechnical mapping of each round. Permanent support was installed during the 6.0 m liner pours using 40 MPa concrete. A lag distance of 12–18 m between the installation of temporary and permanent ground support allowed for stress relief ahead of the permanent liner, ensuring the stability of the shaft.

The Wira Shaft utilised various raise bored hole diameters tailored to the geotechnical conditions of each stage. The first leg, extending 604 m, used a 3.5 m diameter hole. For the second leg, over 304 m, a combination of 2.4 m and 3.5 m diameters were used. For the third leg, 364 m long, a 4.0 m diameter hole was reamed. These dimensions were determined using the McCracken and Stacey (1989) methodology, supplemented by local empirical knowledge and risk-based decision-making. By integrating innovative stage design with comprehensive geotechnical management, the team optimised the hole sizes to maximise project efficiency and minimise risks.

As each leg was completed, a shaft station construction scope was completed before transitioning to the next. Pillars were left in place to separate the legs, allowing the raise boring of subsequent legs to proceed while facilitating the removal of cuttings and mullock from the current leg. This approach enabled efficient progression and continuity throughout the sinking process.

Shaft stations were excavated using conventional lateral development methods ahead of the arrival of the sinking stage. Dimensions of these stations were relatively large, with spans exceeding 10 m and drive heights also exceeding 10 m.

PROJECT EXECUTION

Organisational structure

The sinking contract was structured as a cost-reimbursable alliance, designed to enable rapid early execution and maintain flexibility in project planning before the completion of the permanent IFC (Issued for Construction) designs. This flexible approach allowed the scope, schedule, and cost to evolve over time to align with project requirements and conditions.

Two levels of governance were established to ensure effective management and oversight. At the project management level, the Alliance Management Team oversaw day-to-day operations, while at the corporate level, the Alliance Leadership Team provided strategic direction and support. Figure 4 illustrates the organisational structure utilised during the shaft sinking project.

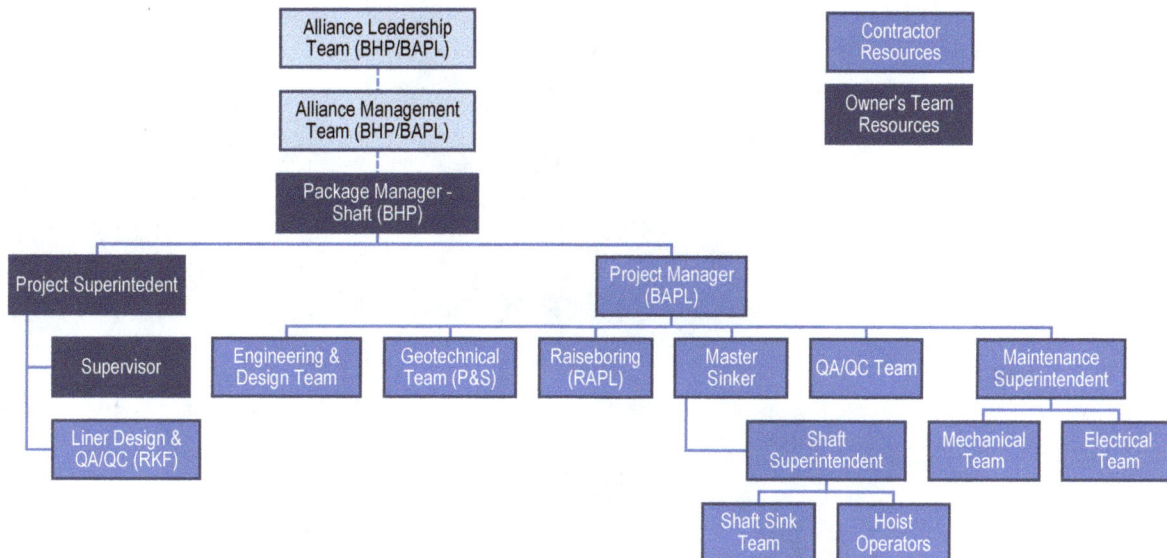

FIG 4 – Organisation structure.

The sinking contractor and liner designer/engineer reported directly to the owner's team Package Manager. The sinking contractor engaged specialised geotechnical, and shaft sink design teams as to support technical services and ensure operational success.

Geotechnical engineering resources were embedded within the sinking contractor's team, forming an integral part of their scope. This arrangement proved highly effective in ensuring the safe and efficient installation of temporary ground support, as well as the implementation of both temporary and permanent instrumentation and the final liner. By aligning these critical functions within the contractor's framework, the project maintained a cohesive approach to geotechnical management, contributing to the overall success of the shaft sinking process.

Raise boring

In early 2023, the first disposal hole for Leg 1 was successfully raise bored from the bottom of the pre-sink to the upper midshaft chamber. The operation utilised a Robbins 123RHc raise boring machine, which was positioned on a surface platform. To ensure stability and accuracy, a collar pipe was installed to support the drill string, running from the pre-sink bottom to the top of the shaft collar platform. This pipe, as shown in Figure 5, was grouted into the shaft face within the basement volcanics and erected using a crane and a personnel basket. The drilling achieved excellent accuracy, with a deviation of less than 0.1 per cent, thanks to the proven MICON-Drilling RVDS system. Reaming was completed without complications, achieving a final diameter of 3.5 m, which was ready ahead of the shaft sinking activities.

For Legs 2 and 3, a single pilot hole was drilled to shaft bottom from the lower midshaft chamber to streamline the process. Along the vertical 703 m long pilot hole, a deviation of only 140 mm (0.02 per cent) was recorded, demonstrating exceptional accuracy. Reaming for these legs was executed in two phases: Leg 2 was predominantly reamed to a 3.5 m diameter, with a smaller section reduced to 2.4 m to address a specific area of poor ground stability near the lower midshaft chamber. By early 2024, this phase was completed.

The sinking of Leg 3 presented additional challenges due to anticipated high *in situ* stresses at depth. The final diameter of Leg 3's disposal hole was determined through a comprehensive risk assessment. There was a compelling case to increase the diameter to 4.0 m to promote early convergence and stress relief, given the elevated stress levels encountered with increasing depth. However, this strategy carried risks, including the potential for raise wall failure and excessive 'dog-earing' (localised overbreak) along the walls. Three specific subsections of the raise were identified as particularly susceptible to material fallout and instability.

FIG 5 – View looking down pre-sink during raise boring of Leg 1.

A nearby 5.0 m diameter return air raise, established at similar elevations, had experienced significant wall breakout and challenging reaming conditions due to the higher stress environment, underscoring the potential risks at depth. After consultation with site geotechnical engineers, the final liner engineer, and the shaft sinking geotechnical team, a decision was made to ream Leg 3 to 4.0 m. This risk-based approach balanced the benefits of stress relief against the risks of raise failure.

Ultimately, this decision highlighted the project's adaptability and commitment to geotechnical management, ensuring that the shaft sinking process remained on track despite the inherent challenges of working in higher stress ground conditions.

Headframe

The Wira Shaft features an open A-frame style headframe with a separate, free-standing Skyshaft (as shown in Figure 1), which houses the skip dumping, conveyance catch gear, and arrestor systems. This design is particularly well-suited to the strip-and-line sinking methodology and provides ample open space above the brace. The absence of a Skyshaft during sinking—brought in after the completion of the sink—further complements the sinking process. Additionally, this design allows for rapid construction, minimising delays. The headframe was successfully erected in August 2023.

During the raise boring phase, the Sheave Deck module was pre-assembled adjacent to the shaft, complete with all necessary sinking sheaves and equipment. The assembly was lifted into position, with each leg of the headframe lifted separately and temporarily supported by guy ropes. This method allowed for fine adjustments to be made to the leg positions during the Sheave Deck lift. The 365 t Sheave Deck was lifted using a 900 t Manitowoc Lattice Boom Crawler crane, and placed atop the four legs, with each joint subsequently welded in place. Following this, the Auxiliary Crane Structure (ACS) was lifted onto the Sheave Deck, as illustrated in Figure 6.

FIG 6 – Wira Shaft headframe erection, sheave deck lift with ACS in foreground.

During shaft sinking, the permanent headframe is used with temporary sinking sheaves placed on both sheave decks for the kibble winder and four stage winders. Upon completion of the shaft sinking, these temporary sheaves will be replaced by a set of permanent sheave clusters for head ropes.

The Sinking Winder Building was positioned to the south of the headframe, whereas the Permanent Winder Building is located to the north, ensuring optimal space utilisation and operational efficiency throughout the construction process.

Commissioning of sinking plant

The Shaft Sinking System was designed in accordance with the latest NSW Technical Reference Guide: Powered Winding Systems (NSW Resources Regulator, 2021), which incorporates several older draft standards, including MDG 33 and EES008. The design philosophy for the Winder Controls and Safety Systems adhered to AS61508 (Standards Australia, 2011), following a functional safety approach supported by a formal Quantitative Risk Assessment.

Off-site construction and commissioning of the sinking stage and winders were generally productive and successful. The winding system integrated legacy equipment from previous projects with new components specifically designed for the Prominent Hill project. Factory Acceptance Testing (FAT) of the system was conducted at the manufacturer's facilities in Newcastle, allowing for energization and testing of drives, motors, and control systems. However, delays in the supply chain for stage winder drive components prevented pre-testing before the equipment's arrival on-site.

Similarly, the stage was fabricated and load-tested to AS/NZS3785 (Standards Australia, 2002) at the structural fabricator's facilities in Parkes, NSW. A significant portion of the electrical, hydraulic, and service reticulation was pre-fitted and tested before delivery to the site.

On-site system commissioning was complicated by delays in the delivery of stage winder components from Canada, extended periods for rope-up, and an underappreciation of the level of verification required to meet the standards of a Functional Safety approach. Siemag Tecberg

Australia (STA) was responsible for safety system design and verification, with Siemag Tecberg USA (STU) acting as the validator.

The widespread adoption of these standards in shaft sinking across Australia highlights the need for a thorough understanding of commissioning requirements. To ensure that scheduling and resourcing are aligned with the project's needs, this understanding must be established during the earliest phases of engineering. This proactive approach is crucial to managing the complexities of system integration and ensuring that all safety and functionality standards are met.

Shaft sink cycle

There are two main processes that take place in the strip-and-line sinking method: the mining cycle; and the final concrete lining cycle. The final concrete liner must be installed at a specific lag distance from the active face (or bench) to meet constraints of maximum residual stress applied by shaft wall rock convergence on the final liner and the capacity of temporary ground support installed after each round is fired. Figure 7 provides a simplified schematic of the full cycle.

The mining cycle begins with drilling and blasting a round 3.0 m in-depth. After the round is blasted, it is blown down using compressed air and water. Next, temporary ground support—typically consisting of a 75 mm thick layer of fibre-reinforced shotcrete—is applied. Another 3.0 m round is then fired, blown down, and supported. This process continues until the required minimum lag distance tolerance is met.

Once the minimum lag distance is reached (typically two rounds), the concrete lining cycle begins. This involves setting up a curb form, which is split from the main form and suspended by hanging rods from the previous liner. The curb is then surveyed into place, and concrete is poured with accelerating admixtures to help it set more quickly. After the curb pour, the main portion of the barrel form is lowered into position and the rest of the concrete pour is completed. Once the concrete has set, mine services like compressed air and water are extended, and the process repeats.

Both the concrete and ground support cycles incorporate the use of a slickline.

As the shaft progresses deeper, liner lag distances are increased to account for increasing *in situ* stresses and areas of poorer rock mass conditions. Several iterations of the mining and concrete lining cycles were trialled throughout the sinking process to determine the most efficient approach. The goal was to complete a full cycle—two 3 m bench rounds blasted, and a 6 m liner poured—within 28 hrs, and this target was achieved multiple times during sinking.

Drilling, charging and muck clearing

A drill pattern consisting of 64 mm drill holes arranged in two to three circular rings was used (depending on the diameter of the disposal hole). The inner rings are drilled to greater depths than the outer rings, creating a conical face profile. This profile is beneficial in facilitating the effective blowing down of the bench after firing. Importantly, no personnel are on the bench during drilling operations (see Figure 8).

During initial drilling, some rounds did not break to the intended depth. This issue was attributed to a combination of uneven blasting from previous rounds, limitations on how closely the stage could be positioned to the face, and the travel distance of the jumbo feed. To address this, modifications were made to extend the drill depth by an additional 300 mm, increasing the total hole depth to 3.7 m. A laser leveling system was also implemented to provide accurate mark-ups to the crews. This adjustment resolved the short round issue and allowed for a maximum of two rounds to be drilled per liner advance cycle.

Given the absence of muck handling equipment on the stage, maintaining an optimal blasted face profile is crucial to minimising residual blasted muck which must be blown down the disposal hole using pressurised air and water from the shotcrete unit.

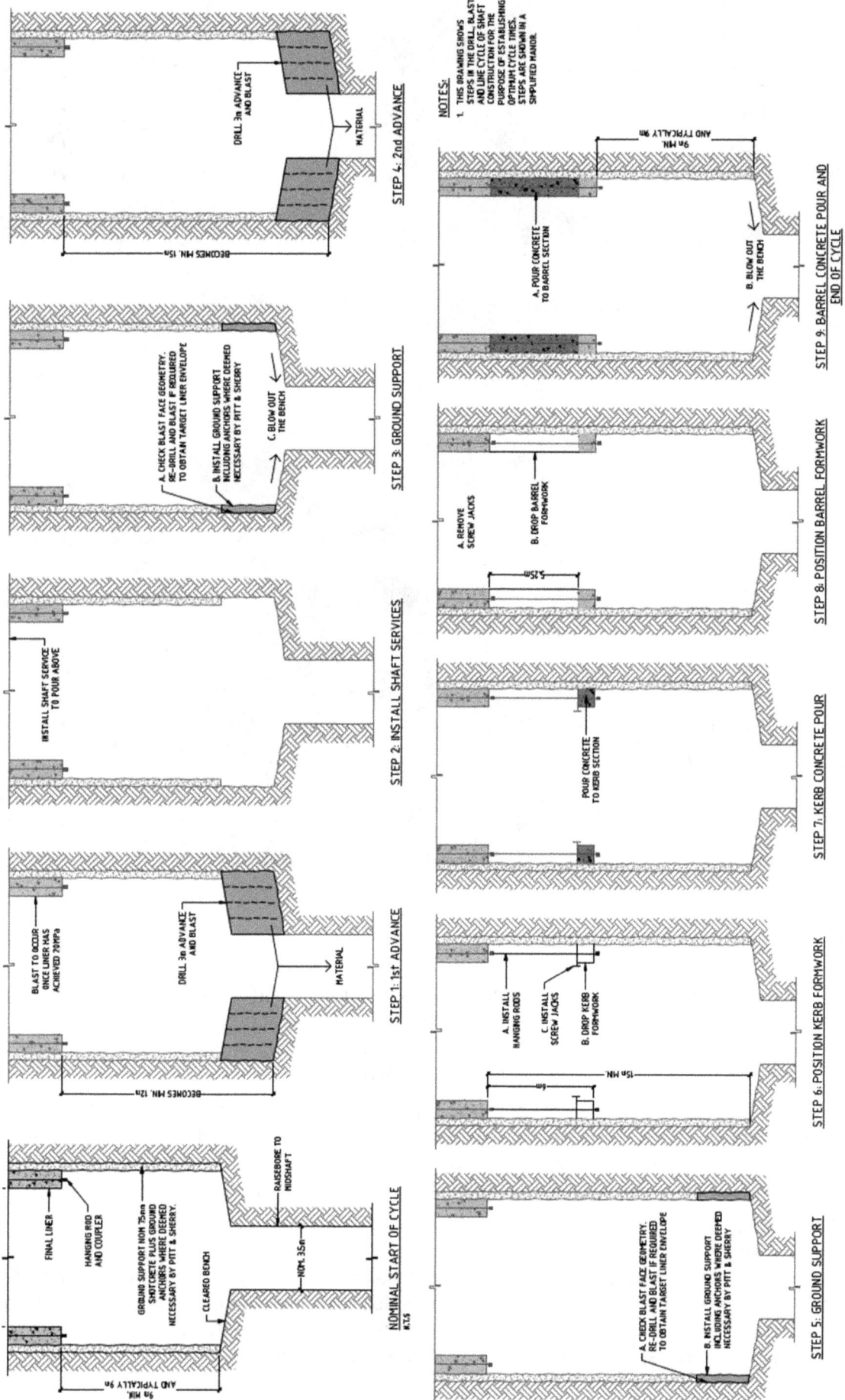

FIG 7 – Simplified strip and line cycle steps (RKF).

FIG 8 – Stage-mounted jumbo drilling bench from E-Deck.

The charging cycle is the only part of the cycle in which personnel work directly on the bench. Initially, the concept of charging holes from the lowest deck (E-Deck) with no personnel on the bench was tested. However, this approach had limited success due to challenges with hole cleaning and difficulties accessing holes because of the bench profile and space constraints on the deck. As a result, it was decided to charge the rounds with personnel working directly on the bench, ensuring that adequate controls were in place for safety and efficiency.

An Orica Miniloader was used to transport and pump emulsion into blastholes, which were initially primed with Orica's Exel™ LP detonators. To improve vibration control and minimise muck remaining on the bench, the initiation system was later upgraded to Orica's i-kon™ electronic detonators. This change resulted in noticeable improvements in wall control and a reduction in blow-over times.

After firing, most of the blasted material falls in the disposal hole, aided by the conical face profile and large disposal hole diameter (>2.4 m). Any remaining material is blown down by compressed air and water from the shotcrete sprayer. There were instances where the blasted face was too flat and resulted in excessive hung-up material, necessitating longer blow-over cycle times. This was quickly rectified with standardised mark-up and drilling practices and was largely resolved for the rest of the sink. Ventilation was in-cast and re-entry times were typically very short.

Extending services

In-shaft services were designed with two primary objectives: to support shaft sinking operations through the provision of compressed air and service water, and to facilitate concrete placement and shotcrete spraying with two slicklines. Some of the pipework installed for sinking purposes will later be repurposed to support life-of-mine operations. For this reason, the pipework was installed in its permanent configuration, enabling a seamless transition to the permanent service arrangements once sinking was completed.

The installation of this pipework (see Figure 9) saw several modifications and improvements throughout the sink. The design proved challenging, requiring Schedule 80 heavy-duty pipes and large high-pressure couplings (Victaulic 808). Although the installation process remained manual, lifting aids were developed to assist with slinging, hanging, aligning, and connecting the pipework, improving the overall efficiency and safety of the task.

FIG 9 – In-shaft services slinging during sinking.

Ground support

Standard temporary ground support during sinking primarily consisted of fibre-reinforced shotcrete (fibrecrete), applied to a minimum thickness of 75 mm (up to 150 mm where required), along with spot bolting as determined by geotechnical staff. The mix was designed to provide sufficient early age strength to allow rapid re-entry after spraying. The objective was to have 1 MPa within one hr, which was achieved in the initial few months of sinking.

Like the final concrete liner, all fibrecrete was delivered via slickline.

The use of fibrecrete as temporary ground support in shaft sinking is not widespread but has been successfully used at other projects like Turquoise Ridge in Nevada (Wu, Dase and Howell, 2022).

Final liner

The Wira Shaft was lined using Super-Workable Concrete (SWC), a high-flowing, self-compacting, and non-segregating concrete mix designed to meet the durability and workability demands of shaft lining. The SWC mix comprised cement, fly ash, 10/7 mm aggregate, sand, various admixtures, and, in some cases, steel fibres. Rigorous testing was conducted both at the surface batch plant and at the shaft collar prior to its transfer down a 150 mm slickline. Upon reaching the sinking stage, the concrete was distributed into form pockets, with vibration provided by pencil vibrators and form-mounted pneumatic vibrators to ensure even distribution and compaction.

The target spread of the SWC mix before slickline transfer was 700–740 mm, ensuring the appropriate flow characteristics. The mix was designed to achieve a minimum compressive strength of 40 MPa at 28 days; actual results consistently exceeded this target. To monitor curing progress, temperature sensors embedded within the concrete provided real-time data, allowing the project team to estimate when the final strength would be attained.

During the sinking process, challenges with overbreak and underbreak posed difficulties for the final liner cycle. Overbreak resulted in excessively thick concrete sections, increasing the risk of thermal cracking due to uneven heat dissipation. Conversely, underbreak required additional drilling and firing of tights, potentially exacerbating overbreak in subsequent cycles. Striking the right balance in the drill-and-blast cycle was critical to maintaining efficiency, reducing costs, and ensuring consistent liner thickness.

Lateral development

Strip-and-line sinking depends on established underground access for materials handling (material dropping from disposal hole) and transitioning between different legs of the shaft. This method involves significant interaction with ongoing underground activities, requiring careful coordination between the shaft sinking teams and the underground development teams. The shaft sinking plan divided the shaft into three legs, with *in situ* rock pillars separating each leg. These pillars were to be excavated in a 'just-in-time' fashion to support the overall progression.

Leg 1 of the shaft sink terminated at the midshaft area (MSA), a chamber approximately 10 m wide, 10 m long, and 12 m high, located at the shaft's midpoint. The MSA was designed to house the head rope change-out facility for hoisting operations. Figures 10 and 11 illustrate the sinking stage breakthrough into the lower MSA chamber and subsequent civil works undertaken in the chamber.

FIG 10 – Sinking stage breakthrough to lower midshaft chamber (looking east).

FIG 11 – Lower midshaft chamber during construction (looking east).

The MSA was excavated ahead of the sinking stage using conventional lateral development for the top bench and longhole drilling for the bottom bench. The pillar separating Leg 1 and Leg 2 was drilled from below and charged/fired from above under the protection of the stage. After firing the pillar to a 2.4 m diameter, sinking proceeded through the pillar to the MSA chamber. Excavating the MSA in advance of the sinking stage saved approximately three weeks of critical path schedule by enabling more efficient ground support installation and completing parallel construction activities.

Sinking paused for three weeks to facilitate construction within the lower MSA chamber. Activities included reinforcing two concrete walls of the shaft station, installing a floor slab for the head rope change-out facility, and setting up shaft stations on the north and south sides for the Rope Reels and Deflection Sheave frames.

Beyond the MSA, the project included another large station at the Workshop Level (between Leg 2 and Leg 3) and two additional shaft stations at the Loading Level and Shaft Bottom. These stations were excavated using two-bench, two-pass lateral development ahead of the sink.

Successes

Continuous improvement

The Wira Shaft Sink Project leveraged a Time Use Model (TUM) to enhance operational efficiency and optimise cycle times during the sinking phase. The Winder Operator played a pivotal role in managing data input, utilising PITRAM software on a tablet mounted in the Winder Control Room. This TUM was specifically developed for the sinking stage to record detailed time data, enabling the project team to track and analyse cycle times, identify areas for improvement, and assess the impact of implemented changes on overall performance.

The implementation of the TUM provided the team with a deeper understanding of cycle time dynamics and actionable insights into optimising the sinking process. As a result, several improvement initiatives were identified and successfully implemented, leading to significant reductions in cycle times. For instance, standardised work methods were introduced, allowing crews to adopt best practices and ensure consistent task execution across the team.

Figure 12 highlights the improvement in boring cycle times for each 6 m liner cycle, showcasing the tangible benefits of the optimisations.

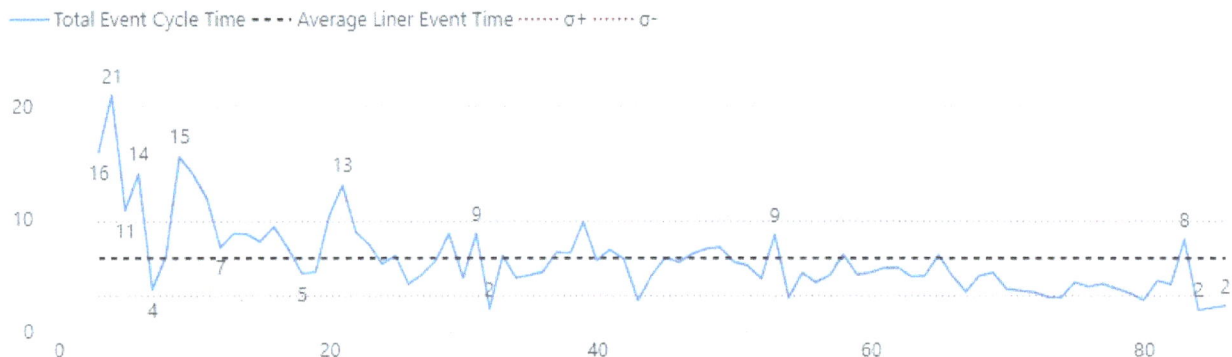

FIG 12 – Actual boring time for each 6 m liner cycle.

Fibre-reinforced shotcrete

This project has successfully demonstrated the efficient use of fibrecrete as a boltless temporary ground support system in mine shaft sinking operations. The success of this approach relied on a thorough understanding of anticipated ground conditions, the development of tailored ground support designs to address varying rock mass characteristics, and the use of a high-quality fibrecrete mix to meet early strength requirements. By adjusting the thickness and strength of the fibrecrete as needed, the system provided adaptable support capacities, ensuring its suitability across the diverse ground conditions encountered. Full-time geotechnical monitoring, analysis, and testing were integral to achieving these outcomes, ensuring ongoing optimisation of the support system.

The positive results of this project suggest that fibrecrete could be considered as both a primary and final ground support method for future shaft sinking projects. While its use as final support remains relatively uncommon in the industry, the successful implementation at Wira demonstrates its potential for broader application. Historical precedents include the use of fibrecrete in a raise bored hoisting shaft at the Renison Mine in Tasmania, as documented by Hexter and Fullelove (1998), and at the South Deep shafts in South Africa. Adoption of this approach should be carefully evaluated, considering geotechnical conditions, ventilation requirements, and shaft design considerations.

Safety and control system design

The project highlights the complexities of today's shaft construction, where each project is unique and subject to different regional legislation and updates to safety standards. These regulatory challenges often prevent direct adoption of previous designs and systems, especially when complying with Functional Safety requirements. This was particularly evident in the development of the shaft's control system, which was designed following the NSW Resources Regulator (2021) and the AS61508 (Standards Australia, 2011) life cycle approach.

The Functional Safety system incorporated advanced technologies and several complex features, including three separate hoisting systems, two separate conveyances, and extensive sensors and instrumentation—some of which had not been previously used in shaft projects. This complexity required careful and early engagement with key stakeholders, including SafeWork SA, equipment OEMs, and Functional Safety Engineers. Early collaboration allowed for the pre-assembly and testing of systems, particularly the sinking stage, which was equipped and tested prior to being installed in the shaft.

Despite some issues and faults encountered during commissioning, the project benefited from the robust processes established during the design and commissioning phases. These processes, along with on-site technical support, fault finding, and change management, enabled the team to address and resolve issues swiftly. This approach minimised disruptions and allowed for rapid modifications to the system, ensuring that the project continued on schedule and met safety standards.

Stage design innovations

The sinking stage incorporates several innovative designs aimed at improving both productivity and safety during shaft sinking operations. A short list of these key innovations includes:

- Messenger Winch and Cable Reel for Stage Power Supply: This system enhances the efficiency and reliability of the power supply to the sinking stage.

- Integration of Stage Mounted Shotcrete Pump and Boom: This integration streamlines the shotcrete application process, improving efficiency and safety in applying ground support.

- 'Scorpion Tail' Remix Chute and Concrete Remix System: This design enables effective handling and remixing of concrete, ensuring consistent quality for shaft lining.

- Kibble Well Hatch Doors: These hatch doors provide overhead protection and improve the ease of disembarkation and materials handling, increasing safety and operational efficiency.

- Leading Form Design: The robust design of the shaft lining form ensures both ease of use and durability throughout the sinking process.

- Stage Winders, First Large Set Compliant to AS4730 (Standards Australia, 2018) in Australia: The incorporation of stage winders compliant with this standard enhances safety and performance for the hoisting system.

- Integrated Load Sensing in Stage Rope Attachments, with Inclinometer Feedback to Winder Drivers: This system provides real-time data on load and inclination, optimising hoisting operations and improving safety by ensuring that equipment operates within safe limits.

However, the most prominent innovation is the use of a rotating lower deck. This design evolves the traditional sinking cycle by integrating the drilling, charging, ground support, and blow-over cycles into a single rotating platform. This innovation significantly reduces worker exposure to the critical 'open hole' risk, a major safety concern in strip and line shaft sinking by removing personnel from the bench. Additionally, it facilitates the possibility of semi or fully remote sinking cycles and supports parallel sinking and lining operations, greatly improving productivity.

Furthermore, this methodology supports the construction of larger diameter shafts in deeper or higher stress environments without compromising advance rates or ground support integrity, making it suitable for more challenging conditions while maintaining operational efficiency.

Limiting worker exposure

The project team adopted a holistic approach to system design, prioritising worker safety and minimising exposure to hazards. This strategy combined rigorous industry standards to control low-frequency, high-consequence risks (eg falling objects, winder control failures) with efforts to eliminate high-frequency manual handling tasks, such as installing ground support with handheld drills and using blow pipes over the bench.

A key driver of this approach was recognising that limited availability of shaft sinking expertise in Australia could increase reliance on lower-level controls (eg administrative measures, PPE), potentially leading to higher safety risks. One critical safety objective was eliminating personnel from working directly on the bench—an area with significant hazards from open holes and ground falls. This objective was achieved for all cycles, though a risk-based decision was made to revert to conventional bench charging for practicality during charge-up.

Equipment availability

Shaft sinking presents one of the harshest environments in mining, characterised by wet, hot conditions and equipment exposed to percussive blasting and heavy mechanical loads. The need for safety-rated electrical systems and machinery capable of withstanding unique stresses adds further complexity to ensuring reliable performance.

To address these challenges, collaboration with operations and maintenance personnel was essential in developing robust, maintainable equipment suited to this demanding environment. An experience-based design and verification team further ensured that equipment was fit for purpose, contributing to high system availability throughout the sinking phase.

This focus on durability, maintainability, and continuous feedback from the operational teams proved critical to the project's success. As shown in Figure 13, the Sinking Stage TUM from June and July 2024 demonstrates the effectiveness of these strategies in achieving continuous operation with minimal downtime and maintenance issues.

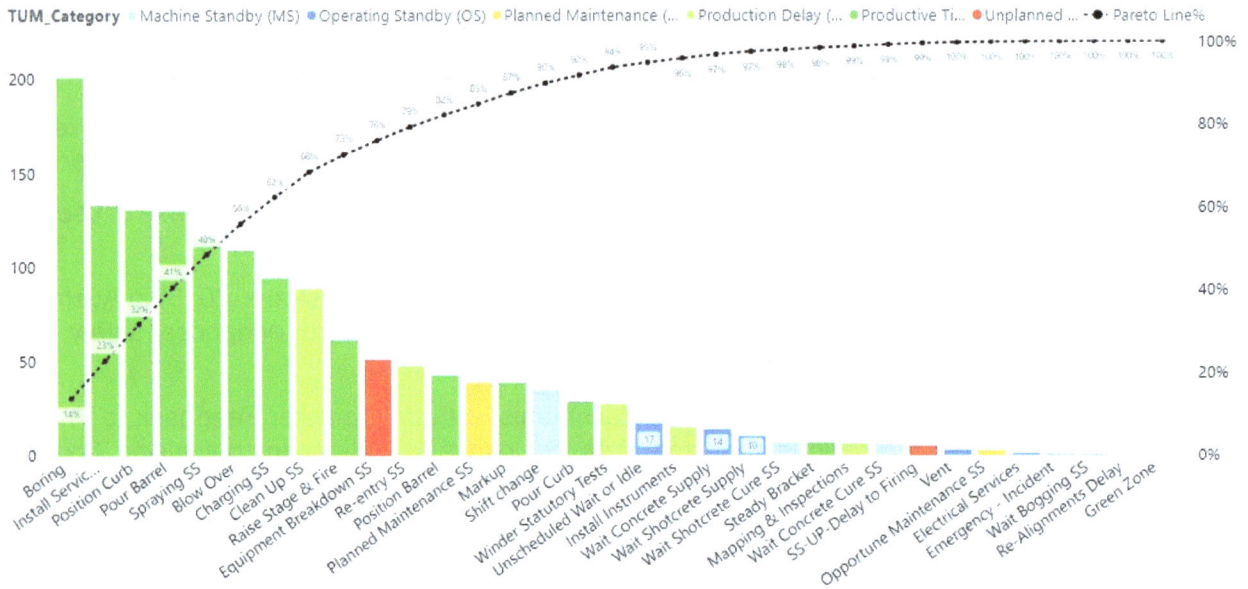

Total Event Statuses

FIG 13 – Example sinking stage time use Pareto Chart (June/July 2024).

Concrete lining

The concrete final liner for the shaft exceeded design expectations in both quality and strength, supported by a robust quality assurance (QA) and quality control (QC) program that minimised non-conformance reports (NCRs) and streamlined slickline operations.

A standout practice was real-time match curing, which used temperature probes embedded in the concrete liner to monitor curing conditions. Surface curing tanks with replicated conditions were used to more accurately test actual concrete strength. This real-time data improved forecasting of early age concrete strength and assisted in optimising blasting near freshly poured concrete, minimising impacts on curing and strength development.

The slickline delivery system also performed reliably, with no major blockages or failures to date. Concrete was delivered to the shaft collar with a high spread (~740 mm), tested and adjusted at the brace, and sent down the slickline. While some challenges with workability arose, the use of high-strength, self-levelling Super-Workable Concrete (SWC) proved effective in meeting performance requirements.

Challenges

Supply chain

The bespoke nature of the equipment needed to construct a large diameter, deep shaft means most of the electronic hardware is not off the shelf, however, normal lead times range from 12–20 weeks. As discussed in Kilkenny, Clark and Hill (2023), commencing a large infrastructure project during a global pandemic is not ideal. In addition to the disruption to global shipping and logistics causing havoc to mobilisation schedules, global semi-conductor manufacturing effectively shutdown. This resulted in exponential increases in lead times for project critical components. Delivery times were being quoted at over six months and in some cases, suppliers stopped taking orders entirely.

Geotechnical instrumentation

During the Early Contractor Involvement (ECI) stage, the project team identified significant ground stresses as a key challenge for the main shaft sinking process. To mitigate these stresses, the planned lag between the active bench and final liner placement was increased, allowing stress-induced convergence to occur before liner installation and reducing residual stress on the liner. This approach balanced long-term liner integrity with the need to minimise temporary ground support exposure and maintain sinking efficiency.

The final liner design incorporated *in situ* stress field data and real-time instrumentation readings. Extensometers near the shaft active bench measured ground convergence as a function of face advance, while concrete stress metres cast into the liner provided stress data. However, installing extensometers near blasting was challenging, and not all instruments were successful.

Below 950 m, a larger-diameter raise bore was used to facilitate early stress relief, paired with an increased bench-to-liner lag to ensure adequate convergence before final lining. These strategies were critical for managing ground stresses and maintaining shaft stability.

Concrete production

Producing Super-Workable Concrete (SWC) is more complex than general-purpose concrete due to the need for more ingredients, tight aggregate and sand specifications, and greater sensitivity to water content and temperature. SWC production requires higher levels of control and expert supervision to maintain performance during transport and placement, ensuring it does not segregate. Initial challenges with achieving the specification were overcome through continuous improvement in material quality and better execution of the mix at the batch plant, which led to a more consistent product and fewer rejected loads.

During the shaft sinking process, the concrete's spread, a measure of workability, was observed to decrease as it moved from the batch plant to the brace and finally to the stage for placement. This effect was attributed to the hot environment, dry materials, and clay content in aggregates. To address this, the team gradually increased the allowable spread at the top end and introduced a pre-conditioning process for the materials to reduce moisture absorption by aggregates from mixing to placement.

Concrete delivery also posed a challenge, with a limited number of agitator trucks initially creating a bottleneck in the cycle. Additional trucks were added to the fleet, which shifted the bottleneck to the batch plant.

Rope-up

Difficulties were encountered during the initial rope-up of the main sink, as well as interruptions to the main sink to re-tension the kibble rope. These difficulties were primarily driven by limited in-country resources (people and equipment) and expertise. Some learnings from these experiences will be applied to the initial rope-up of the permanent hoisting system.

CONCLUSION

Lessons learned

As the Wira Shaft sink is still underway at the time of writing, the knowledge and experience gained by the team is still in progress. However, the following points highlight lessons already captured during the early experience of the sink.

Going slow to go fast

Investment in engineering design, execution preparedness, and training proved invaluable, though unexpected challenges highlighted the need for more thorough pre-planning. Key issues included the initial rope-up of the sinking plant, kibble rope re-tensioning, surface/brace arrangements, in-cycle QA/QC design, concrete handling on stage, and communication protocols between surface and underground.

The strip-and-line cycle presented multiple sequencing options for mining, concrete, and maintenance tasks. Early experimentation established baseline cycle times and clarified task dependencies, such as firing rounds during concrete curing and scheduling preventative maintenance off critical path. Multi-disciplinary working groups provided significant value by focusing on activities that directly advanced shaft sinking progress, with rapid feedback driving improvements.

Focus on logistics

During the sink, the surface/brace area layout evolved to better support shaft sinking cycles and enable smoother transitions between cycle steps. For instance, gear kibbles for materials and tools were prepared in advance, ensuring readiness for the next cycle step.

Concrete delivery is significant portion of the cycle, with considerable focus placed on increasing delivery capacity and rate. Placing the batch plant adjacent to or closer to the shaft would be a highly effective strategy; but may be a challenge for some brownfield projects.

From a materials handling perspective, bogging of shaft rounds had to occur shortly after each round was fired. This required a responsive underground team to manage bogging the waste. Developing clear operational protocols for this process was crucial to avoid delays in firing the next round, preventing issues such as choking the brow or disrupting ventilation in the downcast shaft.

Front-end engineering

Shaft sinking projects require significant front-end loaded engineering and planning, which demands substantial time and financial investment. This phase should be prioritised for optimal project execution. Key benefits of front-end loaded engineering include:

- Clear project definition: In-depth review of designs in relation to available construction methodologies and equipment. Establishment of a fixed interface for permanent hoisting system design and execution.
- Cost control: Improved accuracy in time and cost estimation, with design improvements to enhance sinking rates and support logistics.
- Risk Management: Sufficient time for management to plan and allocate resources effectively, especially given the limited pool of local expertise.
- Project execution: Reduced risk of re-work and changes, and a clearer understanding of risks and opportunities and sufficient time to develop effective controls.

While the extended time spent on front-end engineering might initially seem to delay project completion, it ultimately helps control execution risks, which are typically higher for shaft construction compared to more common mine infrastructure projects. This thorough planning phase can prevent costly delays and rework, ultimately leading to better project outcomes.

Project outcomes

Safety performance

At the time of writing, the shaft project was averaging an injury rate of approximately half that of the broader Prominent Hill Operations site. This is credited to several factors, including team focus on hazard assessment and risk control, stage design, and mechanisation.

Overall sinking performance

Recent shaft sinking projects in Australia have excavated to depths ranging from just over 800 m to approximately 1500 m. However, shafts of similar diameter, design, and sinking method have generally not exceeded depths of around 1100 m. The strip and line sinking method is a commonly used technique in Australia, with notable examples being the Olympic Dam Clark Shaft, Telfer Shaft, and Ernest Henry Shaft. Some of these projects have also utilised a combination of blind sinking and strip and line methods to meet the design requirements of each specific project.

Sinking rates in these projects have varied (see Figure 14), with peak monthly rates reaching as high as 5.5 m per day (m/d). However, typical sinking rates from the start of the project to completion tend to fall within the range of 2.0 m per day to 3.0 m per day – inclusive of shaft station development. Sinking rates are influenced by several factors, including the shaft design (particularly the number and design of shaft stations), the chosen sinking method, sinking stage capabilities, and the overall design of the sink plan. Each of these factors plays a significant role in determining how quickly the shaft can be excavated.

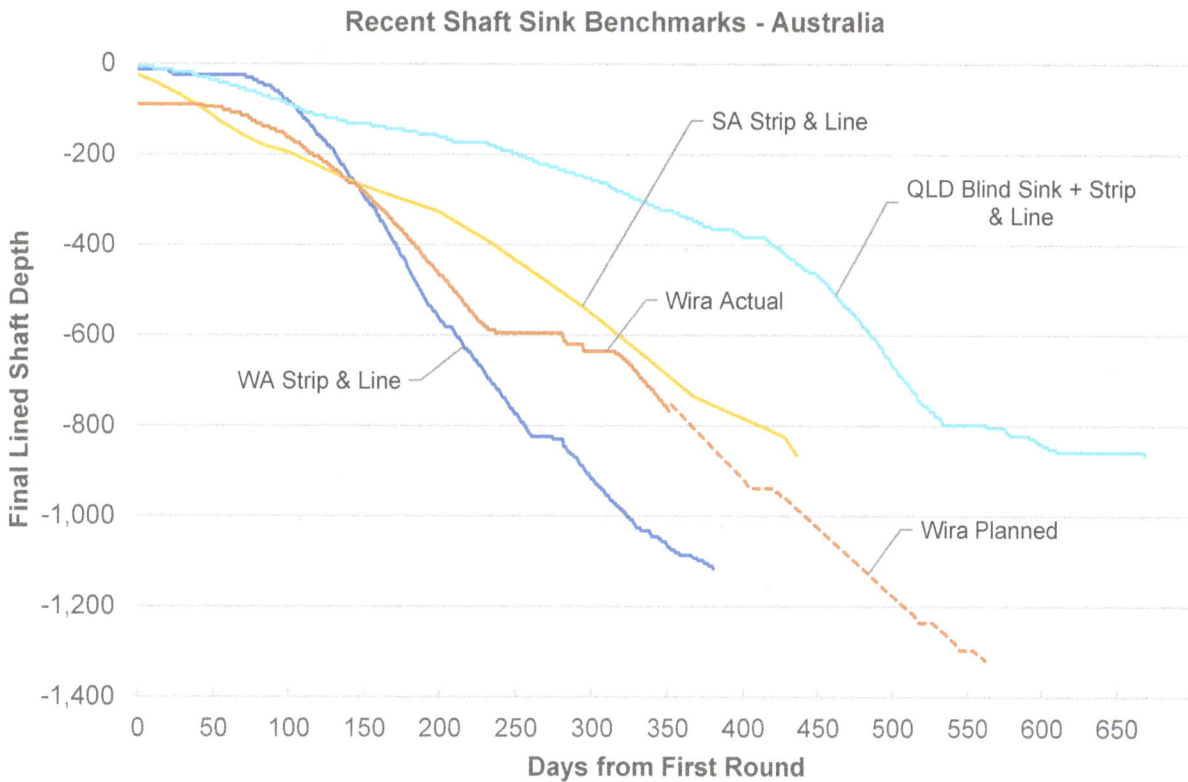

FIG 14 – Recent Australian shaft sink performance, excluding pre-sink.

Furthermore, while peak sinking rates give an indication of the potential for excavation rates, the overall progress and sinking rates for the entire project will also depend on a combination of geological conditions, operational efficiency, and equipment and people performance.

The Wira Shaft sink has experienced a slower than anticipated ramp-up in sinking rate. Despite this, the project has achieved cycle times as low as 28 hrs for full 6.0 m of advance (liner to liner), equivalent to a 5.1 m/d sinking rate. Overall, the Wira Shaft is expected to achieve a final sinking rate of approximately 2.4 m per day (m/d) from start to finish. This rate accounts for various planned and unplanned interruptions during the sinking process.

ACKNOWLEDGEMENTS

The authors wish to express their gratitude to BHP and Byrnecut Australia Pty Ltd for granting permission to publish this paper. Special thanks are extended to the Prominent Hill Expansion Owner's Team, particularly Eric Downing and Charlie Fearon, for their contribution to the project. Equally special thanks are also given to the Byrnecut Shaft Sink team for their dedication, hard work, and commitment to safety and continuous improvement. Further appreciation is extended to project contract partners including RKF Engineering Services, Pitt & Sherry, Siemag Tecberg, Raising Australia, Jetcrete, Hallet Concrete, and others, for their valuable contributions to the success of the project.

REFERENCES

AMC Consultants, 2021. Prominent Hill Shaft Design, AMC Consultants, internal report.

Hexter, J and Fullelove, N, 1998. Renison Internal Shaft Materials Handling System, in *AusIMM, 1998 – The Mining Cycle*, Mount Isa (The Australasian Institute of Mining and Metallurgy: Melbourne).

Hills, P B, O'Toole, D A, Kilkenny, D J and Seah, K L, 2014. Exploring the use of fibre reinforced shotcrete in large diameter shafts, in *Proceedings of the Third Australasian Ground Control in Mining Conference*, pp 169–178 (The Australasian Institute of Mining and Metallurgy: Melbourne).

Kilkenny, D J and Dennis, J M, 2011. Mine shafts – Planning, optimising and constructing, in *Proceedings of the 11th Underground Operators' Conference*, pp 115–126 (The Australasian Institute of Mining and Metallurgy: Melbourne).

Kilkenny, D J, Clark, J and Hill, C, 2023. No entry mechanised pre-sink of the Wira Hoisting Shaft, in *Proceedings of the Underground Operators Conference*, pp 380–395 (The Australasian Institute of Mining and Metallurgy: Melbourne).

McCracken, A and Stacey, T R, 1989. Geotechnical risk assessment of large-diameter raisebored shafts, in *Shaft Engineering,* pp 309–316 (Institute of Mining and Metallurgy).

Norwegian Geophysical Institute (NGI), 2022. *Using the Q-System: Rock Mass Classification and Support Design* (2nd ed), Norwegian Geophysical Institute.

NSW Resources Regulator, 2021. Technical Reference Guide: Powered Winding Systems, NSW Resources Regulator.

Standards Australia, 2002. AS/NZS3785.4:2002 Underground Mining – Shaft Equipment, Part 4: Conveyances for vertical shafts, Sydney, Australia.

Standards Australia, 2011. AS61508.1–2011: Functional safety of electrical/electronic/programmable electronic safety-related systems, Part 1: General Requirements, Sydney, Australia.

Standards Australia, 2018. AS4730.2:2018 – Mining – Winding equipment – Braking systems, Sydney, Australia.

Wu, R, Dase, J and Howell, R, 2022. Using fiber shotcrete as temporary wall support at Barrick Turquoise Ridge #3 shaft sinking, SME Annual Meeting, Salt Lake City.

Considerations for paste fill inrush exclusion zones

L Maia[1], J Percy[2] and B Barsanti[3]

1. Underground Geotechnical Engineer, Norton Goldfields, Kalgoorlie WA 6430.
   Email: leandro.maia@padgold.com.au
2. Project Lead: Backfill, Operational Geotechs, Perth WA 6000.
   Email: jed@operationalgeotechs.com.au
3. Principal Geotechnical Engineer and Director, Operational Geotechs, Vancouver BC.
   Email: ben@operationalgeotechs.ca

ABSTRACT

Cemented Paste Backfill (CPB) has become integral to the mining cycle in many underground mines in Australia. However, there are inherent safety concerns associated with CPB, as it introduces the hazard of inrush into operational areas during backfill operations in the event of a paste wall failure. Typical operational controls to prevent the consequences of an unwanted inrush event involve the establishment of inrush exclusion zones. The mine design layout and sequence can have a significant impact on the implementation of these, which can cause interactions with other production activities, leading to disruptions if not scheduled and designed adequately. These issues are particularly prevalent in bulk stoping mines, where paste fill volumes and interactions on production levels are much higher compared to narrow vein mining.

Exclusion zone design is undertaken using various methods, each with its own advantages and disadvantages. The selection of the appropriate methods can significantly impact the safety and efficiency of mining operations. This review assesses the typical methods used in Australian mines and outlines suitable approaches for different stope types, providing considerations for exclusion zones design to support safe and efficient operations.

INTRODUCTION

Paste fill has become an important tool for underground mining operations due to its capacity of stabilising voids and improving general geotechnical stability. There are many other benefits from improving ore recovery to providing flexibility to mining strategies and disposing of tailings underground to increase tailings storage facilities life cycle.

This technology is popular due to its rheological characteristics making it commonly transported via gravity through reticulation systems; its capacity of retain water behaving as a paste; and the ability to achieve high strength results in a few days allowing optimised mining cycles.

Although the benefits of implementing this backfill technique to a mine site are of great interest, the paste fill operation introduces a principal mining hazard as it is susceptible to inrush scenarios in the case of paste wall failures. The establishment of exclusion zones is one of the main controls to prevent the consequences of a wall failure and inrush, however it introduces an extra delay in the mining cycle during the curing periods of plug pours, before exclusion zones are removed, and production activities can continue. Other risks such as dilution, fall of ground, pipe bursts, must also be risk assessed when implementing paste fill systems.

Exclusion zones are perceived as necessary however operationally restrictive. This paper addresses the establishment of concepts, definitions, methodologies and best practices on paste inrush exclusion design and aims to provide a technical basis for future works that could lead to reducing exclusion zones.

Terminology

Although terminology on paste fill inrush exclusion zones is straightforward, definitions should be made to avoid confusion and clarify concepts.

Inrush exclusion zones or exclusion zones

Area exposed to the flow path of the inrush of paste in which no personnel or equipment is allowed. Also known as containment zones.

Containment bunds

Geotechnical structures designed to be physical barriers that delimitate the exclusion zone with the purpose of containing/catching the volume of fill egress in an inrush scenario.

Containment capacity or catchment capacity

The volume that the containment bunds are capable of retain/catch inside the exclusion zone in an inrush scenario.

Paste wall

A barricade, a geotechnical structure, designed and engineered to retain paste fill inside the stope allowing water drainage. Also known as barricade or bulkhead.

Plug pour

A pour designed to achieve a sufficient height to cover the paste wall usually rested and cured to a sufficient strength that prevents the next pour (body pour) applying load to the paste wall.

Body pour

The volume of fill placed on top of the plug after the plug is finished.

Open hole bund

A rock bund 1.5 m high positioned 3 m away from the stope brow to protect equipment and personnel from rockfalls and fall from height into open holes.

Point of inrush

The exact point in which the fill egress is inrushing from used to assess paste flow path considering a specific rill angle.

Rill angle

The deposition angle of the inrush material to the horizontal.

Stope footprint

The area of the bottom of the stope in which the fill accumulates first.

Rate of rise

Meters of paste level achieved per hour calculated according to paste plant throughput.

Void to fill solid

A 3D solid of the empty stope to be paste filled combined with the drive connecting the stope void shape to the paste walls.

Figure 1 shows main terminology used for Paste Inrush Exclusion Zones.

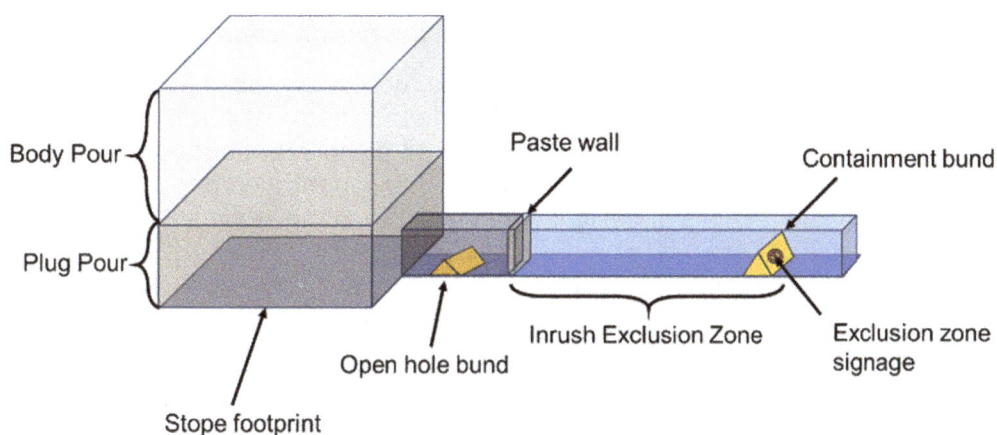

FIG 1 – Schematics of a general paste fill inrush exclusion zone.

BACKGROUND

Inrush and wall failure

According to the Western Australian Work Health and Safety (Mines) Regulations 2022, an inrush scenario is considered a principal mining hazard as it has the potential to result in multiple fatalities in a single occurrence. Therefore, the risk of inrush of a paste fill mass must be managed, including:

- The source of the inrush material.
- Potential failures of paste walls.
- The flow path the inrush can take.
- Areas likely to be affected.
- Additional controls to avoid inrushes.

Source of inrush material

During the stage of designing the filling strategy of a specific excavation, it is important to understand the presence of groundwater while understanding the additional pressure this applies to the paste wall, the decrease in containment capacity, and reduction in paste fill strength generation.

The paste fill mix design performed at the paste plant is also a source of inrush material, especially when the specifications are not met. Quality control on components materials of paste and in the final product must be done throughout the operation and trigger action response plans created for any non-conformity scenario. An out of specification paste fill could result in longer curing periods or in poor strength development which let the paste wall susceptible to overpressure as paste is not strong enough to withstand the pressure from paste poured above, which in turn can lead to over pressurisation of walls, failure, and consequent inrush.

Wall failure

Wall failures are the triggers of inrush in a paste fill operation. Paste walls are considered Geotechnical Structures and must comply with regulations. The potential of failure is managed considering:

- Influence of geological features and hydrological environment.
- Design, location, construction and integrity.
- Properties of materials being used.
- Operational factors in the construction process and the influence of operational activities in the surroundings.
- Quality control and quality assurance.

- Inspections and the competence of personal from the design stage to final construction.

Paste walls can fail due to poor design, poor construction practices, damages to wall, and overpressure. Recent wall failure reports show incidents occurs mostly in overpressure scenarios where high and non-expected paste rate of rise occurred along with poor practices on tight filling. Revell and Sainsbury (2007) and Thompson *et al* (2024) have published papers on wall failures and recommendations on best practices of design, construction and monitoring.

Flow path of inrush and areas affected

The flow path that a paste inrush can take is usually assessed when designing a stope filling strategy through the creation of exclusion zones. When establishing the area within the exclusion zone, the design must account for all openings, boreholes, and excavation connections that can provide possible paths to the inrush mass to guarantee the physical barrier installed (containment bunds), will capture the fill mass egress from the wall failure and there is minimum exposure of personnel, equipment and infrastructure.

Additional controls

Additional controls to avoid paste inrush are related to:

- The filling strategy itself by using higher cement contents for the first volumes poured and dividing the filling strategy in a plug-body pour filling strategy.

- Collection of samples to perform Uniaxial Compressive Strength (UCS) tests allow understanding of paste strength development.

- Monitoring of paste walls using cameras, water/paste level with weep holes, installation of pressure sensors and all the related TARPs.

Purpose of establishing an inrush exclusion zone

In the initial stage of stope filling, the paste is in direct contact with the wall and has not yet become strong enough leading to pressure being exerted on the paste wall. To prevent wall failure due to overpressure, paste runs are limited to a fill 'plug' poured to a few metres above the stope brow. The plug must then cure to sufficient strength to resist the driving forces of the paste deposited above, preventing failure through plug-induced pressure when the main pour begins. Insufficient CPB plug strength can result in wall failure, leading to fatalities, equipment damage, and production loss (Grabinsky, Bawden and Thompson, 2021).

The most noticeable incident of wall failure and inrush was at Bronzewing, which sadly led to three fatalities. Since then, standards on hydraulic fill operations have been raised starting from the implementation of cemented fills and exclusion zones, going through monitoring, 3D modelling, and tighter quality controls. It is understood that the Bronzewing incident was the trigger to implement Inrush Exclusion Zones over Australia.

A series of papers published and technical reports on walls, stope instrumentation, and 3D modelling of paste walls to understand loads and capacities when pouring paste suggest pouring approaches might be very conservative.

Continuous curing approaches in stopes with large footprint, where the paste achieve curing requirements before production reaches containment capacity of exclusion zones, also have been proved as safe to do in terms of paste wall pressures. However, there are several factors that might change the conditions of the void to fill and the paste wall, therefore additional care must be taken to these approaches.

As mentioned in the previous sections, Paste Inrush is a principal mining hazard which could end up in multiple fatalities in a single incident, therefore the risk must be managed. When the risk of a fill wall is assessed to the worst practicable scenario, there are only two controls that guarantee the inrush will be contained and no personal, equipment or infrastructure is exposed to the fill mass egress. These are controlling volumes poured through plug-body lift strategies using continuous curing and establishing inrush exclusion zones.

Interactions with mining cycle

Establishing an exclusion zone, especially in dynamic and high production mine sites, can bring a considerable number of interactions and disruption of activities, as containment bunds often block main access and ore drives which will require pausing development heads, vertical drilling sites, blasting, and block access to critical infrastructure. This is more critical the closer the stope and wall are to level accesses and other active ore drives.

It is therefore critical to rely on optimised exclusion zone designs using appropriate design considerations and achieving requirements to lift exclusion zones as quickly as safely possible through improved strength development and robust quality control practices.

CURRENT DESIGN METHODS FOR PASTE INRUSH EXCLUSION ZONES

Operations throughout Australia use different tools and methods to estimate inrush exclusion zone capacity often with no clear guidelines or procedures from design to construction. Four design methods were assessed in this paper based on authors practical experience using a practical case to provide insights on each of these.

Flat surface

The flat surface methodology consists of using a 3D CAD software to split the void to fill and the development drive solids in a flat surface at a chosen containment bund position to the standard height (2.5 m). The volume of the bottom slice is the containment capacity of the exclusion zone. Additional iterations are performed to optimise bunds positions and containment capacity. Bunds can be design inside the exclusion zone to increase the capacity. Figure 2 shows an example of Flat Surface method on CAD software.

Pros:

- Simple to use.

- Allows additional bunds (double bunds and open hole bunds).

- Can be performed with any 3D CAD software.

- Uses drives and stopes actual shapes allowing accuracy on volumes estimation.

- Often one of the most used for its simplicity and conservative safety factor.

- Design based on containment bund standard height.

Cons:

- Cannot apply rill angle.

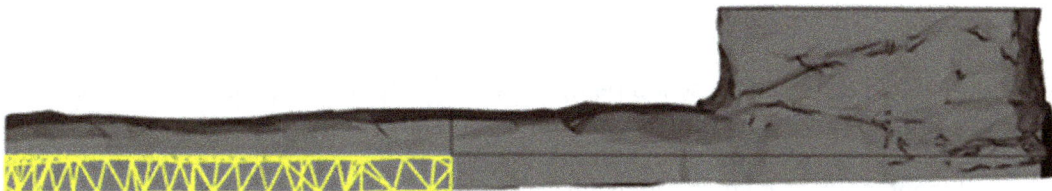

FIG 2 – Example of Flat Surface method on CAD software. Yellow portion on bottom left side highlights a portion of volume representing containment capacity.

Deswik backfill tool

The Deswik backfill tool calculates the inrush volumes by creating a 3D cone slice from a chosen point of inrush and splitting the chosen 3D solid at the rill angle input.

The volume of inrush and the point of inrush are inputted by the user and the software provides the flow path through the chosen solids. Bunds are added manually by splitting solids. The height of the inrush material to the bund can then be measured. Multiple iterations are performed to optimise containment capacity as the inrush volume input can be higher or lower than bund standard height.

Applying additional bunds to increase containment capacity in this methodology requires several manual additional steps.

This is the second most used tool as many operations uses the same software for underground mine design. When a rill angle of zero is applied, this tool provides the same result as the Flat surface method. Figure 3 shows an example of Deswik Backfill Tool calculating containment capacity.

Pros:

- Software available in most mine sites.

- Allows applying rill angle.

- Allows designing by a chosen volume.

- Uses drives and stopes actual shapes allowing accuracy on volumes estimation.

Cons:

- Many iterations trialling different volumes to achieve bund standard height.

- Does not allow additional bunds (double bunds and open hole bunds).

- Rill angle associated with paste wall failures has little to no literature available.

- Choosing point of inrush might lead to errors on estimation.

FIG 3 – Example of Deswik Backfill Tool calculating containment capacity. Red mark on top left indicates filling point. Dark purple is the chosen inrush volume. Query tool on bottom right indicates inrush height on 'DZ'.

mXrap

Creates a block model of the mine solids being assessed. Input is required for the location of both the paste walls and bunds, a beach angle of the paste fill and the volume of paste fill. A simulation is run with the wall being removed which shows the flow of the paste fill into the drive. The beach angle is a significant input as it limits the distance the paste fill will flow. This software can be used to understand how paste would flow-through vertical development such as escapeways or ventilation if required. Figure 4 shows an example of inrush capacity using mXrap software.

Pros:

- Few iterations to perform exclusion zone design specially for positioning bunds.

- Multiple bunds and barricades can be installed into the model quickly easily identifying the limiting bund.

- Shows potential inrush path relative to drives and other voids.

Cons:

- Block modelling of the as-built solids requires adjustment on resolution to avoid simplifying the geometry which can cause minor discrepancies in the estimated volumes.

- Many iterations trialling different volumes to achieve bund standard height.

FIG 4 – Example of Inrush capacity using mXrap. Light green star indicates the fill point. Purple barriers at both sides are containment bunds. Coloured layers are the chosen inrush volumes.

Centreline

Analytical calculation based on triangle prism that uses length of centrelines available in level for containment, design drive profile of the drives, height of containment bund, stope volume up to bund height to calculate the containment volume. This methodology does not consider changes in gradient of the drives, under and overbreak of drives changing the volume.

Pros:

- Simple calculation with just a few inputs.

Cons:

- Does not consider drives and stope actual shapes and floor gradient variations, leading to volumes discrepancies.

- Does not allow additional bunds (double bunds and open hole bunds).

PRACTICAL CASE – EXCLUSION ZONE DESIGN FOR DIFFERENT STOPE TYPES

The four methods described on last section were assessed considering containment capacity results, ability to use rill angles, ability to apply additional bunds, and practicality in a simulation of inrush exclusion zone design considering three different stope shapes: longitudinal, transverse, and bulk (Table 1). Figure 5 shows schematics of distances considered on practical case.

TABLE 1

Stope types assessed in the practical case.

Stope type	Footprint	Constrain
Longitudinal	5.5 × 15 m	Mined in the trans drive. Bund 70 m away.
Transverse	20 × 25 m	Mined in the trans drive. Bund 120 m away.
Bulk	28 × 20 m	Mined in the ore drive towards trans drive. Bund 70 m away.

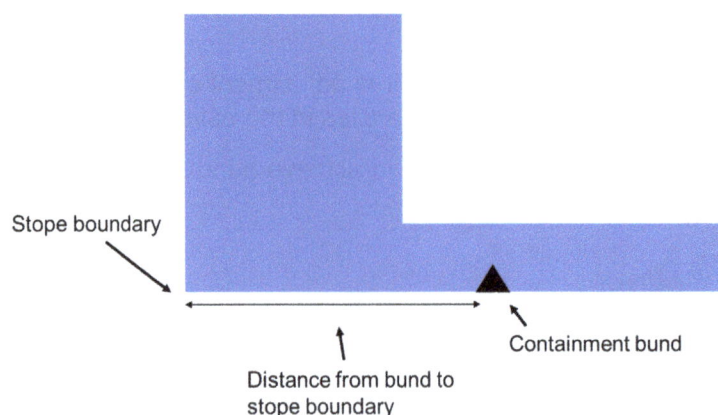

FIG 5 – Schematics of distance considered on practical case.

The three stope types were compared in three different scenarios: No rill angle, 3° rill angle, Double bunds with no rill angle. Bund standard height was considered as 2.5 m. An open hole bund of 1.5 m high at 3 m away from stope brow was also considered for each stope type in this simulation. Additional bunds were positioned halfway from stope boundary to the furthest bund.

Comparison on current design methodologies results

Table 2 summarises and compares the current design methods for inrush exclusion zone calculations within each stope type.

TABLE 2

Comparison of current design methodologies results.

	Stope type	Capacity no rill angle (m³)	Capacity 3° rill angle (m³)	Doubled bunds capacity no rill angle (m³)
Flat line	Longitudinal	828	-	918 (**)
	Transverse	1478	-	2015
	Bulk	1222	-	1435
Deswik backfill tool	Longitudinal	822	1155	-
	Transverse	901 (*)	4770	-
	Bulk	1208	2004	-
mXrap	Longitudinal	940	1500	930 (**)
	Transverse	1520	4450	2120
	Bulk	1100	2050	1470
Centreline	Longitudinal	962	-	-
	Transverse	2029	-	-
	Bulk	1145	-	-

The lower capacity for transverse stopes from the Deswik backfill tool (*) compared to flat line and mXrap in the no rill angle scenario is due to the unpracticality on the Deswik tool to design an open hole bund. Capacity when using double bunds for longitudinal stope type is not increased due to a flat portion of the drive on the chosen shape (**).

The use of rill angles is a feature available only for the Deswik backfill tool and mXrap, enabling significantly increases the containment capacity. However, further site-specific investigation into the flow characteristics during a wall failure is required to allow selecting a rill angle. Based on the authors' experience, adopting a rill angle of zero provides a more conservative yet suitable approach.

Double bunds increased capacity with significant improvement to the transverse stope type as the furthest bund is positioned further away than for longitudinal and bulk stopes types (120 m × 70 m). This implies the benefit is enhanced if the first bund position is already optimised. For bunds closer to the stope boundary (<50 m) the increase in capacity is reduced.

The centreline numbers over and underestimate volumes when compared to the other methods, due to not consider drive profile and stope shapes, indicating a less accurate approach.

In general, the numbers are comparable through the different methods. However, users must evaluate the associated risks and limitations of each method when considering using these approaches.

The following section summarises general observations on exclusion zones capacity behaviour.

Capacity behaviour for different stope types

Capacity of exclusion zone for longitudinal stopes increases as the distance of bund to the stope boundaries increases and stabilises in a maximum volume when bund is moved further in a single drive. Additional bunds are optimised when positioned halfway between furthest bund and paste wall, leading to higher design capacities.

For transverse and bulk stopes, a similar capacity behaviour is expected when positioning bunds. When inserting additional bunds, a maximum capacity is achieved when positioning a bund in front of the wall as the stope footprint increases the containment capacity.

Positioning an additional bund at halfway the distance between stope boundaries and furthest bund also gives maximum design capacity for transverse and bulk stopes. This is more practical if there are constrains or concerns when building bunds close to paste walls.

Figures 6 and 7 are schematic diagrams to summarise the containment capacity of exclusion zones behaviour when positioning bunds and adding extra bunds for longitudinal, transverse and bulk stopes, providing a general idea on how to maximise capacity. The diagrams have six points of interest: (A) is the catchment capacity provided by the brow bund, (B) arbitrarily selected bund position, (C) maximum containment capacity using a single bund. (D) optimum position for the doubled bund inside first bund area to increase exclusion zone capacity, (E) is the increase of capacity to maximum when adding extra bund close to stope wall in traverse and bulk stopes. Diagrams are not to scale.

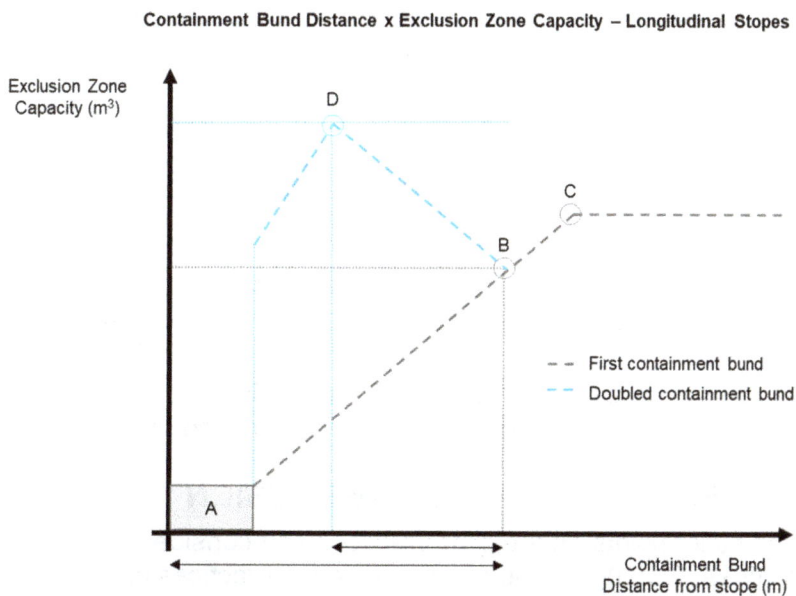

Containment Bund Distance x Exclusion Zone Capacity – Longitudinal Stopes

FIG 6 – Containment capacity of exclusion zones according to bund distance from stope for Longitudinal stope type in a single drive.

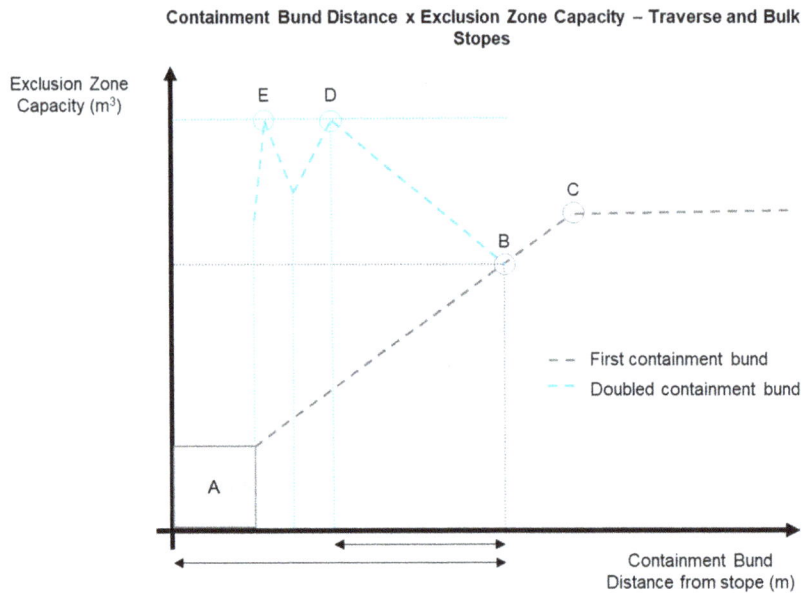

Containment Bund Distance x Exclusion Zone Capacity – Traverse and Bulk Stopes

FIG 7 – Containment capacity of exclusion zones according to bunds distance from stope for Transverse and Bulk stope types in a single drive.

Significant increased capacities are achieved when bunds are moved to multiple access drives. However, this often blocks the level access, disrupting activities in surrounding stopes, and can also disrupt access to critical mine infrastructure. The increase is more pronounced for bulk stope types, where there are multiple drives in which paste can inrush to. Figure 8 represents the abovementioned scenario.

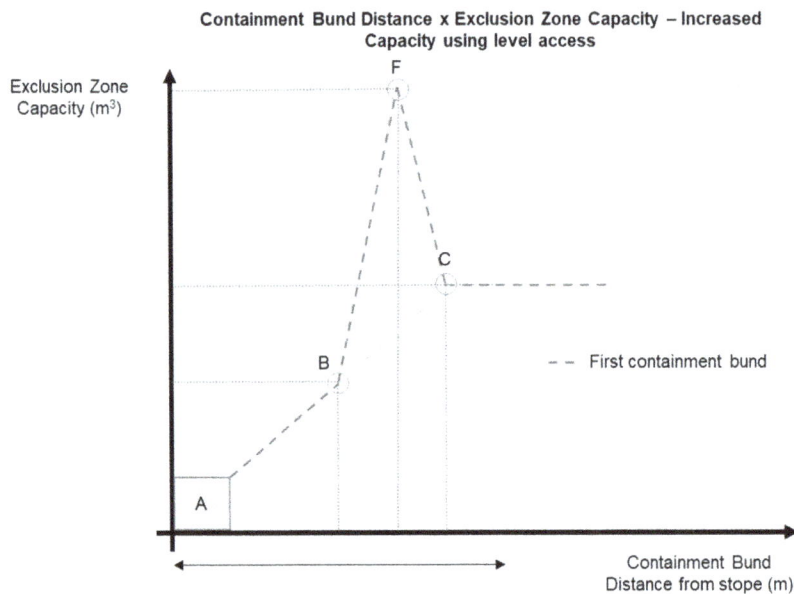

Containment Bund Distance x Exclusion Zone Capacity – Increased Capacity using level access

FIG 8 – Containment capacity of exclusion zones according to bunds distance from stope. (F) is the increased volume when bunds are moved to level access.

BEST PRACTICES FOR EXCLUSION ZONES DESIGN

This section provides bullet points on the best practices and considerations when designing paste fill inrush exclusion zones. Although some of the mentioned practices are not always feasible or are constrained, this is a snapshot of what the authors recommend and have experienced as standard and optimised.

- Large stope dimensions have high volume plug pours, consequently requiring exclusion zones with significant capacity. This can result in multiple stages of pours, multiple bunds, lengthy interactions and delays.

- When planning level designs, consider the distances of level accesses, hanging wall and footwall drives to orebody for backfilling requirements to provide sufficient containment capacity leading to less production interactions.

- Where possible, planning for multiple mining areas significantly reduces the effects of interactions when backfilling. It is particularly important when exclusion zones effect level accesses or restrict access to other stopes on the same level. Allowing multiple levels to be accessed is critical when backfill interactions and exclusion zones are restrictive.

- Exclusion zones for stopes close to the level access or hanging wall/footwall drive often block the access to adjacent ore drives, disrupting activities on surrounding stopes, and access to critical mine infrastructure (sub stations, sumps, vent raises, escapeways).

- Exclusion Zone design must consider interactions with boreholes, vertical openings, infrastructure, vent walls, escapeways, and plans for protecting these must be issued and completed before establishing the exclusion zone. Inspections must be done prior to design.

- Open hole bunds must be accounted for in containment capacity. This often provides optimised capacity when the furthest containment bund is 100 m away from the open hole bund.

- Capacity of exclusion zone increases as the distance of the bund to the stope boundaries increases and stabilises in a maximum volume when bund is moved through a single drive.

- For transverse and bulk stopes, positioning a containment bund in front of a paste wall and a second bund further away provides increased capacity due to increased use of stope footprint.

- For transverse and bulk stopes, the pour rate and rate of rise often allow achieving curing requirements before reaching containment capacity and pour plug to be performed in a single run. Additional controls and risk assessments are required to ensure paste volumes are cured.

- Progressive removal of bunds is permitted when curing requirements are achieved in a continuous curing strategy.

- A second bund for increased capacity is best positioned in the middle distance between furthest bund and stope brow.

- The increase in capacity of a second bund is limited if the furthest bund is already close to stope (<50 m).

- If a plug pour is split due to inadequate containment zone, the capacity provided by the stope footprint up to the height of the bund must be reconsidered for subsequent pours.

- Significant increased capacities are achieved when bunds are moved to multiple access drives. However, significantly more interactions and disruptions are expected.

- Information on plug volumes, bund positioning, and curing days for the plug and removal of bunds should be communicated and contained in both the monthly and weekly schedule.

- Memorandums with stakeholders sign off sheet on exclusion zone establishment and removal is to be shared with all personnel on mine site. Information on exclusion zones must also be available on shift sheets.

- Equipment interaction with walls or infrastructure when building bunds must also be considered by the operational team.

- Appropriate signage for exclusion zone is required. A 'No Unauthorised Entry' sign is often positioned in front of any accessible containment zone or escapeway which interacts with an exclusion zone. No personnel or equipment is allowed to go past signage before requirements for removal have been achieved and communicated.

- Quality control on bund construction and exclusion zone establishment through inspections is required to validate design in the field.

- Using a rill angle on paste has a significant impact on increasing capacity of catchment. However, it is more reasonable to assume a zero degrees rill angle when planning for the worst-case scenario in terms of inrush volume due to lack of information on rill angle of failures.

- If using rill angles, the inrush location where paste will rill from is the highest portion of the fill below the fill point.

- In a scenario with two paste walls, the exclusion zone must be designed for the worst case which is when only one wall fails and fill egress flows to one side of the exclusion zone.

FUTURE DIRECTIONS IN REDUCING EXCLUSION ZONES

This paper provides insights into optimised design of paste fill inrush exclusion zones. Reducing or eliminating exclusion zones requires a further technical understanding of paste fill behaviour and risk mitigation strategies. The following topics are suggested as relevant for further investigation:

- Improved curing times for paste plugs by developing optimised mix design using site specific cement and/or admixture to remove exclusion zones earlier and decrease disruptions time.

- Investigation to develop site specific relations between curing times and rill angle to allow using rill angles and a progressive removal of bunds, reducing exclusion zones real estate.

- Development of a mobile expandable physical barrier device capable to retain paste from inrush to be installed in front of the wall and activated in the case of a wall failure, working as a second emergency wall. Exclusion zones sizes could be drastically reduced and plug pours sizes increased.

- Encourage development and improvement of backfill inrush exclusion zone design tools into mine design software to allow for optimised design.

CONCLUSION

Backfill inrush is a principal mining hazard that has resulted in multiple fatalities in the past. Appropriate design considerations are crucial for safely assessing the containment capacity of inrush exclusion zones while minimising impacts and disruptions to mining activities. Despite advancements in backfilling technologies, especially in paste fill monitoring systems, conservative pour strategies are still employed and infrequent monitoring of stopes during paste filling often resulting in conservative estimates of wall pressures, which lead to heightened safety factors. Given the classification of paste fill inrush as a principal mining hazard, it is critical to have effective and appropriate controls for each mine site, however consideration should be given in the planning phase to understand the impact of mine design and stoping strategy on backfill. In this context, exclusion zones play a vital role in preventing exposure of personnel and equipment to the egress of fill mass. Future works are required to provide technical basis for safely reducing exclusion zones and increasing pour sizes.

ACKNOWLEDGEMENTS

The authors would like to acknowledge the contributions of Jeremy Doolan (Operational Geotechs), Peter Fookes (29 Metals), Calvin Gloster (Gold Fields Australia), and Roo Talebi (Gold Fields Australia), for the technical and practical knowledge shared on this paper.

REFERENCES

Grabinsky, M, Bawden, W and Thompson, B, 2021. Required plug strength for continuously poured cemented paste backfill in longhole stopes, *Mining*, 1(1):80–99.

Revell, M B and Sainsbury, D P, 2007. Paste bulkhead failures, in *MineFill 2007*, paper 2472.

Thompson, B D, Veenstra, R, Carmichael, P, Counter, D and Grabinsky, M W, 2024. Lessons learned from tight or blind filling induced barricade failures, presented at *MineFill 2024*.

Western Australia. Work Health and Safety (Mines) Regulations 2022, s 612.

Managing large open voids throughout the production cycle at Olympic Dam

G Mungur[1], C Alberto[2] and B Stevenson[3]

1. Superintendent Geotechnical Operations, BHP Olympic Dam, Roxby Downs SA 5725.
   Email: glenton.mungur2@bhp.com
2. Geotechnical Engineer, BHP Olympic Dam, Roxby Downs SA 5725.
   Email: cassandra.alberto2@bhp.com
3. Superintendent Load and Haul Production, BHP Olympic Dam, Roxby Downs SA 5725.
   Email: ben.stevenson@bhp.com

ABSTRACT

Olympic Dam undertakes a sublevel open stoping (SLOS) mining method. Stopes range from single lift (30 m tall) to multi-lift (up to 260 m tall) with an average footprint of 30 m × 30 m. To meet production targets, Olympic Dam mine approximately 50 stopes per annum. The mining cycle of a stope (undercut firing to stope backfilling) creates large voids that need to be managed.

Voids, if not managed appropriately, can pose a significant risk to workers due to the potential of failure or collapse including impacts to reliable production through access closures and increased scheduled activities.

This paper describes the process of managing stope voids at Olympic Dam, including effective design and engineering, methods of void monitoring and use of Trigger Action Response Plans. These proactive applications ensure higher levels of controls are in place and ultimately reduce the risk to personnel and major unplanned changes to the mine schedule throughout a stope's production cycle.

INTRODUCTION

Olympic Dam (OD) mine is situated approximately 570 km NNW of Adelaide, South Australia. The orebody is an iron oxide copper gold deposit with copper being its primary commodity. Gold, silver, and uranium are the other commodities produced.

The underground mine has a length of approximately 5 km and width of 2.5 km with more than 700 km of accessible development (Figure 1).

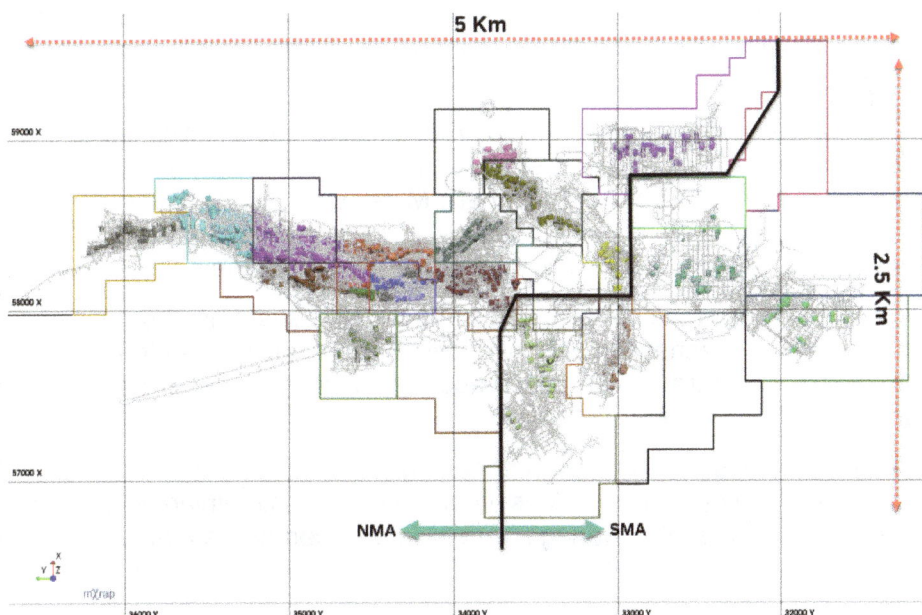

FIG 1 – Plan view of current Olympic Dam underground mine.

The mining method at Olympic dam is a mechanised sub-level open stoping, with cemented aggregate backfill (CAF). Figure 2 shows a ring design and blast packets of a typical longhole open stoping method. In this method, the development has a drill drive access with a perpendicular expansion slot drive. Depending on the height of the ore zone, the number of drilling horizons may vary between 2 to 6 levels, spaced 30 to 60 m apart vertically, to suit local orebody geometry, geotechnical conditions, and the capabilities of the longhole drilling equipment.

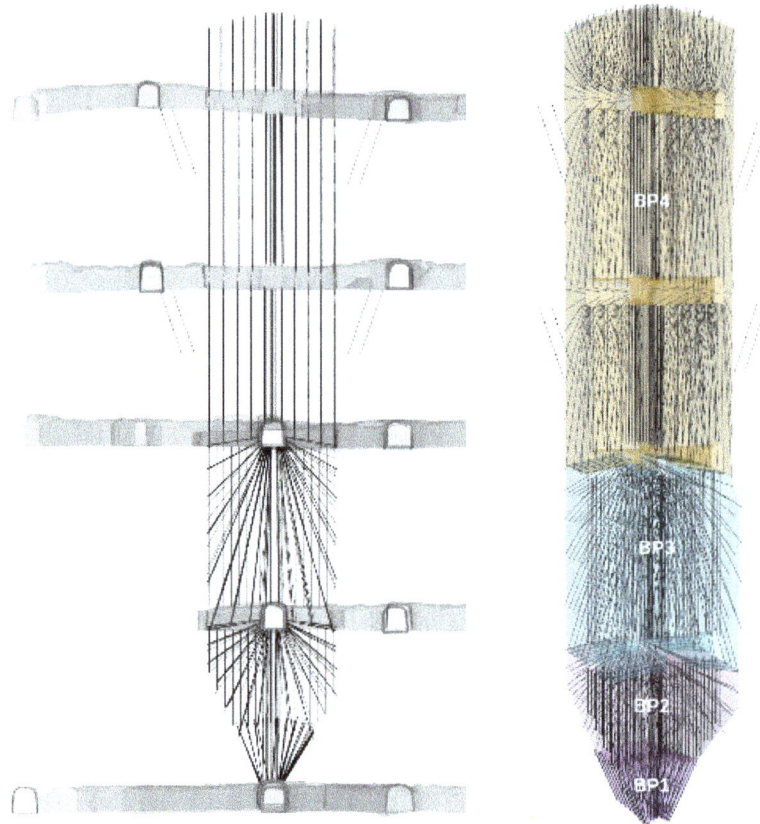

FIG 2 – Ring and blast packet design for a longhole open stope.

The intra-stope extraction sequencing is bottom-up level by level, in a series of blast packets. This method minimises exposure near an open void, improves ventilation usage and enables improved blasting techniques. The blasted ore is extracted by loaders via drawpoints and stopes are backfilled with CAF once emptied.

Over the years, the average stope sizes have decreased from 300 kT (max stope > 1.1 MT) in the early 2000s to current average stope sizes of approximately 200 kT (max stope ± 600 kT). This is due to the progression of mining from the Northern Mine Area (NMA) to the Southern Mine Area (SMA), which is accompanied by a higher degree of geological complexity. Figure 3 show the distribution of ore tonnes from the NMA/SMA as a percentage over the past nine years.

The geology in the SMA is more complex driven by larger fault systems and an increase of weaker geological units. The failure mechanism is gravity driven (unravelling) due to lower tensile strength and cohesion. However, structural driven failure accompanied by higher stress has been observed in the deeper areas within the SMA.

With an increased number of stopes and greater geological complexity, the risk of exposure to large open voids increases. Managing this risk is imperative to maintaining a safe and productive workplace. This paper describes the management of large open stopes throughout the production cycle at Olympic Dam.

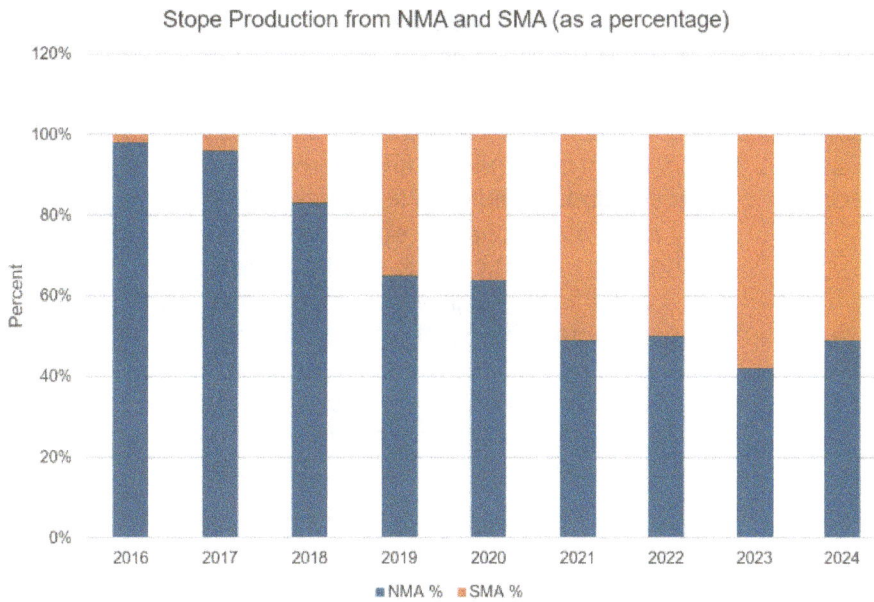

Stope Production from NMA and SMA (as a percentage)

FIG 3 – Distribution of ore tonnes as a percentage between NMA and SMA.

STOPE VOID EXPOSURES

Throughout the life cycle of a stope, multiple activities occur which expose personnel and equipment to the stope void. Table 1 describes these activities and the risk of void exposure, and Figure 4 illustrates where the activity takes place relative to the stope void.

TABLE 1

Activities near a stope void during the production cycle.

	Activity	Description	Risk scenario
1	Stope prep and charge, Re-drills	Prior to each blast packet firing, production holes are cleaned out and filled with explosives ready for firing. Access is required over the void (interim crown) created from the previous blast packet firing. An interim crown is an arch shape temporary backs of a fired blast packet. The pillar thickness ranges between 20 m to 40 m depending on the blast packet. Whilst the interim crown is un-supported, the arch shape and pillar thickness leads to increased stability.	Interim crown failure of the previous blast packet exposing prep and charge crew to the void below.
2	Clean-up, Push-off, stope bombing and operating drones	With each blast packet firing, some blasted material gets deposited into development access. Loader cleanup and stope push-off is required to ensure drive is clear for barricade building, bombing activities and drone scanning. Bombing activity is the process of placing an explosive package from the access development over the edge of the stope and fired to shake down any potential loose material on stope walls or broken ore standing up in the stope.	Failure of the stope wall encompassing (undercutting) the access development exposing personnel and equipment to the stope void.
3	General access (adjacent)	Travelling past an open stope as part of other mining activity.	Failure of the stope wall encompassing (undercutting) the adjacent development exposing personnel and equipment to the stope void.
4	Backfill with underground rock fill (URF) using trucks	Stopes with no future adjacencies are primarily filled with waste (URF). Trucking waste to the edge of the stope and tipping into the stope.	Failure of the stope wall undercutting the tip point exposing personnel and equipment to the stope void.
5	Access (level above stope crown)	Travelling along levels above the stope footprint	Stope crown failure exposing personnel and equipment accessing above crown footprint.
6	Backfill barricade building	Barricades are built to restrict backfill flowing outside the stope area. The barricades are built as close as possible to the stope edge to reduce the volume of fill and reduce future CAF development.	Stope failure exposing near stope edge operations.

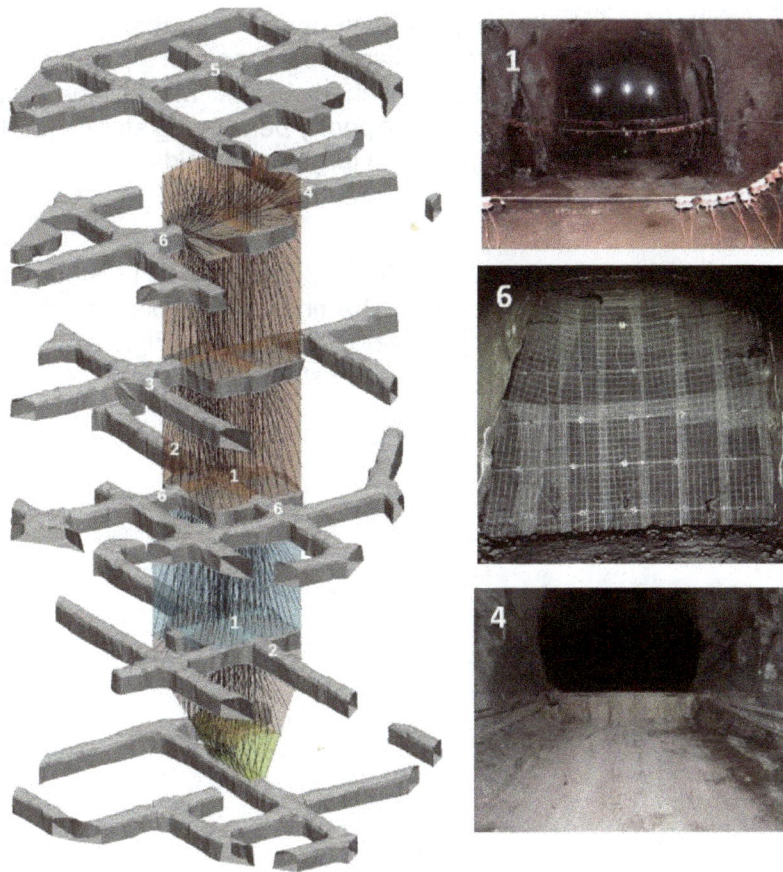

FIG 4 – Typical stope showing multiple stope void exposures (numbers correspond to activity in Table 1).

STOPE PERFORMANCES

Stope performance, quantified as overbreak, have typically ranged between 10–15 per cent for areas where rock mass is considered fair to good (Q' > 4). In weaker rock mass conditions (Q' < 4) increased overbreak is observed, up to 25 per cent. Figure 5 show the historical stope overbreak at Olympic Dam.

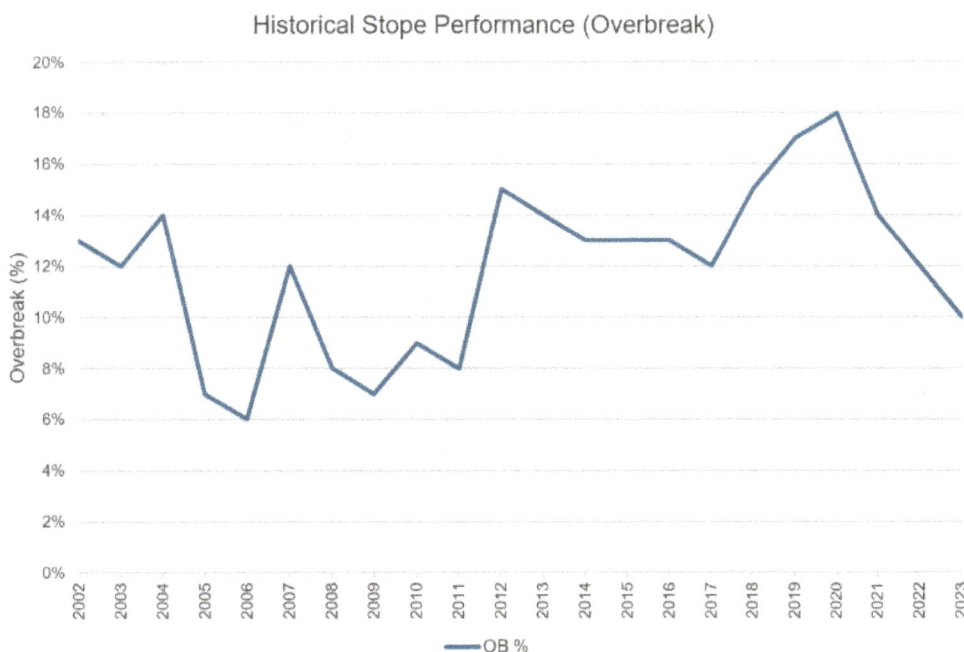

FIG 5 – Historical stope performance as a function of overbreak.

As the stope performance does not always go to plan, it is crucial that a stope void management plan is in place to reduce the risk to both personnel and operations. In a paper presented at the 2023 Underground Operators Conference by Capes, Lachenicht and Mitkas, the authors describe the failure of a stope (LIM 290). Below is a description of poorly performing stopes (LIM290 and TAN422) that potentially exposed personnel and equipment to large open voids.

LIM 290 stope

LIM 290 was designed as a 50 m tall stope with a 30 m × 30 m crown. The stope had caved vertically approximately 120 m above the designed crown. The final stope void was determined through a series of diamond drill probe holes from nearby development drives (Figure 6). At the time of mining, there had not been any development directly above the stope footprint.

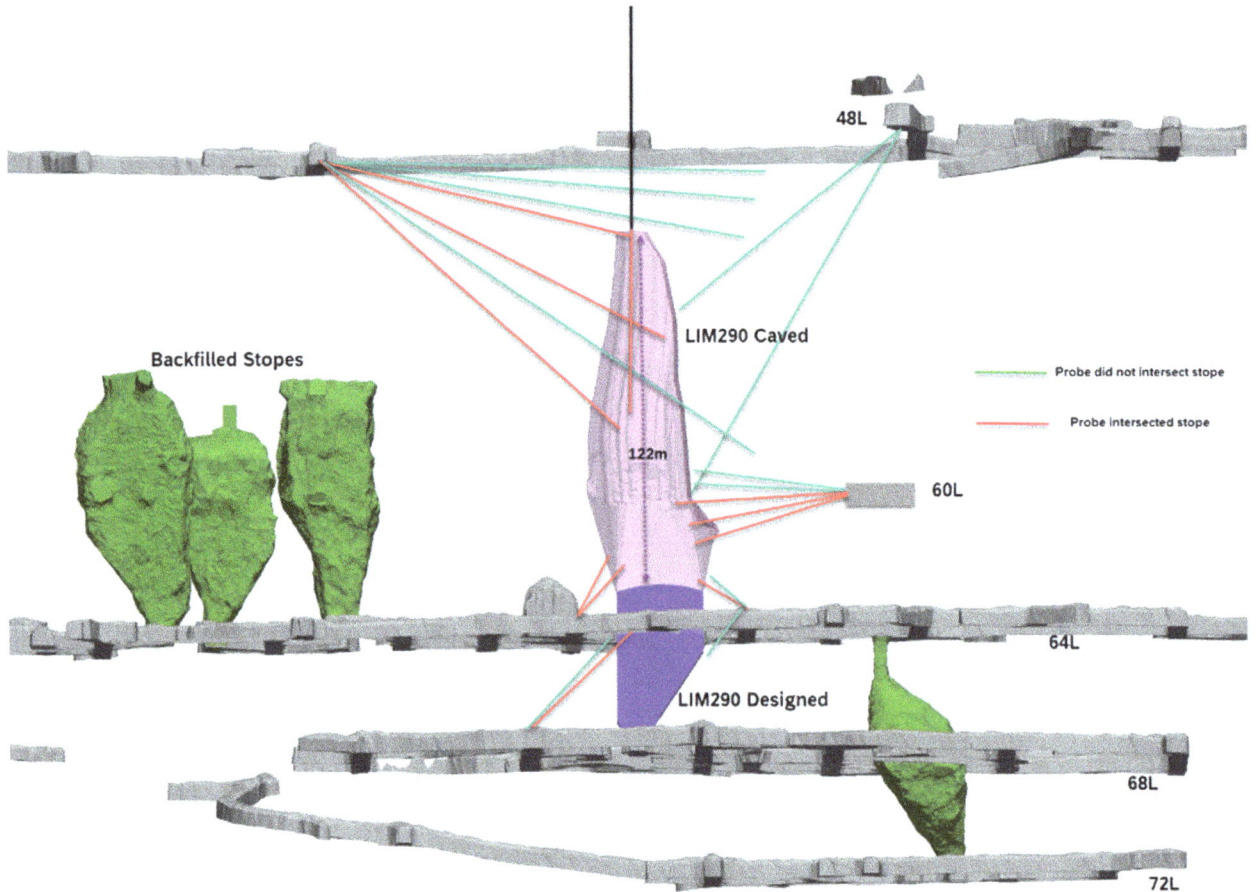

FIG 6 – LIM 290 stope failure.

TAN 422 stope

The stope was design with the following dimensions (28 m L, 26 m W and 85 m H). This stope had over-broken 60 m horizontally (Figure 7) driven primarily by a weak, altered and structured rock mass. The void had progressed within a few metres of the adjacent access development. Void management processes and TARPS were used to prevent any personnel exposure to the unplanned void. These processes are described in the next section.

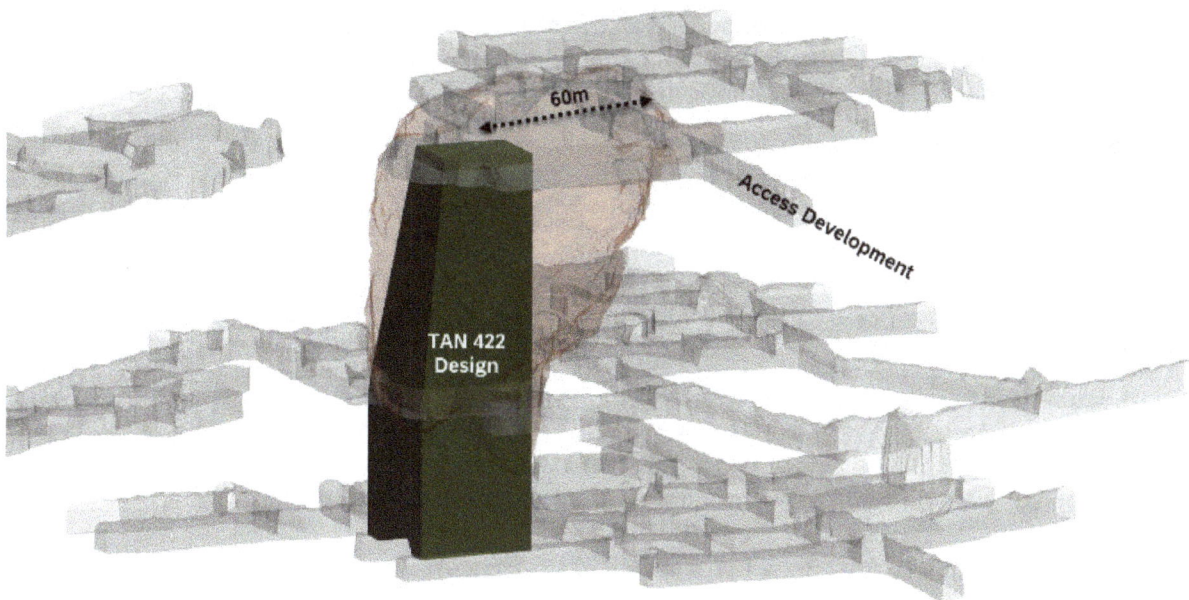

FIG 7 – TAN 422 stope failure.

VOID MANAGEMENT – CONTROLS FOR STOPE FAILURE

Void management processes are used to manage the Stope Failure risk at Olympic Dam. Stope Failure is identified as a material risk (potential for fatality) and is described as any area where personnel may be exposed to a failed stope (crown failure and/or overbreak into adjacent drives). The stope is considered failed when it has over-broken to a point where geotechnical assessment indicates it is no longer safe for personnel to access the crown or adjacent drives. The risk is removed once the stope has been backfilled and the void is no longer available to facilitate a failure. For the Stope Failure material risk, seven critical controls have been identified as shown in Table 2.

TABLE 2

Stope failure critical controls.

Critical control	Control type
Geotechnical design model	Characterisation/Design
Stope design	Design
Earth bunds, signage, and tipples	Separation/Monitoring
Hazard zones – exclusion and separation	Separation
Monitoring – install buckets	Monitoring
Blasthole dipping	Monitoring
Backfill reconciliation	Monitoring

MANAGEMENT OF STOPE VOIDS – IMPROVEMENT CYCLE

Through understanding changed conditions (TARPs and Hazard reporting systems) the mine has been proactively strengthening its critical controls using the PDCA process. The application of the improvement cycle is to move up the hierarchy of controls, ie from administrative through to engineering controls, substitution and elimination.

The improvement cycle for the management of stope voids include:

- **Planning:** The geotechnical design model aims at having the rock mass characterisation with sufficient confidence (quality and quantity) to enable improved and reliable mine designs (eliminate potential instabilities). The stope design process ensures that crowns (interim and final) and adjacent development have been assessed for stability and redesigned if required.

- **Execution:** This relates to the execution of the stope, ie procedures and work instructions are followed; void protection and separation controls are in place and risk assessments associated with non-standard practices are undertaken.

- **Monitoring and analysis:** To ensure stope performance meets the expected design, sufficient monitoring and analysis is required throughout the life cycle of the stope. Monitoring includes the use of LiDAR scanning, dip-hole measurements to validate crown pillar thickness and 'bucket' monitoring to alert of any changed condition during the operational stage. Trigger action response plans (TARPs) are used to define responses to observed changed conditions through monitoring. Near stope edge activities require additional geotechnical assessment to be undertaken.

- **Improvement:** The reconciliation process compares the as-built to the design and identifies gaps in understanding or data. These are used to improve and support future stope designs. External reviews such as Geotechnical Review Board provide external subject matter to verify existing systems and processes and recommend leading practices to implement.

CONTINUOUS IMPROVEMENT THROUGH DESIGN AND TECHNOLOGY

The following examples demonstrate using the continuous improvement process to manage stope void exposure.

Stope design – engineering controls

Capes, Lachenicht and Mitkas (2023) showed that through a change in stope design, the performance of stopes can be improved in poor ground conditions. This was achieved through numerical and empirical back analysis and an improved understanding of rock mass conditions. Stope shapes were changed to a chamfered back design and a few stopes were mined demonstrating the improved performance and restriction of un-controlled caving of stopes. Figure 8 shows the performance of the subsequent stope (068_LIM_468_489_STP) mined following the LIM 290 caving failure. Both stopes were mined in similar geological conditions but different design methodologies had been applied.

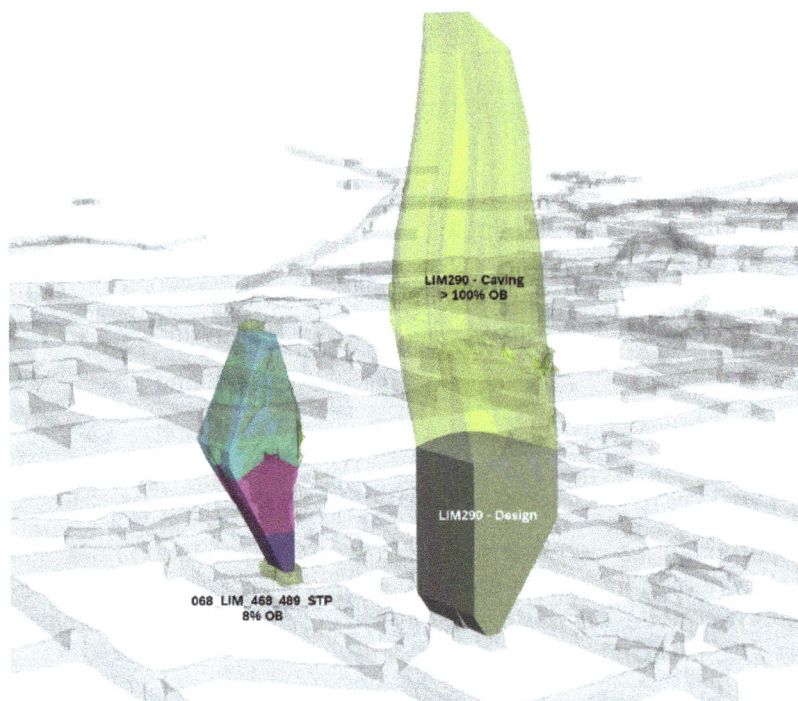

FIG 8 – Comparison in performance of two similar LIM's stopes through modification of design.

Stope design – elimination of stope edge truck tips

Backfilling of stopes can take the form of cemented aggregate fill (CAF) and surface rock fill (SRF) poured from surface or underground rock fill (URF). For URF stopes, a truck tip (on the crown level) is designed to allow trucks to access near the open stope to fill the stope (Figure 9). The stope is designed with a slide and has protection of a stop block and 'matty' chimes including monitoring in the form of probe-holes and 'buckets' (discussed later in the paper).

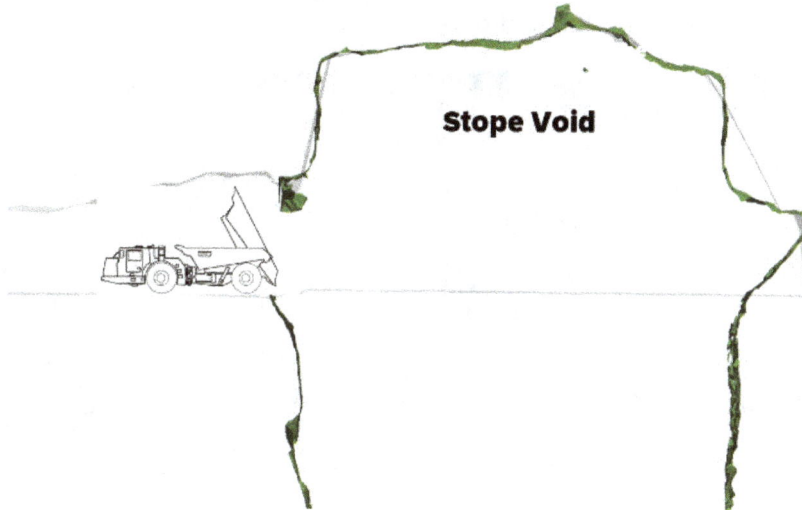

FIG 9 – Truck tip showing proximity of truck to open stope.

To reduce the exposure of personnel and equipment from the stope edge during the filling activity, truck tips are now designed on the level above and offset from the footprint with a longhole raise (LHR) drilled into the top of the stope as shown in Figure 10. Similarly, stop block and probe hole checks exist to manage any changed conditions.

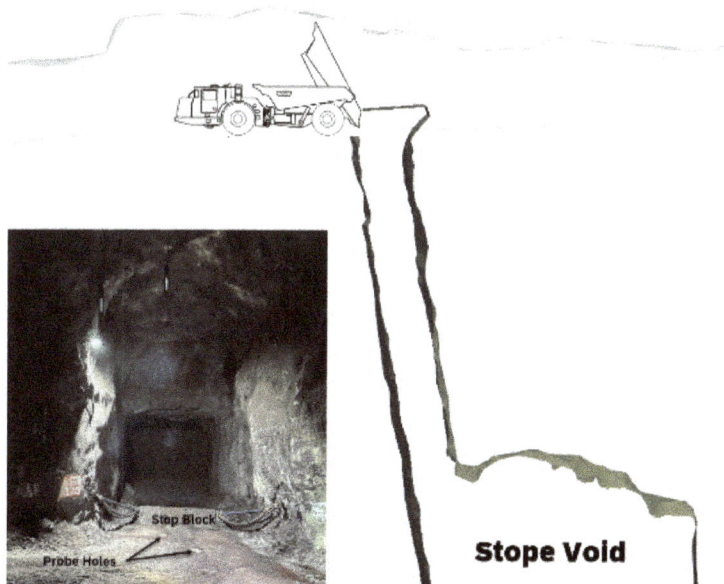

FIG 10 – Section view showing tip point above stope with LHR into stope.

Stope scanning – substitution/elimination

Monitoring is important to understand the performance of the stopes during its production cycle. Monitoring typically takes the form of scanning the void which provides vital information on overbreak, access undercut and stock levels. Scanning in the past was undertaken by surveyors using a cavity monitoring system (CMS). This system required the surveyors to go close to a stope edge, typically in a basket on an integrated tool-carrier (IT), with a pole extended into the void and the CMS attached to the pole. This method exposed personnel and equipment to the open stope through potential undercutting of the access drive.

Drone and LiDAR technology have greatly improved and potentially eliminated the risk of personnel undertaking scanning of large open voids. Drones with LiDAR attachments can be flown with line-of sight or around corners at a distance away from the stope edge. This technology improvement has also enabled improved scans (shadowing resulting from standard CMS scanning) and the ability to execute the task at a greater frequency. Figure 11 shows the process of undertaking a stope scan at the edge of the stope with a CMS and a drone scan that eliminates personnel from the stope edge.

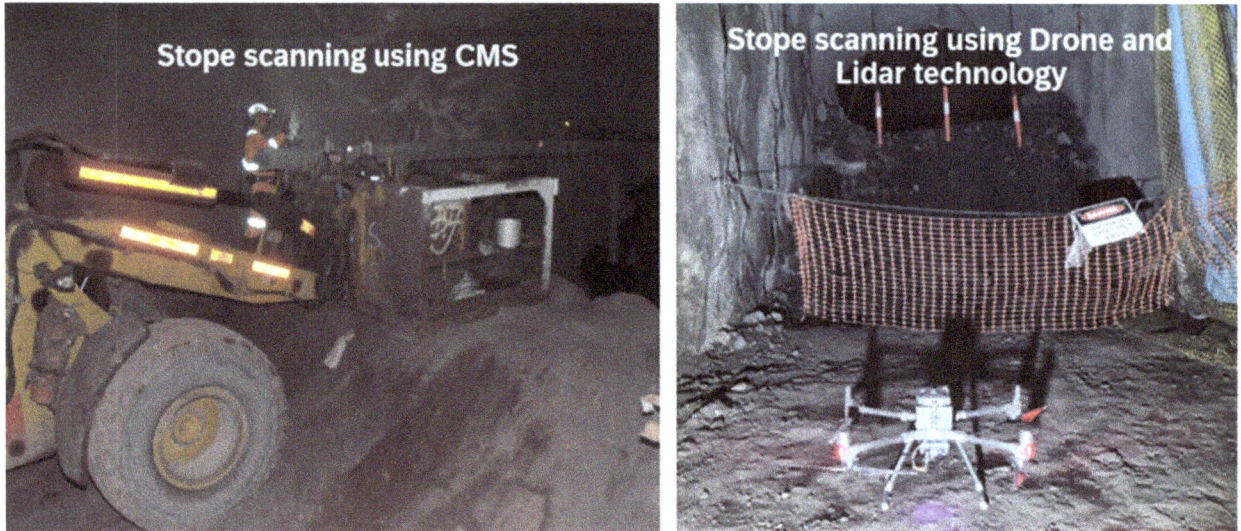

FIG 11 – Stope scanning highlighting the elimination of personnel and equipment from a stope edge using Drone and LiDAR technology.

MANAGEMENT OF STOPE VOIDS – TACTICAL APPROACHES

Table 3 illustrates the tactical approaches applied to manage the exposure during risk scenarios identified previously (Table 1) for the various activities that occur during the stope production cycle.

TABLE 3

Tactical approaches applied manage void exposure.

	Activity	Risk scenario	Control type	Tactical approaches
1	Stope prep and charge, Re-drills	Interim crown failure of the previous blast packet exposing prep and charge crew to the void below.	Monitoring	• 'Bucket' monitoring • Production hole dipping
2	Clean-up, Push-off, stope bombing and operating drones	Failure of the stope wall encompassing (undercutting) the access development exposing personnel and equipment to the stope void.	Separation Monitoring	• 'Push-off' probe dipping • Hazard zones • LiDAR drone scanning • Geotechnical assessments • Void protection
3	General access (adjacent)	Failure of the stope wall encompassing (undercutting) the adjacent development exposing personnel and equipment to the stope void.	Separation Monitoring	• 'Bucket' monitoring • Hazard zones • LiDAR drone scanning •
4	Backfill with underground rock fill (URF) using trucks	Failure of the stope wall undercutting the tip point exposing personnel and equipment to the stope void.	Design Separation Monitoring	• LiDAR drone scanning • Void protection and tipples
5	Access (level above stope crown)	Stope crown failure exposing personnel and equipment accessing above crown footprint.	Separation Monitoring	• Hazard zones • LiDAR drone scanning
6	Backfill barricade building	Stope failure exposing near stope edge operations.	Separation Monitoring	• Hazard zone • LiDAR drone scanning • Geotechnical Assessment

Stope hazard zones

A hazard zone (Figure 12) represents a zone (typically 15 m) around a designated stope shape. The hazard zone can be increased based on expected rock mass conditions and expected stope performance. The hazard zone has three levels of access control namely:

- Red Hazard Zone – No access is allowed within this area, and it is demarcated with 'No-Traveller' signage.

- Yellow Hazard Zone – Access allowed with authority, 'Authorised Entry Only,' with information pertaining to access requirements on the signage.

- Green Hazard Zone – Access allowed and no signage underground. However, the hazard zone is represented on mine plans which prevents any mining (development) to occur within the zone. Extraction levels have a green hazard zone as there is no void below the level.

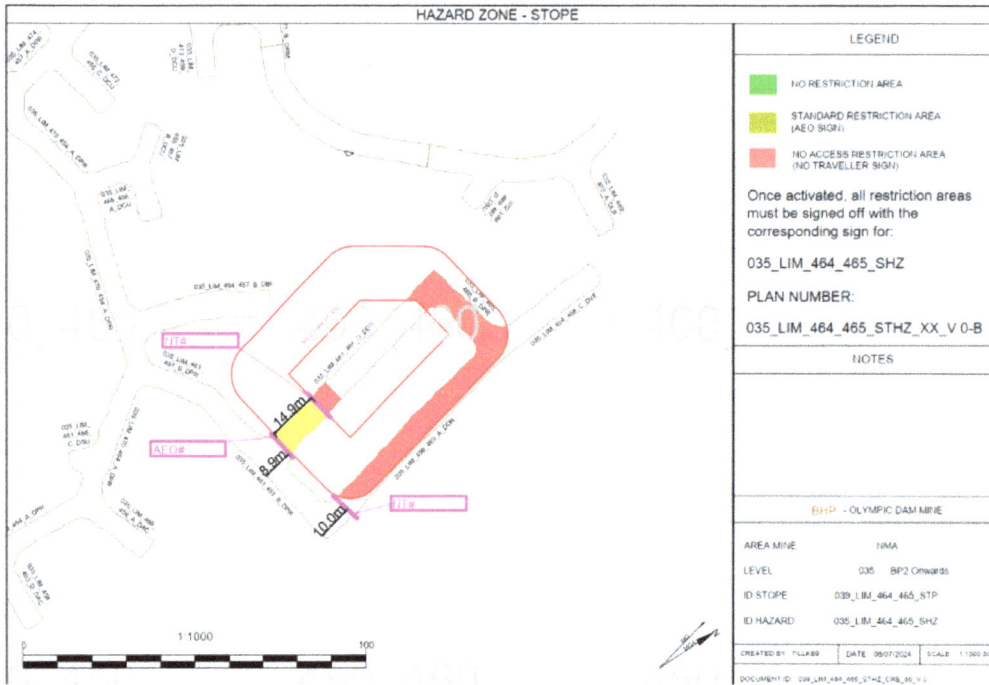

FIG 12 – Level plan hazard zone showing access level restrictions.

Void protection and 'push-off' probes

Void protection (Figure 13) is installed at all accesses where there is a risk for personnel to fall into the stope void. It consists of a 1.5 m high bund wall as a physical barrier separation control, as well as three 'matty chimes' hanging from the backs as a visual indicator. The bund is installed 3–4 m from the design stope edge. All accesses to the stope void have two probe holes drilled towards the void at 12 m and 9 m from the stope edge. These 'push-off' probes (Figure 13) are dipped on entry to verify the access drive is not undercut.

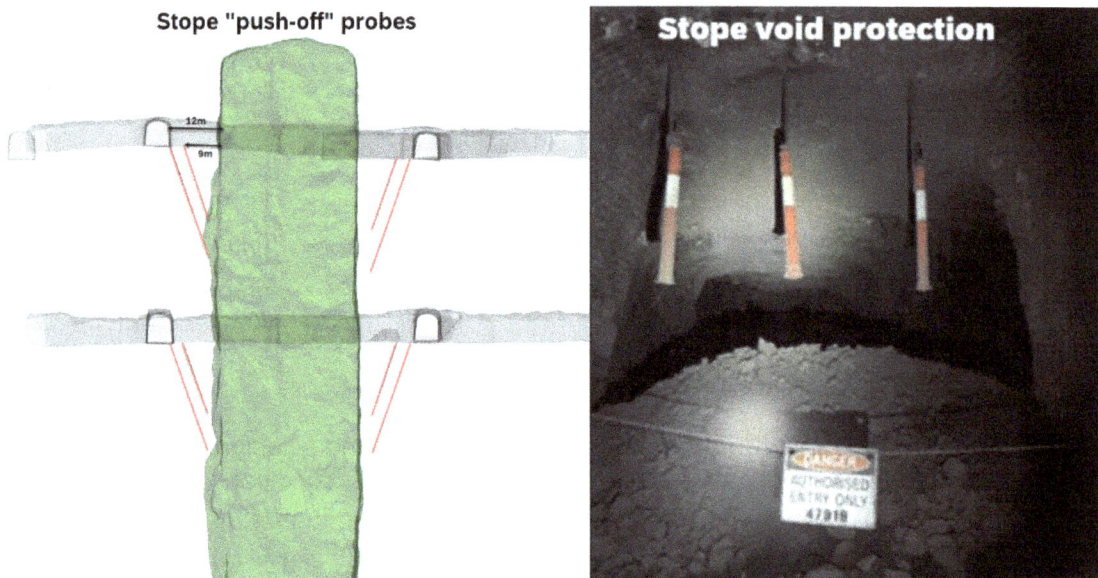

FIG 13 – Void protection and 'push-off' probes installed at all accesses to a stope void.

Monitoring

Monitoring is an important control used to verify that the performance of the stope throughout its production cycle is in line with expected behaviour (as per the design). It also alerts of deviations from the design and allows for corrective measures to be applied to ensure safe production. Deviation triggers identified through monitoring (instrumentation and observations) enables responses through the stope performance TARP. Typical stope void monitoring is described below.

'Bucket' monitoring

Bucket monitoring is a method of monitoring which provide 'real-time' information on stope fall-off. It works on the basis that the bucket will drop from its original position (marked up on the wall) when fall-off in the stope occurs. This triggers a separation control to be installed requiring a scan and a geotechnical assessment. The monitoring method is used primarily above interim stope crowns where access is required to prep and charge the next blast packet. It is also used along access drive when no line of sight to the void is possible. Figure 14 shows a schematic of the bucket monitoring method including underground applications.

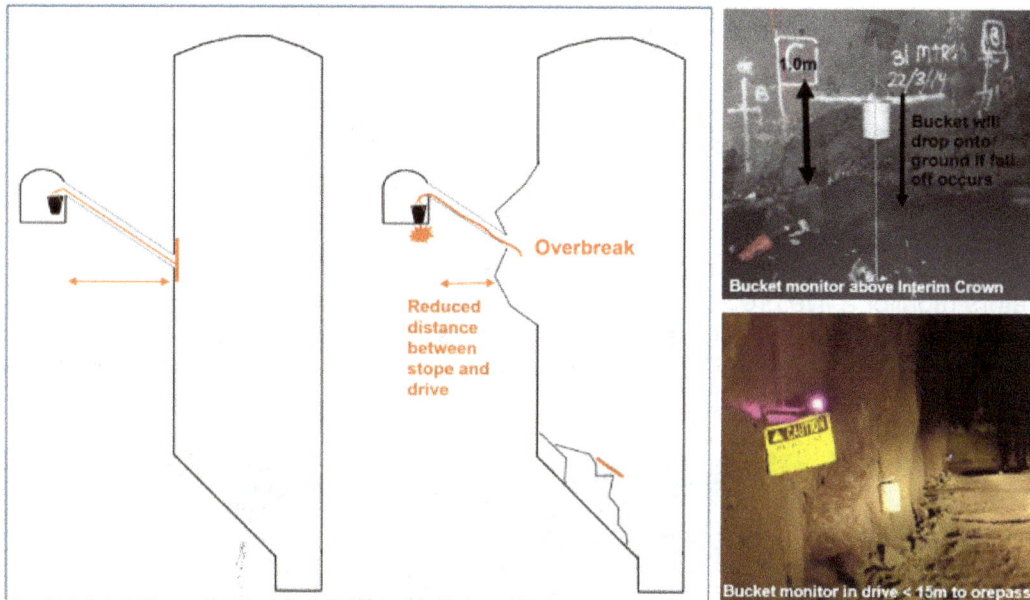

FIG 14 – 'Bucket' monitoring method and underground applications.

Production dip-hole monitoring

Prior to installing bucket monitors above the interim crown, production holes are dipped (Figure 15) on approach and over the interim crown. This is to ensure that there is a sufficient rock pillar beneath them, as the conditions may have changed due to the subsequent blast packet firing. If production holes are less than 15 m or 20 per cent less than what has been designed, access is restricted and further assessment is undertaken.

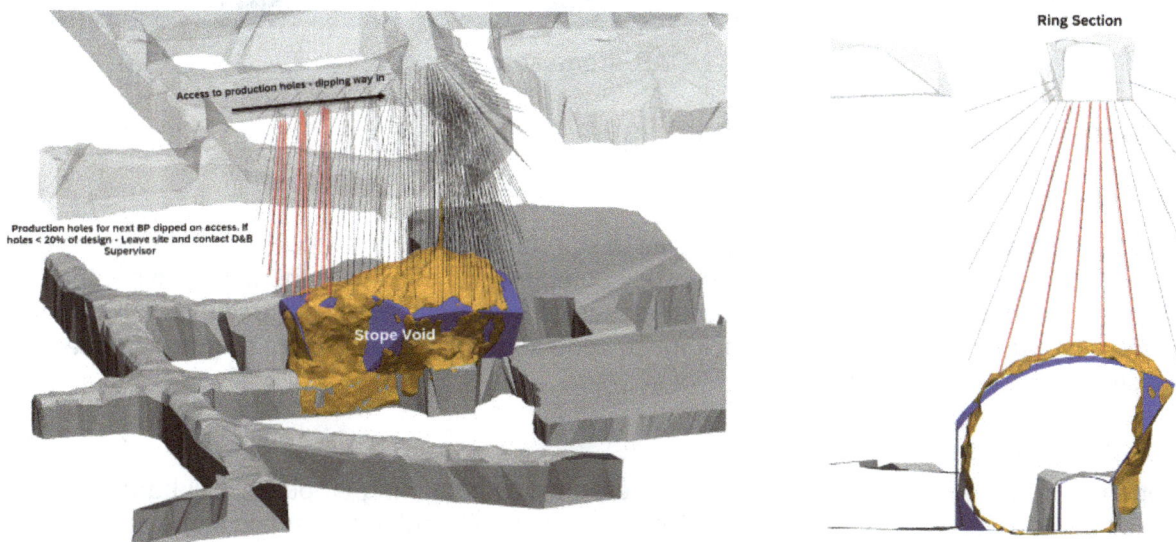

FIG 15 – Production dip-hole monitoring.

LiDAR drone scanning

LiDAR drone scanning is the primary method of monitoring the stope performance. During the production cycle, there are multiple triggers for scanning ranging from operational requirements to instability concerns. The triggers are:

- Stock level checks assisting with scheduling requirements.
- Variation between bogged and planned tonnes exceeds 20 per cent.
- Stope instability – CAF, large rocks or cablebolts reporting to the drawpoints.
- Production holes above interim crown dipped ±20 per cent of design.
- Dropped bucket monitors.
- Geotechnical requirements due to overbreak identified from previous scans or concerns with ground conditions.
- Seismicity.

The LiDAR scanner is mounted on a drone and flown into the stope by certified drone operators. The operators monitor a live preview on a tablet to ensure the void is captured as best as possible. Once the scan is processed, the stope performance is reviewed against the TARP to determine any responses required. Depending on the assessment of the stope void, frequency may be increased (eg weekly). Figure 16 shows the LiDAR drone scan process.

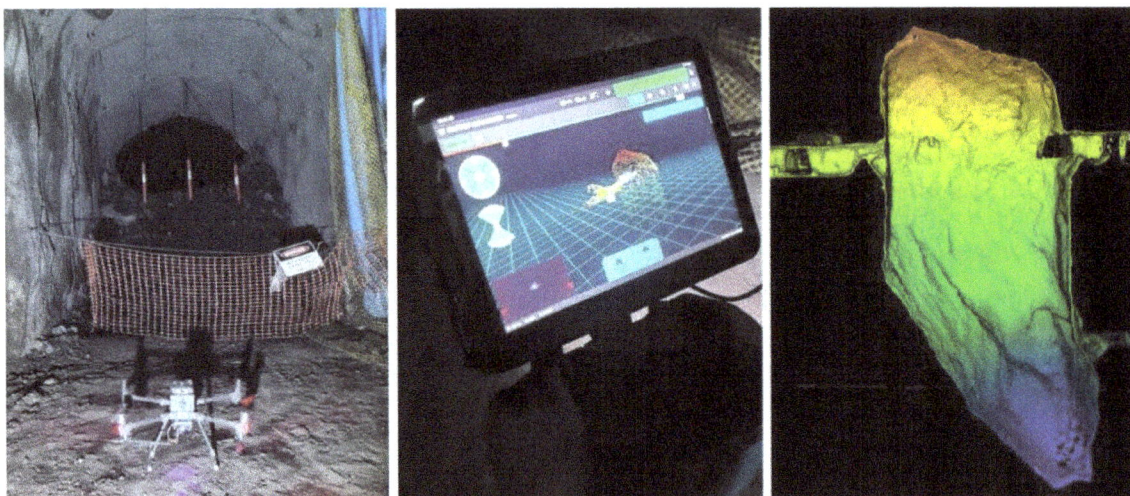

FIG 16 – LiDAR drone scan process.

Geotechnical assessments – checklists

Near stope activity checklist

The near stope edge activity checklist is a way to standardise how the stope void is being assessed and ensure the scheduled activity can be safely completed. It considers the activity type, the stope performance to date, performance of the wall closest to where the activity is, and the expected stope void stability based on geometry and rock mass conditions. Stope scans are required to enable the assessment to be undertaken.

Stope barricades build checklist

A stope performance review checklist is completed for all post-firing stope backfill barricade builds within a hazard zone. To assess the risk of CAF or rock engulfment, air blast and stope void instability at the barricade locations, the checklist considers the amount of lateral underbreak and overbreak, overall stope performance and the location of the barricade to be built. Stope scans are required to enable the assessment to be undertaken.

Hazard zone reduction or removal checklist

Hazard zones are removed from all levels as the stope is being backfilled. The exception is the crown or the level immediately above the crown. Due to potential void in the crown resulting from poor tight-filling, probes are required to verify whether the stope is tight-filled prior to any access or development is done. Probe results are required to enable the assessment to be undertaken.

CASE STUDY – 068_SCA_497_546_STP

This case study describes the identification of changed conditions during the production cycle of the 068_SCA_497_546_STP stope, the process to manage personnel exposure to the large stope void and the successful execution of the stope to the design.

The stope – 068_SCA_497_546

068_SCA_497_546_STP was a standard stope consisting of four blast packets with a tonnage of 415 Kt. Prior to production, three stopes were previously mined near the studied stope and each performed close to design (Figure 17).

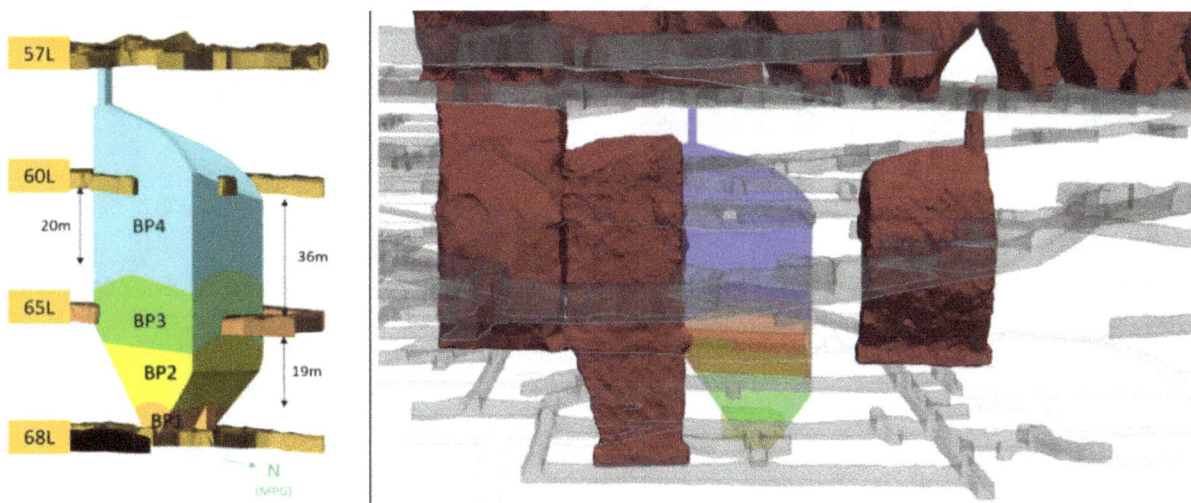

FIG 17 – 068_SCA_497_546_STP blast packet design and location relative to other mined scarlet stopes.

Identification of changed conditions

On approach to prep and charge BP4, following push-off probe hole dipping, the first ring was dipped and identified to be shorter than 20 per cent of the design length. This triggered a separation control and a geotechnical assessment requirement. A drone scan was undertaken showing a 11 m pillar at the centre of the stope and the overbreak was arched shape. At the time of the scan, 200 Kt was still to be fired. Figure 18 shows the interim crown failure, pillar thickness and production prep hole results.

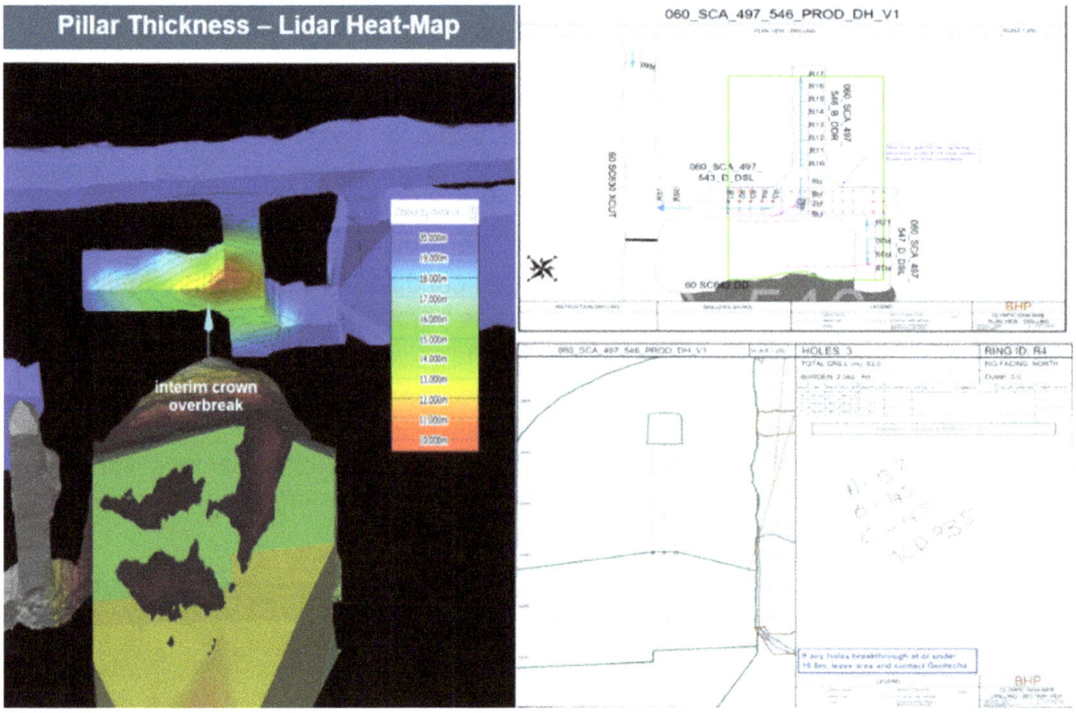

FIG 18 – LiDAR heat map and production hole dipping results.

Stope recovery

Following a geotechnical assessment and stakeholder engagement, it was deemed unacceptable to access over the stope footprint to prep and charge BP4. A decision was made to split the final blast packet into two firings. Figure 19 shows the recovery plan for the stope.

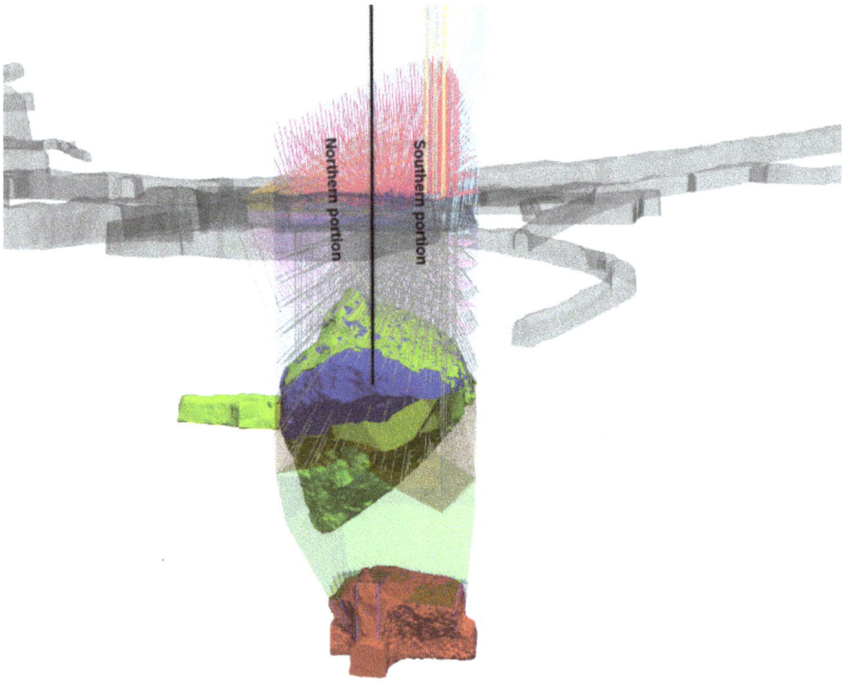

FIG 19 – 068_SCA_497_546_STP recovery plan.

Southern portion (BP4)

The southern portion of the interim crown had to be accessed with new development through the adjacent CAF filled stope which was 20 m from the void. The controls used to manage this development were:

- Regular drone scans to monitor stope void.

- Stopped bogging stope to maintain confinement to the CAF adjacent stope.
- Installed in-cycle buckets with development to actively monitor the void.

The southern portion was charged out of a Franna box using a Franna crane utilising the heat map to guide which rings could be accessed over.

Northern portion (BP5)

The northern portion was drilled from outside of the stope footprint and fired separately. Monitoring during this phase of recovery included 'bucket' monitoring during recovery drilling and an increase in the scan frequency.

Outcome

The remaining 200 Kt was extracted safely and near design. The void management controls used throughout the stope production were:

- Dipping production holes (20 per cent).
- LiDAR drone scans.
- Stope performance TARP.
- Bucket monitoring.
- HZ modification risk assessments.
- Changed design (split BP4 into two firings).

CONCLUSION AND FORWARD WORK

There are many potential failure scenarios existing around large open stope voids. Critical controls are designed to prevent such failures or alert of change conditions throughout the stope production cycle. The collectiveness of the Stope Failure Material Risk controls defines the stope void management plan. They are based on characterisation, design, monitoring, and separation. Through continuous improvement and advancement in technology, the risk is managed with higher order controls. This is evident in the design change to manage stope performance, separation from stope edges, drone scanning and tactical approaches. A safer, more productive environment is created through the management of stope voids.

The future work looks at building on the continuous improvement cycle moving ultimately to elimination through automation and technology.

REFERENCE

Capes, G, Lachenicht, R and Mitkas, N, 2023. Chamfered Backs to Improve Open Stope Stability in Poor Ground Conditions, in *Proceedings of the Underground Operators Conference 2023*, pp 399–406 (The Australasian Institute of Mining and Metallurgy: Melbourne).

Hydropower energy recovery in mines using pump as turbine technology – three case studies

D Novara[1], M Crespo Chacon[2], M Pedley[3], A McNabola[4], P Coughlan[5] and J Warren[6]

1. CEO, Easy Hydro Ltd, Dublin 2, Ireland. Email: novarad@easyhydrosolutions.com
2. CTO, Easy Hydro Ltd, Dublin 2, Ireland.
3. Commercial Director, Easy Hydro Ltd, Dublin 2, Ireland.
4. Dep. Dean International, School of Engineering, RMIT University, Melbourne Vic 3000.
5. Professor in Operations Management, Trinity College Dublin, Dublin 2, Ireland.
6. Senior Project Engineer, Cornish Metals, TR15 3QT, UK.

ABSTRACT

Mining is one of the industries with the highest energy and water consumption. In Australia, the sector can account for around 10 per cent of the nation's total energy use. Reducing the level and impact of this energy use is imperative to improving sustainability metrics and reducing costs. Most underground and open cut mines feature complex hydraulic networks with water used for processes such as drilling, air conditioning, dust abatement, leaching, flotation, tailings transportation. In many sections of a mine's hydraulic network there are opportunities to recover excess hydraulic energy and convert it to electricity. The renewable power generated can either be fed directly into the mine's internal power grid for self-consumption or can be used locally for applications such as the charging of Battery Electric Vehicles (BEVs). This article presents three recent case studies undertaken by the authors, describing and comparing the design, installation and operation of Pumps As Turbines (PAT) as hydropower energy recovery devices in different areas of mining infrastructure. The cases illustrate the technology and the collaboration with the mine operators. As a set, these installations can generate around 1 GWh annually, preventing 210 t of CO_2 emissions per annum and generating economic savings through reduced energy purchase. Finally, the potential for industry-wide adoption of this technology is outlined.

INTRODUCTION

The mining industry is a significant consumer of water, which is used within several processes including mineral processing, drilling, dust suppression, equipment cleaning, mineral slurry transport and cooling. At the same time, the treatment and pumping of such large volumes of water require large amounts of electricity (Bangerter and Dixon, 2010; Brown, 2003). In Australia, it has been estimated that 10 per cent of the country's total energy use can be ascribed to the mining sector (SunSHIFT Pty Ltd, 2017). Yet, there is an opportunity to utilise the water towards recovering energy in mining through hydropower. Fundamental to realising this opportunity is an understanding of the concepts underlying hydraulic energy recovery and their application in mining.

Hydraulic energy recovery – general concepts

Hydraulic energy recovery involves utilising the excess pressure or the water level elevation difference within a hydraulic system, which would otherwise be dissipated in non-useful forms, to generate electricity. The hydraulic power that can be recovered is calculated using Equation 1:

$$P = Q * H * \gamma * \eta \tag{1}$$

Where:

Q   is the available flow rate in m^3/s

H   is the net hydraulic head (or pressure differential that can be used by the turbine) in metres

γ is the specific weight of the fluid in kN/m^3 (9.81 for water)

η is the turbine efficiency

P   is the power in kW

Typically, a hydraulic energy recovery installation will have the following elements as depicted in Figure 1: a turbine inlet valve, the turbine itself and an automated bypass valve. The valves typically feature either an Electric or an Hydraulic actuator (marked E/H in Figure 1), depending on the operational requirements of the system.

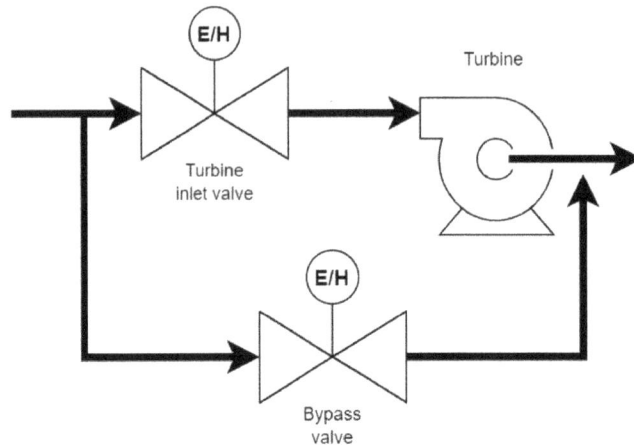

FIG 1 – Typical layout of a hydraulic energy recovery facility.

Hydraulic energy recovery – the potential in mines

Most mines, both open cut and underground, have extensive infrastructure in place for water supply and wastewater treatment and discharge. Often, there are locations within these pipes and channels where the water experiences a drop in elevation or in pressure which can also be used to generate hydroelectric power. Even in the presence of a modest elevation/pressure drop, having a substantial and consistent flow of water can result in a significant power generation potential.

As an example, cooling is one of the primary operational costs in deep underground mines. It is typically achieved through piping systems that transport cold water to the necessary depth, thus increasing the water's pressure which is then dissipated through Pressure Reducing Valves (PRVs) or Break-Pressure Tanks (BPTs). Numerous studies have explored methods to enhance the energy efficiency of mining operations by optimising the internal water distribution systems of mines (Van Antwerpen and Greyvenstein, 2005; Van Niekerk, Uys and Van Rensburg, 2014). Similarly to what has been widely proposed for drinking water networks, also in underground mines it is possible to reduce the excessive pressure build-up via hydraulic turbines rather than dissipating it as heat and noise (Gaius-Obaseki, 2010; McNabola et al, 2014; Tabrizi and Markoski, 2021).

Given that the recoverable hydroelectric potential in mines generally falls within the micro or small hydro range (less than 1 MW), utilising low-cost, standardised turbine technologies is crucial for project feasibility. The use of hydraulic turbines derived from conventional water pumps (Pumps-As-Turbines, PATs) offers a lower capital cost and shorter delivery times with respect to conventional hydro turbines due to mass production. PATs have been employed to recover energy from various water-intensive sectors, including potable water treatment networks, water treatment plant inlets, industrial cooling networks and irrigation networks. Several studies have analysed the design and performance of these installations, confirming the technical and economic feasibility of PAT technology for reducing operational costs and environmental impact by improving the overall energy efficiency of the process (Chacón et al, 2021; Novara and McNabola, 2021).

Objectives of the paper

The present paper explores three recent case-studies of hydraulic energy recovery undertaken by the authors in collaboration with the mine operators using PATs as a strategy to improve the mine's energy efficiency, reduce operating costs and lower the carbon footprint. The three sites demonstrate applications in different areas of mining infrastructure: the cold process water distribution system of an underground mine; the outfall of a mine water treatment plant; and the outlet of a clarifier in a mineral processing plant. These installations, having nominal capacities of 30 kW, 88 kW and 12 kW respectively, have resulted in a total combined potential energy saving of 1 GWh/a, preventing 230 t/a of CO_2 emissions and generating economic savings through reduced energy purchase.

CASE STUDIES

The three case studies are described in the following sub-chapters.

Water distribution system of an underground mine

The first case study involves a hydraulic energy recovery pilot plant installation on the cold process water distribution system of an underground lead and zinc mine in Ireland (Tara Mines, operated by New Boliden). Similarly to what happens in many other deep underground mines, the process water/drillwater flows by gravity from the surface down to the lowest level of the facility. Therefore, at regular intervals there are pressure reducing valves installed on the water pipeline in order to maintain the pressure within operating limits. The water flow required by mining operations is relatively stable, and the fluid is primarily used for dust suppression and machinery cooling.

The location of the pilot plant is at a depth of about 300 m, where there was a pre-existing pressure reduction station with three pressure reducing valves installed in series as shown in Figure 2. Thanks to these valves, the incoming water pressure was lowered from 27 bar to 5 bar.

FIG 2 – The three pre-existing pressure reducing valves installed in series.

In parallel with these valves, during 2022 a 30 kW PAT was designed and installed to exploit the 22 bar excess pressure with a flow rate up to 22 L/s. The power generated by the system is fed directly into an adjacent low voltage distribution panel for use by nearby machinery. A schematic view of the system and a photo after construction are displayed in Figure 3.

FIG 3 – A 3D drawing of the installation (left); a photo of the same during the commissioning (right).

The estimated energy production based on historical flow records is 200 MWh/a. However, this figure has yet to be recorded in practice due to the lower than expected production rate of the mine in 2022 and 2023, following which the mine was placed in care and maintenance during 2023. In 2024, a decision was taken to restart full-rate production at the mine, and the turbine is expected to be re-commissioned in early 2025. From an environmental point of view, 200 MWh/a of clean power generation corresponds to a reduction of 66 tCO_2 emissions annually based on the SEAI (Sustainable Energy Authority of Ireland) emissions factor for 2022 of 330.4 gCO_2/kWh (SEAI, 2024).

Whilst the 30 kW power generation from this turbine represents a modest energy saving when compared to the overall energy demand of the whole mine, replication at ten other existing pressure reduction stations could increase the installed capacity to around 300 kW, potentially achieving up to 5 per cent energy recovery (and, so, savings) for the whole mine.

Outfall of a mine water treatment plant

The second case study is at the outlet of the Mine Water Treatment Plant (MWTP) of an underground UK tin mine (South Crofty, operated by Cornish Metals). The MWTP processes the water pumped out from the underground works and discharges it into a 55 m deep shaft. At the bottom of the shaft it is released into a sloped adit which eventually leads to the river outfall. A twin-turbine hydro system was designed and installed at the end of 2023 featuring two PATs of different sizes. They are installed at the bottom of the shaft immediately before the discharge point into the adit as seen in Figure 4.

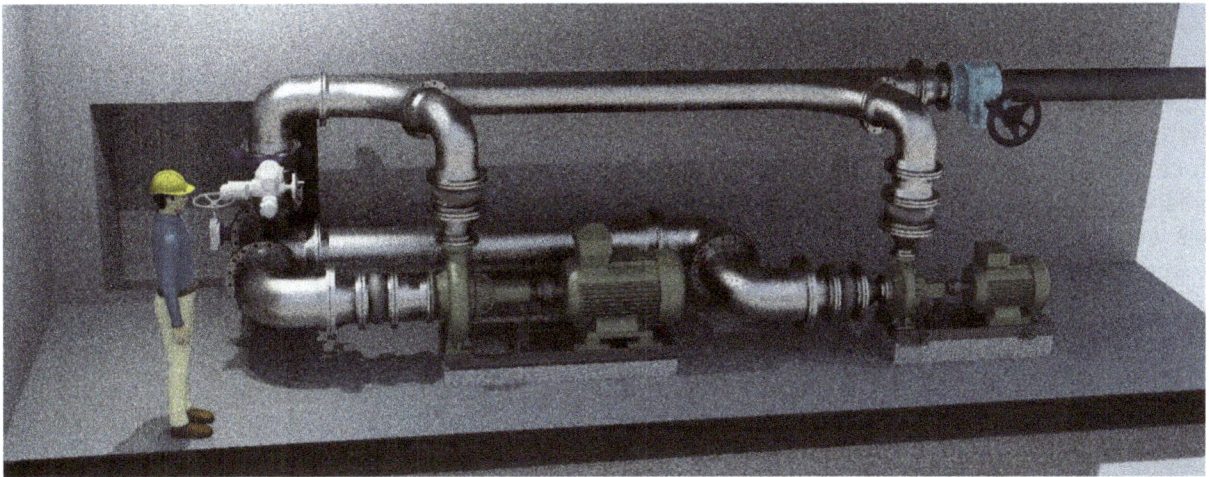

FIG 4 – The twin-PAT system installed at the bottom of the shaft, prior to the discharge of the treated water to the adit.

The system is designed for both short-term and long-term mine operations. In the short-term (14–18 months), the MWTP will process a flow up to around 290 L/s with a design power generation up to 88 kW to allow for a rapid dewatering of the mine. Once the mine reaches full operation, the flow will then reduce to about 75 L/s thus producing an output of around 18.5 kW long-term. In both phases, the electricity generated is self-consumed at the MWTP.

In terms of carbon dioxide savings, the short-term expected power generation (using both PATs at maximum power) is 730 MWh/a while the long-term production (using 1 PAT) is anticipated to be 150 MWh/a. Hence, the annual emission reductions are estimated at 141 tCO_2, based on the UK's emissions factor of 193 gCO_2/kWh (UK Department for Energy Security and Net Zero, 2023).

Outfall of a clarifier in a mineral processing plant

The third case study consists of a PAT installed at the outfall of a clarifier in an Italian mineral processing plant adjacent to an open cut feldspar mine (Fossanova, operated by Sibelco). Once the raw materials are extracted and crushed, they are sent via conveyor belt to the nearby mineral processing plant where large amounts of water are used in a closed loop to wash and sort dolomite, feldspar, and silica. The process terminates with an 11 m high clarifier, after which the clean water falls vertically inside a wide metal column into an underground collection channel prior to its recirculation.

The hydraulic energy recovery facility was designed and installed in 2023 and consists of a PAT placed at the foot of the clarifier in order to exploit the 8–10 m net fall of which two photos are shown in Figure 5. The turbine handles a flow rate of 140–190 L/s and produces an output of up to 11 kW of electricity. The energy generated is used for self-consumption in low voltage by the main 55 kW circulation pump which sends the cleaned water from the collection pond back to the start of the washing cycle.

FIG 5 – An aerial view of the clarifier (left), at whose foot the energy recovery PAT was installed (right).

The facility is expected to generate around 70 MWh of clean electricity annually, which corresponds to a carbon footprint reduction of 26 tCO_2/a using Italy's carbon emissions factor of 372 gCO_2/kWh (Statista, 2024).

CONCLUSIONS

These systems respond to the challenges of water and energy use in mining. They have been designed and installed in different sections of the mining infrastructure in three different countries. Common across the systems is the use of PATs for energy recovery. The results from these three case studies demonstrate the technical and economic viability of using PATs for hydraulic energy recovery in the mining sector. A summary of the key results is provided in Table 1.

TABLE 1

Summary of the results obtained in the three projects.

Case	Flow rate (L/s)	Head (m)	Power (kW)	Energy (MWh/a)	Emissions reduction (tCO$_2$/a)*
Water distribution system of an underground mine	18–22	220	15–30	200	66
Outfall of a mine water treatment plant	75–290	40–50	18.5–88	730	141
Outfall of a clarifier in a mineral processing plant	140–190	8–10	7–11	70	26

* – based on the gCO$_2$/kWh carbon intensity figure of each country.

This paper has presented three case studies where hydraulic energy recovery systems using PATs have been designed and successfully implemented in mining-related infrastructures, delivering both economic and environmental benefits for operators. The combined electrical capacity of these facilities is approximately 130 kW, leading to an annual energy saving of up to 1 GWh and avoiding around 230 tCO$_2$ in emissions per annum. These projects also demonstrate the flexibility of the PAT technology in dealing with low flow/high head (the first site) as well as with high flow/low head (the last site).

These three situations are not unique in mining. The presence of excess pressure and of small hydraulic falls in the water systems of mines presents an opportunity for hydraulic energy recovery, offering an opportunity to enhance energy efficiency while reducing operating costs and environmental impacts.

There is clear potential for industry-wide adoption of this technology. The systems, though installed in different operating contexts, are based on common design principles, include common components, and face manageable challenges in installation and commissioning. Further, given the scope of individual mines, there may be opportunities for multiple replications at a single site, so leveraging the experience gained in operation and scaling the benefits to significant levels.

ACKNOWLEDGEMENTS

The authors wish to thank the mining companies involved: New Boliden, Cornish Metals and Sibelco. This activity has received funding from the European Institute of Innovation and Technology (EIT), a body of the European Union under Horizon 2020, the EU Framework Programme for Research and Innovation.

REFERENCES

Bangerter, P and Dixon, R, 2010. Improving overall usage of water in mining – A sustainable development approach, in Proceedings of the 2nd International Congress on Water Management.

Brown, E, 2003. Water for a sustainable minerals industry – a review, in Proceedings of Water in Mining.

Chacón, M C, Rodríguez Díaz, J A, Morillo, J and McNabola, A, 2021. Evaluation of the design and performance of a micro hydropower plant in a pressurized irrigation network: Real world application at farm-level in Southern Spain, Renewable Energy, (16).

Gaius-Obaseki, T, 2010. Hydropower opportunities in the water industry, International Journal of Environmental Sciences, 1(3).

McNabola, A, Coughlan, P, Corcoran, L, Power, C, Williams, A, Harris, I and Styles, D, 2014. Energy recovery in the water industry using micro-hydropower: An opportunity to improve sustainability, Water Policy, 1(16):168–183.

Novara, D and McNabola, A, 2021. Design and Year-Long Performance Evaluation of a Pump as Turbine (PAT) Pico-Hydropower Energy Recovery Device in a Water Network, Water, 21(13).

Statista, 2024. Carbon intensity of the power sector in Italy from 2000 to 2023, Emissions, Energy & Environment, Statista. Available from: <https://www.statista.com/statistics/1290244/carbon-intensity-power-sector-italy/>

SunSHIFT Pty Ltd, 2017. Renewable Energy in the Australian Mining Sector, white paper, SunSHIFT Pty Ltd and the Australian Renewable Energy Agency (ARENA). Available from: <https://arena.gov.au/assets/2017/11/renewable-energy-in-the-australian-mining-sector.pdf>

Sustainable Energy Authority of Ireland (SEAI), 2024. Conversion factors, Ireland's energy statistics, Energy data and insights, SEAI. Available from: <https://www.seai.ie/data-and-insights/seai-statistics/conversion-factors/>

Tabrizi, A B and Markoski, R, 2021. Small Hydropower, Another Source of Renewables for Australian Underground Mining?, *Insights Min Sci Technol,* 4(2).

UK Department for Energy Security and Net Zero, 2023. Green Book supplementary guidance: valuation of energy use and greenhouse gas emissions for appraisal, Energy and climate change: evidence and analysis, UK Department for Energy Security and Net Zero. Available from: <https://www.gov.uk/government/publications/valuation-of-energy-use-and-greenhouse-gas-emissions-for-appraisal>

Van Antwerpen, H and Greyvenstein, G, 2005. Use of turbines for simultaneous pressure regulation and recovery in secondary cooling water systems in deep mines, *Energy Conversion and Management,* 46(4):563–575.

Van Niekerk, A, Uys, D and Van Rensburg, J, 2014. Implementing DSM interventions on water reticulation systems of marginal deep level mines, in Proceedings of the Conference on the Industrial and Commercial Use of Energy, ICUE.

Managing operator fatigue underground – the importance of management of change

J Savit[1], A Heieis[2] and M Quintela[3]

1. Principal Advisor, Hexagon, Tucson Arizona 85701, USA. Email: joshua.savit@hexagon.com
2. OAS Product Manager, Hexagon, Vancouver British Columbia V6E 3Z3, Canada.
   Email: adrian.heieis@hexagon.com
3. Portfolio Manager – Underground, Hexagon, Vancouver British Columbia V6E 3Z3, Canada.
   Email: mateus.quintela@hexagon.com

ABSTRACT

Traditionally accepted as standard in surface mining, Hexagon and MMG partnered to deploy Australia's first application of Hexagon's Operator Alertness System (OAS) to 11 underground mining trucks at MMG's Rosebery mine in Tasmania, Australia.

MMG wanted a solution that would help detect fatigue in their operators, with the challenge of this being at a complex underground environment. Hexagon's OAS works in both light and dark conditions to provide real-time notifications to operators, empowering them to act should they experience fatigue or distraction.

Operators could rely on a single solution in the cab for fatigue and distraction monitoring, from underground to when the trucks come back up to the surface.

MMG was able to easily integrate OAS into its systems to provide its operators with an additional safety check during their 12-hour shifts.

Hexagon and MMG worked together through the course of the deployment to ensure the project was introduced seamlessly, from procurement to user acceptance testing and implementation. Over a year later, the project is successful at reducing risks and further strengthen operator safety at Rosebery. The success of the project was a combination of an adaptive technology, and an appropriate level of guided change management that helped re-set the safety culture on-site.

This presentation is a case study that will illustrate:

- The demonstrated level of fatigue and distraction faced in the underground environment and the subsequent improvements.

- How data from OAS allowed for rotations and break times to be unified for a stronger safety culture.

- The Management of Change (MoC) principals applied throughout the OAS deployment that allowed for the operators to be heard and continue to build a stronger and safer work environment.

INTRODUCTION

Underground mines are characterised by narrow tunnels and restricted spaces, making material transportation a challenging task. In such confined environments, specialised equipment is essential for navigating tight spaces, often requiring manual handling and precise operation. The physical demands of this type of work contribute to the accumulation of fatigue among workers (Smith, 2020).

The underground mining environment is harsh, often with high temperatures, humidity, and inadequate ventilation being common issues. These conditions exacerbate the physical strain on miners, leading to quicker onset of fatigue. Studies have shown that exposure to extreme temperatures can significantly impair cognitive and physical performance, thereby increasing the risk of accidents (Jones, Smith and Walker, 2019).

Operating heavy machinery in an underground setting further demands high levels of physical exertion, especially when dealing with equipment that generates significant vibration and noise.

Prolonged exposure to these factors not only contributes to physical fatigue but also adversely impacts hearing and overall health (Brown and Davis, 2018).

Moreover, the design of underground mining equipment and work environments often fail to prioritise ergonomics. As a result, miners frequently adopt awkward postures and perform repetitive motions, leading to musculoskeletal disorders. The lack of proper ergonomic support only intensifies the physical fatigue experienced by workers (Wang and Li, 2021).

Fatigue monitoring solutions have been available to the mining industry for a long time but were largely designed for surface operations. The challenges of underground mining environments were not well understood nor tested for. It must be noted that solutions designed for one and not both types of mining operations will always be deficient; vehicles and personnel working underground will eventually come up to the surface and therefore it is critical to have a purposed-built design that is applicable to detecting driver fatigue and distraction in both surface and underground environments.

The journey for combating fatigue management at MMG's Rosebery site began around 2020. At that time, Rosebery had decided to peruse underground fatigue management technologies after a series of incidents resulted not only in equipment damage but also highlighted the issue of personnel safety caused by fatigue and distraction incidents.

Rosebery's selection criteria were not strictly on the solution's features and capabilities. Rather, through a prior trial with an OEM-incumbent monitoring technology, it further cemented their opinion that engaging a vendor that was willing to collaborate on the change management aspects of the project, was just as critical, if not the lynchpin to a successful project roll-out. It was important to Rosebery that the technology deployment journey was smooth but more so that the stakeholder engagement piece was handled well. They are not interested in deploying from a top-down, forced approach.

TECHNOLOGY OVERVIEW

Hexagon's Operator Alertness System (OAS) is a fatigue and distraction management solution, that helps operators of heavy and light vehicles to maintain the level of attention necessary for long driving hours and monotonous tasks. In the event of fatigue and/or distraction events, the system has real-time audible and vibratory alerts for operators to prevent accidents.

The system is purpose-built for harsh, rugged, environments including underground mines. In the entire time OAS has been deployed at Rosebery, which is over two years now, there has only been one component failure which was a faulty camera hardware that was replaced in the very early days of the project. There has been no other hardware and system faults since.

Hexagon OAS's product design has proven to work equally well in both underground and surface mining environments. There are currently over 10 000 vehicles globally with OAS installed, and the Hexagon team have reviewed over 1 million fatigue and/or distraction events. On average, the Hexagon team reviews about five generated events per vehicle daily. Note these encompass validated fatigue and/or distraction events, as well as false positives. Most of the vehicles installed with OAS are in operation in open pit mines. The Rosebery site is the first underground mine to have successfully run OAS for their underground fleet.

Hexagon OAS has an integrated infrared illuminator in the cameras which enables the system to work in low-light environments like underground mines. The integrated illuminators use a 950 nm wavelength, well above the visible light spectrum of 350–750 nm. The system often works better in low-light environments because there is less contrast between the lighting in the cab and the surrounding environment. This gives the camera a better-quality image when monitoring for unsafe events. This also means that the system works well with tinted safety glasses as infrared light has a higher transmittance relative to visible light.

A study was conducted at the University of Texas to evaluate the efficacy of ultraviolet blocking glasses (400–450 nm) and in doing so have shown that the increased transmittance of infrared light through tinted eyewear (Giannos *et al*, 2019).

Figure 1 shows the transmittance of light at different wavelengths for seven different retail sunglass models, and in most cases, there was a higher transmittance of infrared light than visible light.

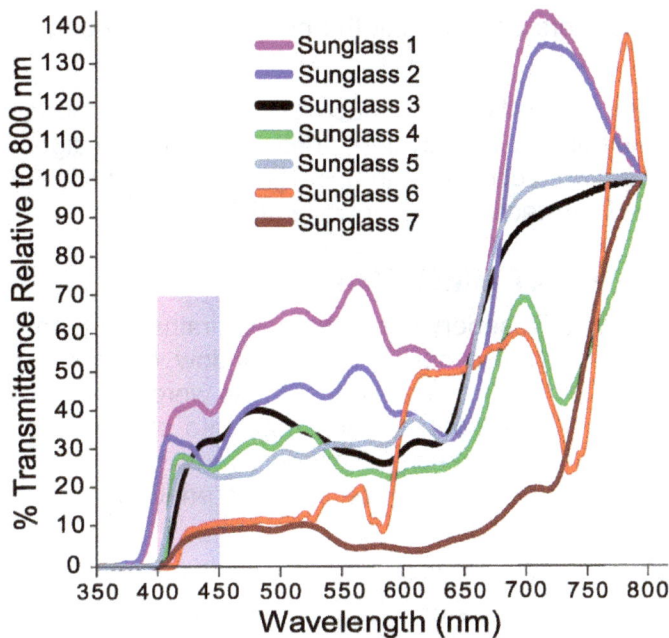

FIG 1 – Transmittance of light at different wavelengths for retail sunglass models.

In 2020, Hexagon conducted a test where different safety glasses were evaluated for their impact on the performance and ability for OAS to detect the percentage of eyelid closure over the pupil, over time (PERCLOS). PERCLOS measures slow, eyelid closures as opposed to normalised blinks.

PERCLOS is determined to be one of the more reliable and valid assessments of driver alertness (US Department of Transport, 1998). Over 30 different models of safety glasses that were tested, 90 per cent were approved by Hexagon for use with its OAS. Figure 2 shows how even tinted safety glasses were compatible as seen by the corresponding infrared camera images.

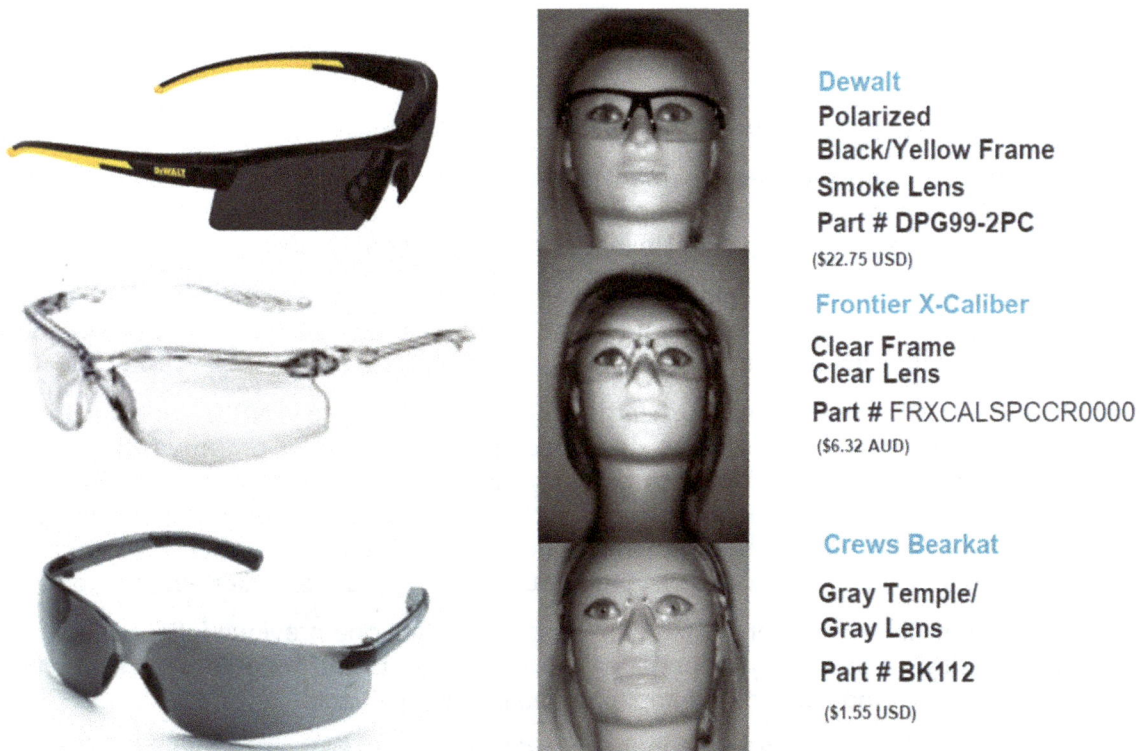

Dewalt
Polarized
Black/Yellow Frame
Smoke Lens
Part # DPG99-2PC
($22.75 USD)

Frontier X-Caliber
Clear Frame
Clear Lens
Part # FRXCALSPCCR0000
($6.32 AUD)

Crews Bearkat
Gray Temple/
Gray Lens
Part # BK112
($1.55 USD)

FIG 2 – Compatibility test of tinted safety glasses with Hexagon OAS.

The study proves that OAS can seamlessly transition from monitoring operator fatigue underground to surface operations. It works equally well in both operating environments and will continue functioning optimally as operators enter or exit the portal to go between surface and underground haul routes.

It is recognised that operator trust is key to the success of fatigue management programs. We know this, and we know what can happen if the system does not work. The algorithm is proven to work in circumstances that no others do, such as in an underground mine where the operator is wearing a hardhat, safety glasses, and half-face respirator while driving.

PROJECT DEPLOYMENT AT MMG ROSEBERY

The severity of the conditions at Rosebery was witnessed firsthand recently by Hexagon's Product Manager for OAS. Standing in the office looking out the window, one can see the haul trucks coming out from the portal. It was obvious from the make that these were meant to be yellow trucks, but it is almost impossible to see the yellow coat anymore because of the environmental factors at the location. Any equipment or hardware used at Rosebery must be robust enough to withstand the moisture levels and low temperatures, and yet be sensitive enough to accurately monitor and track eye movements in very low light.

The search for a fatigue management solution commenced in 2020 as part of Rosebery's 'Fitness to Work' program. In addition to the parent company, MMG's, mandate for engineering controls, Rosebery sought to have additional administrative controls in place to help manage personnel fatigue. Like most mining operations at that time who have not implemented fatigue management technologies, Rosebery relied on several engineering and administrative controls to manage the fatigue issue on-site including reducing shift durations, adjusting shift schedules, implementing more breaks and worker education on how to identify signs of fatigue.

Roseberry had initially started with an OEM-incumbent fatigue management technology but the initial trial of the OEM-owned fatigue management solution uncovered many user issues which included:

- tracking reliability
- product effectiveness and accuracy
- burden on operators as a wearable device
- pricing
- lack of product support from vendor.

However it was also due to low engagement and collaboration from the vendor, that ultimately impacted stakeholder engagement and adoption on-site. This resulted in a search for an alternative solution which led to the partnership with Hexagon. The project go-live was in 2022.

Hexagon was successful in the selection process because at every step of the technology deployment process, Hexagon had demonstrated willingness to come to site and work through the successes and failures with the team at Rosebery.

Hexagon had previous success with a much larger operation – Glencore in NSW – and had facilitated a site visit for Rosebery stakeholders to see the solution in use. The visiting group was representative of the project stakeholder group, which included the project lead responsible for the deployment, a truck operator and a maintenance person. They could see firsthand how the solution was in use and could provide a realistic update back to their peers at Rosebery. The site visit enabled the stakeholder group to dispel misconceptions and untruths and help drive earlier adoption on-site.

It was important that value is demonstrated early, especially with one previous fatigue management project failure already. To be able to understand the key constraints and benefits of the system meant learning from someone who had already implemented it was invaluable.

OAS was implemented at site for a small underground fleet of 11 trucks and a total of 130 operators are currently trained in the system.

Project implementation was in two phases – the first phase went for eight weeks which was largely the hardware install. Whilst installation was performed by Hexagon, they were also simultaneously training the maintenance team. The system was up and running, without live notifications to the operators. This gave the site the opportunity to review all baseline data and weed out false positives. It was crucial to ensure when operators and monitoring centre were getting notified for fatigue events that they were real. Nuisance alarms would simply dilute the system's credibility. During this period, only Hexagon remote monitoring personal had access to the system with clear rules about intervention. The next phase was go-live, which meant all alerts and alarms are turned on.

The recordings from OAS are objective and indisputable. Validated reports are available from Hexagon including footage from the in-cab camera and from the external camera. The major behavioural change in the crew has been the level of self-awareness. They realise now what they are doing outside of work absolutely impacts how they are performing during their shift.

When the operators are now asked what they think about OAS that has been implemented, they say the trust it because they know it works. This is a complete transformation from their original opinion of fatigue management solutions.

CHANGE MANAGEMENT APPROACH

Organisational change is ultimately inevitable and necessary for company growth; companies that fail to manage change in relation to new technology projects often experience employee morale decline and these employees would 'be denied the chance to develop skills and competencies needed in Industry 4.0' (Stobierski, 2020).

In early conversations with the Rosebery site team, discussions were around the systems accuracy, ability to communicate on a very limited network and Change Management.

Preparing workers for the change was a massive undertaking. This is in addition to the scepticism from one failed fatigue management solution.

Operator engagement was slow to take effect. The mine has been in operation for many years with deep rooted behaviours and methodologies. The culture at Rosebery is unique in that many have been at the operation a very long time. Two employees have clocked over 50 years at the site. The challenge to get the crew to accept new systems, technologies and processes cannot be underestimated.

At go-live, Hexagon ensure our safety solutions lead attended four prestart sessions allowing access to all operators. This visit was crucial as it gave both the operators and management at the time the opportunity to speak freely about concerns that they had. It was clear that there was a disconnect regarding the expectations of the Management team that was created by miscommunications, not only about what OAS would bring but also about overall expectations. When putting in a disruptive technology system, especially one that can be seen as an invasion of privacy, there will be concerns and distrust. Prior to the meetings, a Trigger Action Response Plan (TARP) was created allowing for detailed processes and controls on how the system will be used and who will have access. These conversations included operations, HR, and management. Alignment was key as we discovered issues that were not related to the system but were directly due to the camera being installed.

This was key as it enabled the next step. Operators were then presented with the facts of the system, that it would be recording them. It would create events when thresholds of eye closure occurred, and when the head had moved out of the designated yaw percentage. It was also made clear that though the system did not stream video it did record, though no sound. That recording had access controls and could not be downloaded without permissions. That recording would also be overwritten as part of the system configurations.

Once this was agreed upon, then the conversations were had with the union. During these conversations issues related to policy were brought up. Fortunately, the work on the TARP meant all sides were able to agree and see the value of the system for what it was meant to be. These clearly defined parameters were key in this process.

Ensuring all site stakeholders are kept informed with project updates and what is to come became important in helping them embrace the change and diminish scepticism.

Strategically placing of fact sheets and material to create convenient and constant reminders to the operators of what was discussed. Rosebery management also made it easy and was open for people to come back with any questions. The OAS Hardware was also made available for them to see and touch – the tangibility was helpful for operators to understand what was going into the cab with them.

Myth-busting via education on the fatigue and distraction was also critical as it was relatable to every operator. Open and honest conversations about experiences were shared by the Hexagon team as well as the operators. Discussions about driving home while tired, mistakes made while distracted and the impact.

RESULTS AND IMPACT

The success of this project was not just in detecting fatigue events but supporting a culture to address the risk of operator fatigue and alertness systematically to continuously reduce this risk to people and operations. We see the short- and long-term benefit of this initiative by looking at the frequency and severity of fatigue related events at Rosebery.

In the first 30 days of the system being deployed, there were 690 validated eye-closure events. These are PERCLOS events which were picked up by the system and subsequently validated by a remote monitor. As part of the validation process, remote monitors not only confirm the validity of the event, but also rate the severity of the eye-closure event as it corresponds to fatigue. As shown in Figure 3, there was a consistent reduction in validated eye closure events over the first three months of the project, equating to a 65 per cent reduction in eye closure events from Month 1 to Month 3.

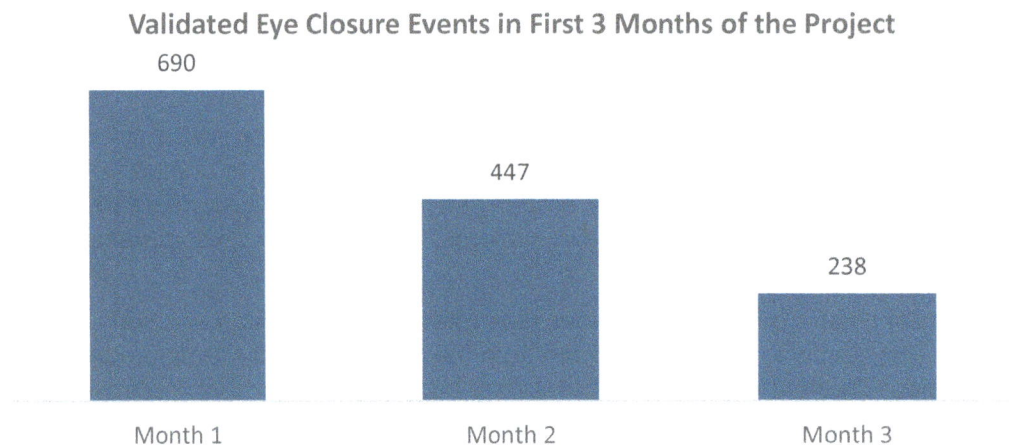

FIG 3 – Reduction in validated eye-closure events over three months.

What is more important to assess than the count of total events is the frequency of validated eye closure events with higher severity. The system cannot prevent fatigue, but alerts when an operator is fatiguing and then it is up to the operation to take the appropriate reactionary steps. This is why culture and change management are so important. This can be observed by the percentage of validated events which were moderate or critical compared to low. What would be expected with a successful fatigue management project, is a reduction of the percentage over time. This is observed at Rosebery, as shown below in Figure 4, where the percentage of eye closure events validated as moderate or critical compared to low, was reduced from 40 per cent to 15 per cent from Month 1 to Month 3.

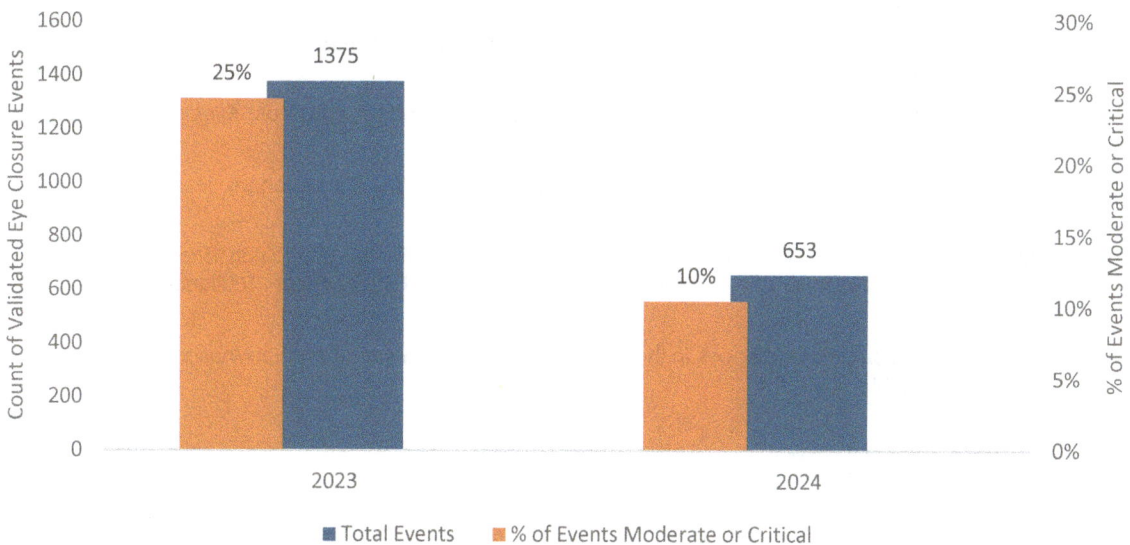

% of Validated Eye Closure Events Moderate or Critical in First 3 Months of OAS Deployment

FIG 4 – Reduction in validated eye-closure events considered moderate or critical over three months.

We see this trend continue when you compare the first three months of deployment to the same three-month period one year later as shown in Figure 5. In the first three months of the project, there were a total of 1375 validated eye closure events, 25 per cent of which were moderate or critical. In the same three-month period the following year, there were 653 validated eye closure events, only 10 per cent of which were moderate or critical. This is equivalent to a 53 per cent reduction in event frequency, and 81 per cent reduction in events which were moderate or critical severity.

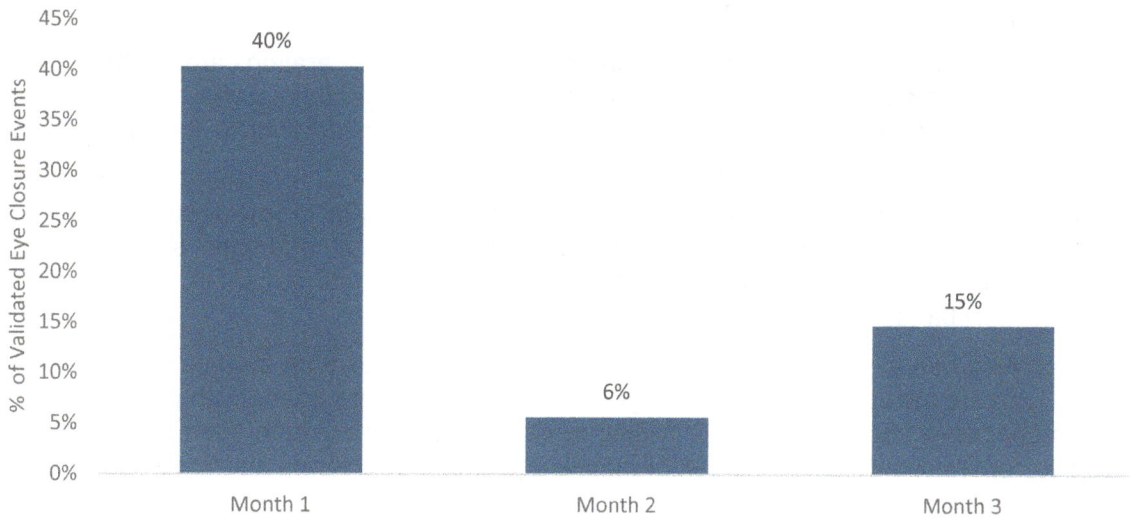

Validated Eye Closure Event Frequency First 3 Months of Year 1 vs Year 2

FIG 5 – Comparison year-on-year reduction in moderate or critical eye-closure events.

Reasons for this observed reduction trend:

- Practical changes resulting from the data observed – rotations and breaktimes, implementation of fatigue break facilities.

- Trust and transparency in the way camera footage is handled and how fatigue and distraction events are managed.

- Employees see a greater level of care by the company when further health assistance is provided to help employees overcome fatigue issues.

FUTURE DEVELOPMENT FOR OAS

There is an increasing interest in OAS adoption from underground operations around the globe, and we are committed to supporting our customers as they go through this journey. This includes the partnership and change management, but there are also enhancements to the product we are planning to make so the system is even more effective underground.

More near term, this means ensuring the system performs optimally in low-network environments where GPS is not available. For example, using ground speed as an additional source of machine speed for low-speed event suppression or intentionally throttling data transmission rates to lessen the impact on low bandwidth networks.

OAS overall also has a very exciting roadmap which will apply to both surface and underground operations. We are always improving the algorithm, focusing on integration with our broader safety and fleet management solutions, and providing more insights that will help operations make use of their data to continuously reduce the risk of mobile equipment operations and keep people safe.

CONCLUSION

Technology is not a silver bullet; without the Management of Change piece to compliment the project, operations will never achieve success. The lessons learnt from Rosebery's initial attempt at a fatigue management project is proof enough that by not engaging your stakeholders early and frequently is a recipe to fail. Encouraging stakeholder to embrace new, disruptive technologies require a robust and structured Management of Change framework to address stakeholder concerns and build trust.

REFERENCES

Brown, T and Davis, L, 2018. The impact of heavy machinery vibration and noise on underground miners' health, *Journal of Occupational Health*, 60(3):215–222.

Giannos, S A, Kraft, E R, Lyons, L J and Gupta, PK, 2019. Spectral Evaluation of Eyeglass Blocking Efficiency of Ultraviolet/High-energy Visible Blue Light for Ocular Protection, *Optom Vis Sci*, 96(7):513–522. https://doi.org/10.1097/OPX.0000000000001393

Jones, A, Smith, R and Walker, M, 2019. Environmental stressors and their impact on miners' performance, *Safety Science*, 114:90–98.

Smith, L, 2020. Haulage challenges in underground mining: A review of current practices, *Mining Engineering Review*, 45(6):223–231.

Stobierski, T, 2020. Organizational Change Management: What It Is and Why It's Important, Harvard Business School online.

US Department of Transport, 1998, October. Publication No, FHWA-MCRT-98–006, Tech Brief: PERCLOS: A Valid Psychophysiological Measure of Alertness As Assessed by Psychomotor Vigilance, Federal Highway Administration.

Wang, Y and Li, X, 2021. Ergonomic challenges in the underground mining environment, *International Journal of Industrial Ergonomics*, 77:102980.

Enhancing raise boring stability assessments through back analysis of overbreak

D Sidea[1] and C Scott[2]

1. Senior Principal Geotechnical Engineer, pitt&sherry, Hobart Tas 7000.
   Email: dsidea@pittsh.com.au
2. Principal Geotechnical Engineer, pitt&sherry, Hobart Tas 7000. Email: cscott@pittsh.com.au

ABSTRACT

Raise boring is a widely used method for drilling shafts in mining and civil engineering projects. Maintaining wall stability and safety during back reaming remains challenging. Increasingly, there is a requirement for shafts, raises and orepasses to be excavated deeper where increasing stresses may present stability issues. Stress induced overbreak complicates raise excavation impacting safety, schedule, and lifespan. A thorough geotechnical investigation remains the starting point in understanding the geotechnical risks to raise bore projects.

Empirical raise boring stability assessments are widely used in the industry and form the starting point in shaft investigations. These methods have been enhanced by the addition of raise bore performance data in recent years due to the improvement in post excavation scanning technologies.

In this study, the authors have back analysed the stability of several raise bored shafts that have experienced overbreak and added to previously published raise bore design charts. The methodology integrates geotechnical review of the initial assessment parameters and *in situ* conditions, material properties and the post excavation performance of the raise bored shaft.

Scans of raise bore walls have been used to compare the measured depth of failure to the widely used empirical relationship developed by Martin, Kaiser and McCreath (1999). This data has been used to increase the understanding of different rock masses during and after raise boring excavation and adds to the confidence in assessment of reliability (probability of failure/success) of raise bored shafts.

INTRODUCTION

In underground mining, raise boring remains the preferred excavation method for ventilation shafts, escapeways and orepasses. There is increased need for deep vertical excavations in civil applications such as pumped hydro projects. Raise boring is being investigated as an option for several planned projects in Australia.

With increased mining depth, there has been an increase in the incidence of stress related issues during raise boring. This has resulted in difficulties during reaming with increased incidence of equipment damage and slow reaming rates. The empirical relationship relating depth of failure and maximum tangential stress developed by Martin, Kaiser and McCreath (1999) is commonly used during the assessment phase of raise boring projects. This paper comprises a review of case studies where overbreak was observed on post reaming laser scans. The aim of the study is to gain a better understanding of the rock mass behaviour under increasing stress conditions in raise bored excavations.

The geotechnical and *in situ* stress conditions were reviewed and the data has been used to add to raise bore design charts relating the ratio of maximum tangential stress and uniaxial compressive strength (UCS) to raise bore diameter as originally published by Edelbro *et al* (2019).

This paper assumes a general understanding of the input parameters for the Q-system for rock mass classification (Barton, Lien and Lunde, 1974; NGI, 2015). The McCracken and Stacey (1989) geotechnical assessment for large diameter raise bored shafts has developed a modified parameter (Q_R) based on the Q-system to assess the classification of rock mass quality for raise boring. The above two methods are not repeated in this paper, readers are directed to the references at the end of this paper for more information about the Q system and the empirical raise bore assessment methodology. The terms 'shaft' and 'raise' are used interchangeably to describe a vertical underground opening created by a raise boring machine.

STRESS INDUCED FRACTURING

Stress-induced fracturing is a function of the *in situ* stress magnitudes and the characteristics of the rock mass intact rock strength and fracture characteristics and distribution. At low *in situ* stress magnitudes, rock mass behaviour in moderately fractured rock masses is controlled by falling or sliding blocks or wedges. The size of blocks and wedges is a function of the continuity, spacing and defect friction.

As *in situ* stress magnitudes increase, stress induced fracturing becomes the dominant failure process around excavations. Martin, Kaiser and McCreath (1999) notes that this fracturing is typically referred to as brittle failure. The relationship between the depth of failure (D_f/a) and maximum tangential stress for circular excavations as established by Martin, Kaiser and McCreath (1999) is defined as follows:

$$\frac{D_f}{a}, = 0.49 \pm 0.1 + 1.25\frac{\sigma_{max}}{\sigma_c}$$

Where:

D_f = Depth of failure

a = Radius of circular excavation

σ_{max} = Maximum tangential stress defined as $3\sigma_1 - \sigma_3$

σ_c = Unconfined Compressive Strength (UCS)

The relationship between the depth of failure and σ_{max}/σ_c is shown in Figure 1.

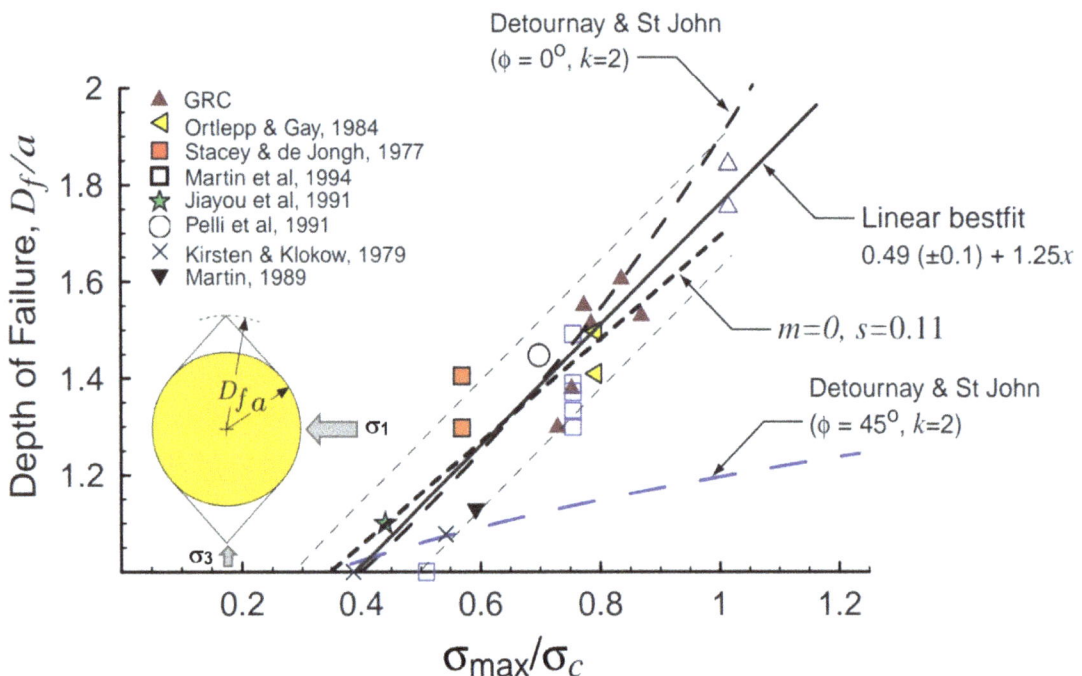

FIG 1 – Relationship between depth of failure (D_f/a) and maximum tangential stress at boundary of a circular opening) (after Martin, Kaiser and McCreath (1999) as presented by Martin, Christiansson and Soderhall (2001)).

The purpose of this study is to understand how back analysis of wall overbreak in raise bored excavations (failure) compares with the empirical relationship in Figure 1. In addition, the tangential stress relationship has been added to the existing raise boring design chart developed by Edelbro *et al* (2019).

DATA SOURCES AND METHODOLOGY

The data used in this study was sourced from various projects in Australia, all of which the authors have directly participated in. The raises studied range in length from 130 m to 890 m and diameters between 4.0 m and 6.3 m. While the study focuses on metalliferous mines, it also includes a review of a case study from a coal shaft with the results discussed later in the paper.

For each site studied, the following process was followed:

- **Review initial geotechnical parameters:** The initial geotechnical rock mass assessment parameters were examined, with a focus on *in situ* stress and the UCS of the intact rock.

- **Evaluate empirical raise bore stability parameters:** The empirical input parameters for raise bore stability assessments were reviewed to ensure consistency across the case studies.

- **Assess relative block size:** Rock Quality Designation (RQD)/J_n (Joint set number). The RQD/J_n ratio has been used to classify rocks into three classes:

  - **'Poor':** RQD/J_n<8 (based on McCracken and Stacey (1989) raise bore classes).

  - **'Jointed':** RQD/J_n>8<30 (based on classification by Peck (2000) for Australian conditions.

  - **'Massive':** RQD/J_n>30, (based on classification by Peck (2000).

- **Analysis of lidar scan data:** Lidar scans of the final raise walls were utilised to measure the average depth of failure within the geotechnical domain intervals used in the initial stability assessment. This was undertaken by measuring the overbreak in horizontal slices of the lidar scans.

- **Assign failure type:** The type of failure observed was classified using three different categories:

  - Structural: type of failure observed due to interaction of main defects.

  - Intact rock: type of failure observed due to stress induced fracturing.

  - Combination: type of failure observed due to a combination of defects and stress induced fracturing

- **Assign raise stability type:** The overall raise stability was classified using three different categories:

  - Stable: Limited or no overbreak observed in the raise profile in the lidar scan.

  - Overbreak: The raise profile shows limited fallout, and the raise is still functional.

  - Collapse: Refers to significant overbreak with the raise either abandoned or its function significantly reduced.

- **Calculate stress ratios:** The ratio of UCS (σ_c) to major principal stress (σ_1) was calculated (σ_c/σ_1). Laboratory intact rock test results from the shaft investigation drill hole were used where available. This was supplemented by field point load tests and/or field estimates correlated with the site wide testing database.

- **Stress Reduction Factor (SRF) calculation:** For cases where the RQD/J_n is less than 30 (blocky ground conditions), and the compressive strength to major principal stress ratio (σ_c/σ_1) is between 10–1.5, the SRF has been calculated based on Peck's (2000) suggested formulas as follows:

  - If σ_3 is known: SRF = $31 \left(\frac{\sigma_1}{\sigma_3}\right)^{0.3} \left(\frac{\sigma_c}{\sigma_1}\right)^{-1.2}$.

  - If σ_3 is unknown: SRF = $34 \left(\frac{\sigma_c}{\sigma_1}\right)^{-1.2}$.

  The authors acknowledge that there are varying opinions on the calculation of the SRF, the above method was used to standardise the data.

- **Failure zone analysis:** The failure zone that forms around the circular raise excavation is a function of the excavation geometry, the far field stresses, and the strength of the rock mass. The shape of the failure zone around the excavation is controlled by the ratio of the maximum principal stress to the minimum principal stress in the plane of the shaft cross-section. The maximum tangential stress is calculated using the relationship proposed by Martin, Kaiser and McCreath (1999) which is defined as $\sigma_{max} = 3\sigma_1 - \sigma_3$. For the raise excavations reviewed in this study, the maximum and minimum principal stresses in the horizontal plane are the two horizontal principal stresses.

- **Normalisation of the failure depth:** The depth of failure (D_f) has been normalised to the shaft radius (a) and is expressed as D_f/a as illustrated in Figure 1.

RESULTS

The back analysis results are shown on the relationship graph developed by Martin, Kaiser and McCreath (1999), with the data points added to the original graph. The data is segmented based on:

- Blockiness/Jointing.
- McCracken and Stacey maximum unsupported span.
- Failure type.

Blockiness/Jointing

McCracken and Stacey (1989) indicated that problems are expected when raise boring in rock masses characterised by small block size. This is defined when the ratio of RQD/J_n is less than 8 (Very Poor to Poor raise bore class). The impact of rock blockiness has been evaluated in the data set in Figure 2.

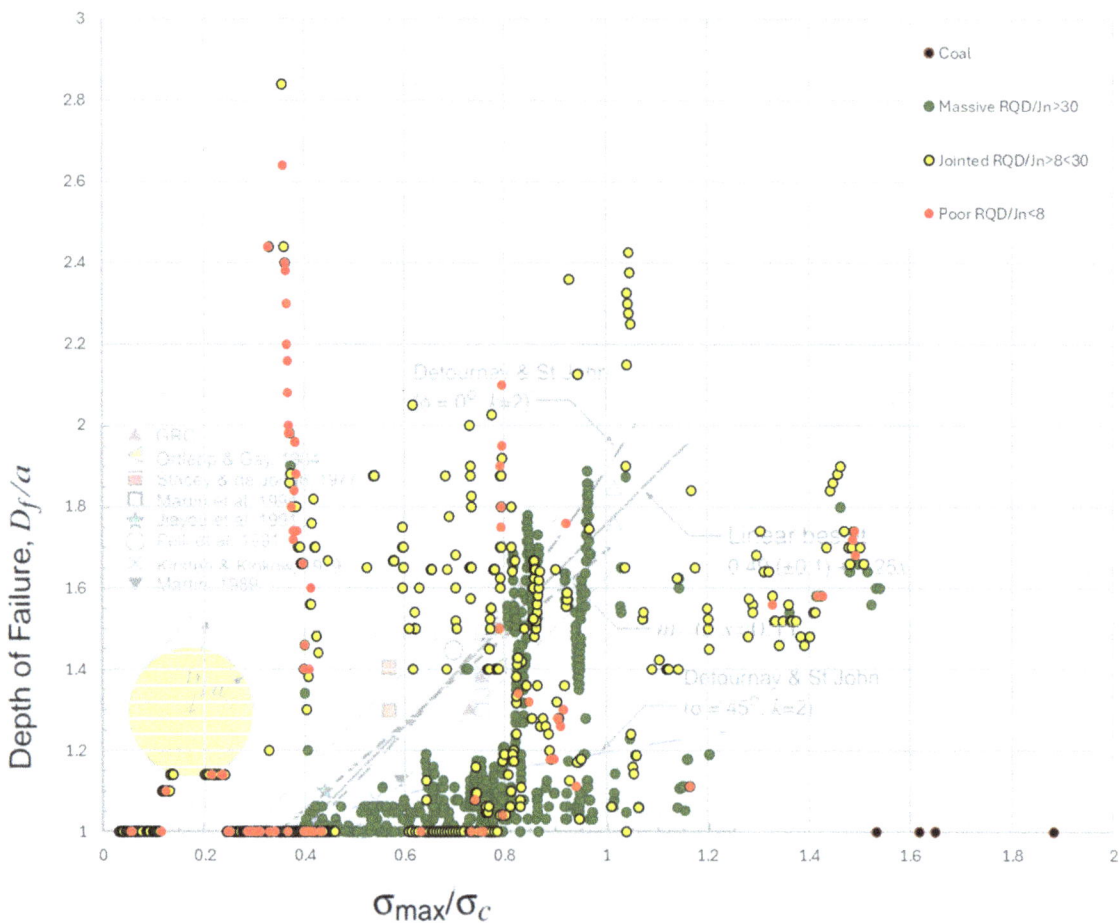

FIG 2 – Depth of failure (D_f/a) versus σ_{max}/σ_c. Data coloured by blockiness. Background graph after Martin, Kaiser and McCreath (1999).

The analysis conducted indicates that there is no discernible relationship between rock masses quantified as 'poor' (RQD/J_n<8) and the depth of failure measured in the raise wall. There is a greater depth of failure recorded from raise walls when RQD/J_n is less than 8 compared to 'massive' rock mass conditions. There does not appear to be a correlation of increasing depth of failure with an increase in σ_{max}/σ_c for poor rock masses.

Rock masses quantified as 'jointed' show significant scatter as well. Although this category contains more data points than 'poor' rock masses, the analysis reveals there is no discernible trend.

For 'massive' rock masses, there appears to be two clusters of data. One data set broadly follows the line of best fit identified by Martin, Christiansson and Soderhall (2001). The second data set is concentrated between σ_{max}/σ_c of 0.4 to ~1.0. Most of the measured depth of failure is limited to 1.2. In massive rock the onset of stress damage can initiate at a σ_{max}/σ_c of 0.4. There is however significant scatter around this point, with the value of σ_{max}/σ_c ranging from 0.1 to approximately 0.8 observed in the back analysis data set.

A small amount of data from coal seams has been included in Figure 2. It can be observed that the behaviour of coal seams does not align with data collected from hard rock. Further data and analysis are required from coal projects to determine if there is any meaningful relationship.

Maximum unsupported span

The measured overbreak is shown in Figure 3 categorised by the empirically assessed maximum unsupported span. This span was calculated using the relationship defined by McCracken and Stacey (1989):

$$span_{max} = 2RSRQ_R^{0.4}$$

Where:

RSR   is a risk term with the Raise bore Stability Ratio (RSR) value set to 1.3

Q_R is the raise bore rock mass quality index

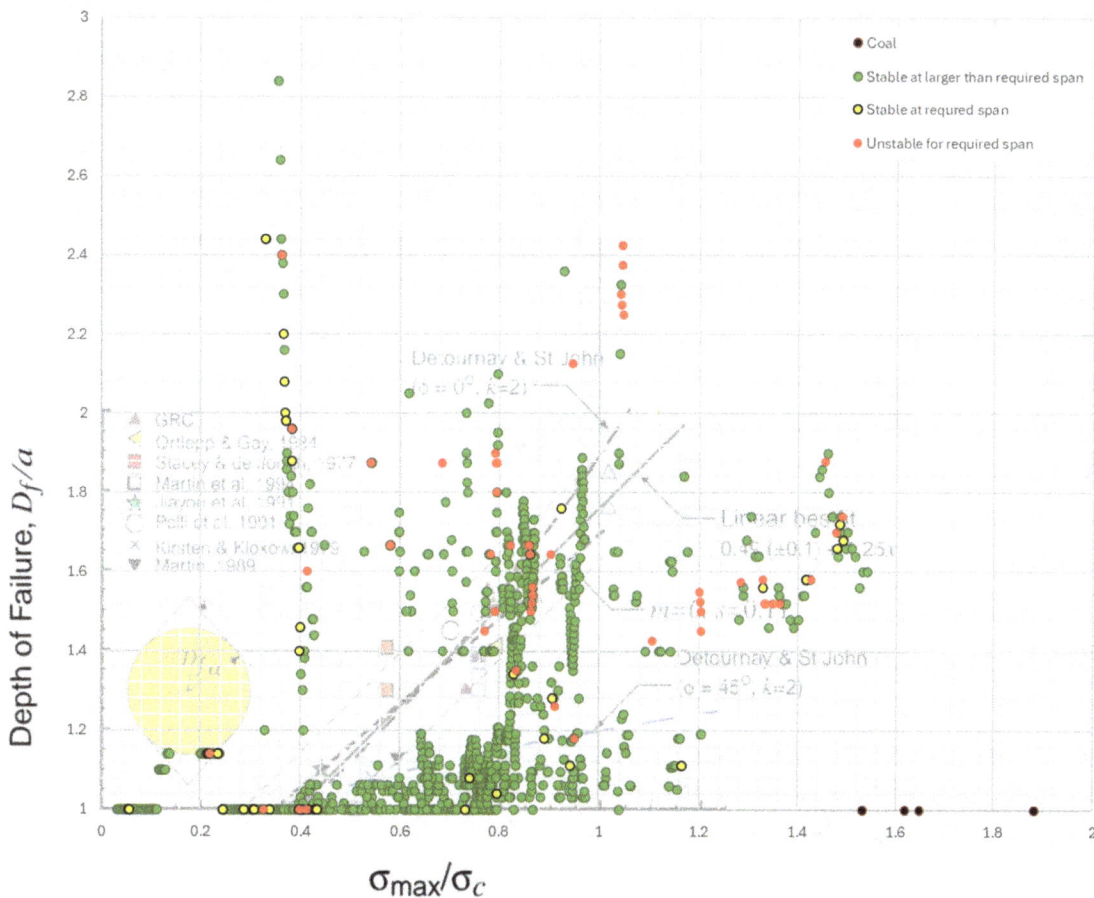

FIG 3 – Depth of failure (D_f/a) versus σ_{max}/σ_c. Data coloured by empirical maximum unsupported spans. Background graph after Martin, Kaiser and McCreath (1999).

The outcome of the back analysis presented in Figure 3 identifies that overbreak did occur in most cases where the McCracken and Stacey (1989) empirical assessment predicted unstable spans. The magnitude of the depth of failure did not follow a clear pattern with increasing stress conditions.

It should be noted that the back analysis has indicated that significant overbreak was measured in zones where the empirical assessment indicated that stable wall spans should be expected.

Failure type

Laser scans enabled the authors to classify the type of failure observed. The process considered the scan outline of the walls, the general rock mass conditions, depth below surface and the *in situ* stress conditions. Some situations or failure types may be difficult to accurately interpret from lidar scans alone.

Figure 4 shows that structural driven failure has the greatest variability in the measured depth of failure for a given set of stress/strength conditions. It highlights the challenges in predicting the maximum depth of failure when failure is structurally controlled. Similarly, where failure has been assessed to be a combination of structure and intact rock failure, the data shows significant variability.

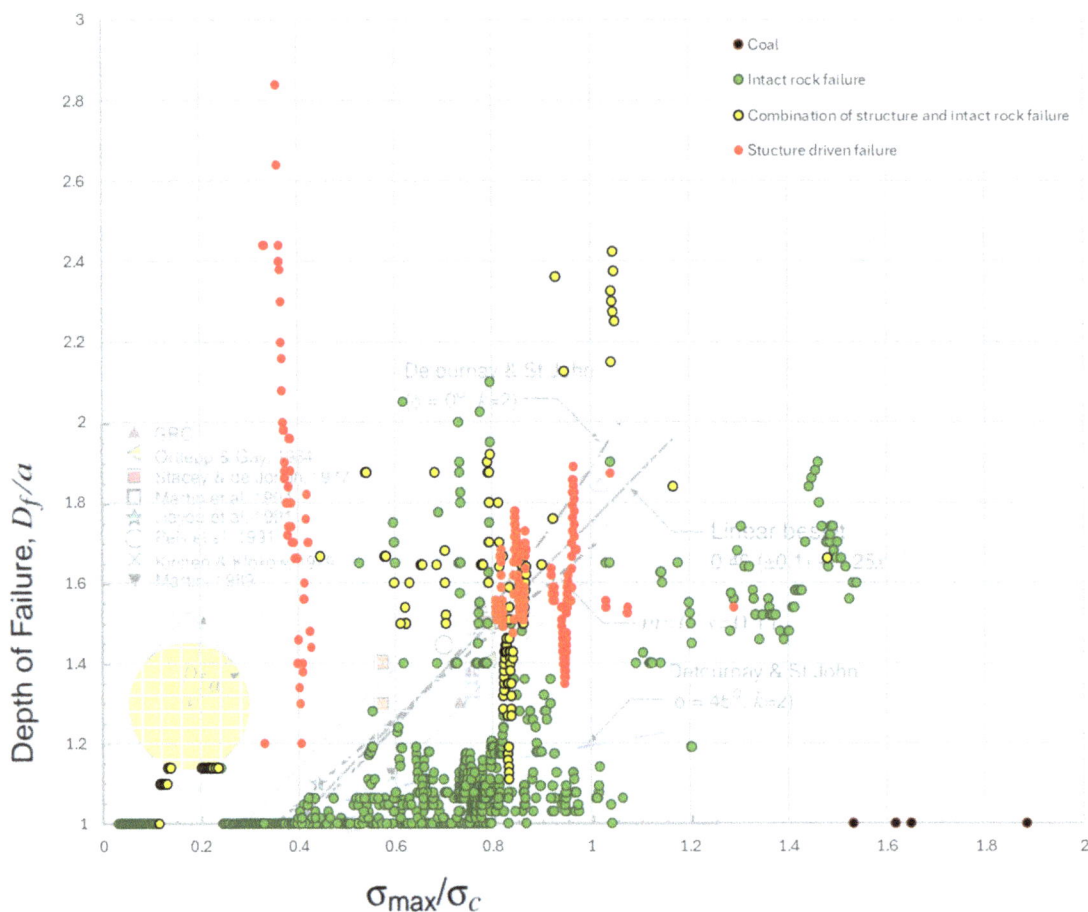

FIG 4 – Depth of failure (D_f/a) *versus* σ_{max}/σ_c. Data coloured by failure type. Background graph after Martin, Kaiser and McCreath (1999).

The majority of the observed overbreak classified as brittle rock failure plots below the line of best fit defined by Martin, Kaiser and McCreath (1999). There appears to be an increase in the depth of failure when σ_{max}/σ_c is greater than 0.8, with another increase in the magnitude of D_f/a observed when σ_{max}/σ_c is greater than 1.2.

Although the original relationship developed by Martin, Kaiser and McCreath (1999) was intended for brittle failure in massive rock, and the results show some scatter, it seems to apply well to rock masses where structure is a contributing or the main assessed cause of failure. Whilst there is significant variability in the initiation of stress failure, the value of the maximum tangential σ_{max}/σ_c of approximately 0.4 remains as the most appropriate point of failure initiation. The value of σ_{max}/σ_c at which failure initiation has been observed in the back analysis data set can however ranges from 0.1 to approximately 0.8.

RAISE BORE DESIGN CHARTS

Edelbro *et al* (2019) developed raise bore design charts relating the ratio of maximum tangential stress (σ_{max}) to UCS. The data from the back analysis has been plotted on the design chart relating raise bore diameter to σ_{max}/σ_c. Figure 5 presents the results.

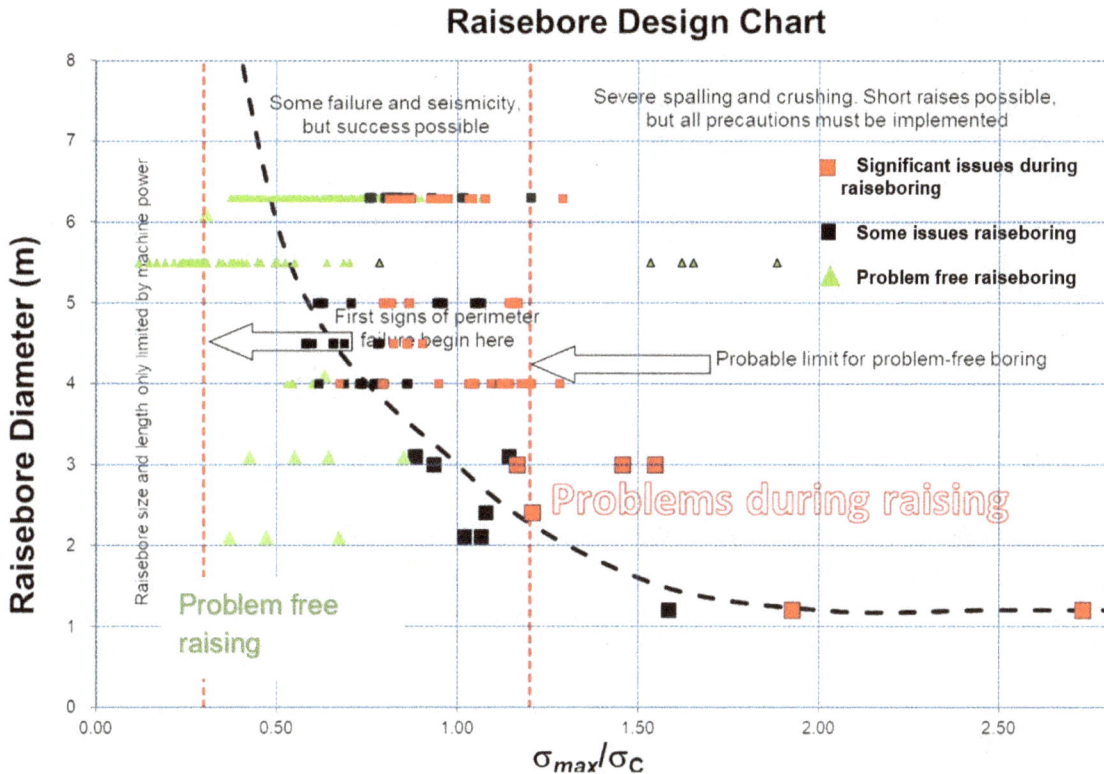

FIG 5 – Raise bore diameter versus σ_{max}/σ_c of back analysis case studies. Background graph after Edelbro *et al* (2019).

The back analysis data covers raise bore diameters from 4.0 m to 6.3 m. There was limited data within this part of the chart in the original Edelbro *et al* (2019) paper. The findings from the study are in broad agreement with the initial chart. For the Australian case studies, it appears that the differentiation between problem free raise boring and first signs of perimeter failure is more likely to occur when σ_{max}/σ_c exceeds approximately 0.7 for larger diameter shafts. This threshold is slightly higher than the initial limit for problem-free raise boring which is represented by the black dashed line in Figure 5.

CONCLUSIONS

Empirical raise bore assessment methods are periodically updated. The relationship between depth of failure (D_f/a) and maximum tangential stress as established by Martin, Kaiser and McCreath (1999) is considered suitable for cylindrical shafts that are not affected by mining induced stresses. This study has back analysed overbreak lidar data to gain a broader understanding of the relationship between depth of failure and maximum tangential stress. The data has been studied in relation to rock mass blockiness, the predicted McCracken and Stacey (1989) empirical stability and interpreted failure mode.

The back analysis of overbreak data from raise bored shaft case studies in Australia has shown that the amount of overbreak and the onset of stress damage value are more variable than the simplified Martin, Kaiser and McCreath (1999) relationship.

The analysis conducted indicates that for rock masses quantified as 'poor' (RQD/J_n<8) there is no discernible relationship between the depth of failure measured in the raise wall and σ_{max}/σ_c. There is a greater depth of failure recorded from raise walls when RQD/J_n<8 compared to 'massive' rock mass conditions. Structural driven failure which occurs in poor to block rock masses has the greatest variability in the measured depth of failure for a given set of stress/strength conditions.

The analysis shows that in massive rock, where brittle failure is predicted to be the main failure type, the existing relationships tend to overestimate the depth of failure. Whilst there is significant variability in the initiation of stress failure in massive rocks, the value of the maximum tangential σ_{max}/σ_c of approximately 0.4 remains as the most appropriate point of failure initiation. There is significant scatter around this point, with the value of σ_{max}/σ_c for initiation of overbreak ranging from 0.1 to approximately 0.8 observed in the back analysis data set.

The McCracken and Stacey (1989) empirical assessment of maximum unsupported stable spans shows that overbreak did occur in most cases where unstable spans were predicted. However, the method is not applicable at predicting stress induced damage. A comprehensive assessment of likely failure methods is important when undertaking raise bore stability assessments.

The high variability observed in the back analysis is attributed to the natural variability in the strength of rock. It is important to understand the geological variability of intersected rock units in a planned raise during the initial assessment and provide results as a guide of the range of potential overbreak outcomes.

The impact of structures to overbreak should be carefully considered in addition to stress induced failure.

Further back analysis of case studies is required to increase confidence in the chart presented in Figure 5. The addition of the author's case studies confirms that raise boring issue were experience when σ_{max}/σ_c exceeds approximately 0.7 for larger diameter raise bores.

McCracken and Stacey (1989) should not be the only method used for raise boring assessments, it should be used in conjunction with a thorough investigation of all potential failure modes and the variance in material properties. The variability should be clearly communicated as a risk to project stakeholders.

REFERENCES

Barton, N, Lien, R and Lunde, J, 1974. Engineering classification of rock masses for the design of tunnel support, *Rock Mechanics*, 6(4):189–236.

Edelbro, C, Brummer, R, Pierce, M, Sandstrom, D and Sjoberg, J, 2019. Raise boring in difficult rock conditions, in *Ground Support 2019: Proceedings of the Ninth International Symposium on Ground Support in Mining and Underground Construction* (eds: J Hadjigeorgiou and M Hudyma), pp 185–198 (Australian Centre for Geomechanics: Perth). https://doi.org/10.36487/ACG_rep/1925_11_Edelbro

Martin, C D, Kaiser, P K and McCreath, D R, 1999. Hoek-Brown parameters for predicting the depth of brittle failure around tunnels, *Canadian Geotechnical Journal*, 36(1):136–151.

Martin, C D, Christiansson, R and Soderhall, J, 2001. Rock stability considerations for siting and constructing a KBS-3 repository: Based on experiences from Äspö HRL, AECL's URL, *Tunnelling and Mining*, technical report, TR-01–38:94.

McCracken, A and Stacey, T R, 1989. Geotechnical risk assessment for large-diameter raise-bored shafts, in Proceedings of the Second International Conference on Shaft Design and Construction (Institute of Materials, Minerals and Mining: London).

NGI, 2015. *Rock Mass Classification and Support Design*, Oslo, May 2015. Original version, Revised and new edition, Oslo, June 2022.

Peck, W A, 2000. Determining the stress reduction factor in highly stressed jointed rock, *Australian Geomechanics*, 35(2).

Remediation of collapsed steel portal tunnel at Yaramoko Mine

C Thomson[1], J Player[2], R Moyle[3] and A Baro[4]

1. Senior Geotechnical Engineer, MineGeoTech, Perth WA 6000.
   Email: colinthomson@minegeotech.com.au
2. Principal Geotechnical Engineer, MineGeoTech, Perth WA 6000.
   Email: johnplayer@minegeotech.com.au
3. Principal Geotechnical Engineer, GGC, West Leederville WA 6007.
   Email: richard.moyle@ggconsultants.com.au
4. Rock Mechanics Engineer, Roxgold SANU, Ouagadougou Burkina Faso.
   Email: abaro@roxgold.com

ABSTRACT

The buried corrugated steel tunnel in the boxcut to the portal of the Z55 underground mine at the Yaramoko project in Burkina Faso unexpectedly collapsed in April 2023 leading to a pause of operations.

The direct cause of the failure was not determined but several bolts pinning two rings of the corrugated steel panels were sheared through with minimal signs of corrosion, which allowed 20 t of backfill material to force its way between the sheets, irreparably damaging the tunnel and blocking the single access decline.

To reopen the mine, the backfill material around the steel tunnel was partially excavated to expose the damaged area but due to the short remaining mine life it was decided not to fully replace the steel tunnel, thereby leaving an open-air gap where the damaged panels were removed.

This necessitated a geotechnical assessment and a plan for remediation works, involving the construction of sumps, diversion walls and earthworks to limit the impact of run-off from the sudden increase in catchment area of the new open boxcut. The designed remedial works were successfully completed, and a prism monitoring program put in place which has so far shown no signs of displacement even through rainy season.

The failure highlights the need for proper installation practices of corrugated steel liners and periodic inspections of the panels which if performed may have caught the shearing and deformation of the bolts early enough to prevent the collapse.

INTRODUCTION

Yaramoko mine site is an underground gold operation in Burkina Faso owned by Roxgold, a Canadian based gold operator. The Z55 underground mine at Yaramoko is accessed by a single decline through a corrugated steel tunnel portal which was constructed in 2015.

In April 2023 the corrugated steel tunnel unexpectedly failed, allowing approximately 20 t of backfill material to enter the decline, blocking traffic from passing through and creating a hazard in the main access, effectively shuttering the mine until it could be rectified.

The backfill material around the corrugated steel tunnel had to be excavated and the damaged portions removed. Due to the short mine life remaining at the operation, the decision was made to leave the damaged section of the boxcut to the portal exposed, requiring remedial works around the perimeter of the boxcut to ensure safe access.

This paper will present the details of the initial portal construction, the failure of the corrugated steel tunnel, the remedial actions taken to secure the portal and the ongoing monitoring put in place to provide a case study in operating practice for unexpected geotechnical hazards at an underground mine.

YARAMOKO PORTAL

The corrugated steel tunnel was designed to access the Z55 underground mine via a boxcut excavation that was later backfilled around the steel structure. The steel archway was designed as approximately 144 m long, 5.9 m high and 6.8 m wide with a 15 per cent gradient resulting in a 15 m height difference from ground level to the roof of the decline tunnel at the deepest point (Roxgold, 2024). The steel tunnel was installed in 2015.

The tunnel was constructed using rings of five steel plates, with alternating footing plate lengths to offset the seams as shown in Figure 1. Plates were fastened together using steel 19.1 mm multiplate bolts. The base plates were mounted into footings which were then covered by a 1 m high brick wall on the inside to protect the footings from vehicle damage.

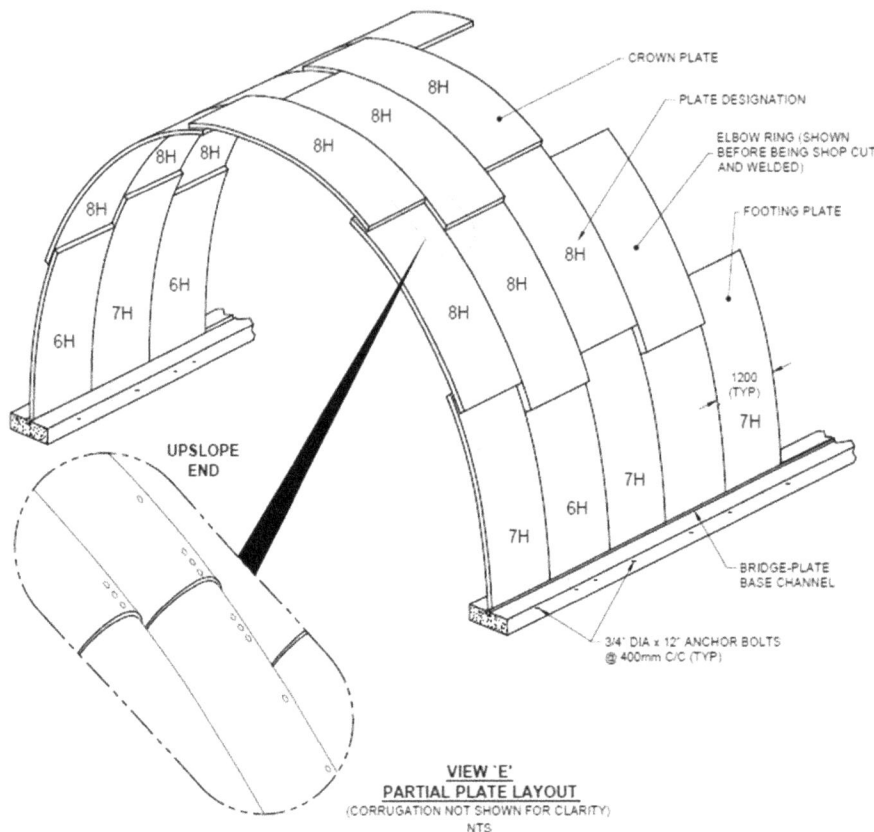

FIG 1 – Corrugated steel tunnel construction (Roxgold, 2024).

Once the tunnel was installed, the boxcut excavation was backfilled with waste material. Finer material was placed around the immediate outside of the tunnel to protect it from damage, with a limitation of no particle greater than 75 mm to be placed within 300 mm of the steel (Roxgold, 2024). The construction guidelines recommended that the fine material be compacted to a 95 per cent Proctor density. Compaction was anecdotally completed using a hand compactor but it is not clear if the material was tested to check if the required density was achieved. Coarser material was then placed on top and backfilled up to original ground level.

TUNNEL FAILURE

On 3 April 2023 the corrugated steel tunnel unexpectedly failed approximately 12–15 m from the bottom of the tunnel (at a depth of ~10 m below surface to the top of the tunnel) allowing backfill material to rill into the decline. The failure is shown in Figure 2.

FIG 2 – Corrugated steel tunnel failure.

Several of the 19.1 mm bolts were found to have sheared through (Figure 3), allowing the material to push open and pass through between two adjacent rings of steel sheets. The movement of the material then displaced additional foot plates, damaging the 1 m barrier wall on the inside of the tunnel.

FIG 3 – Sheared 19.1 mm bolt.

In some instances, the bolt itself failed in shear, and in others the bolt pulled through the steel sheet sideways, allowing the plates to move and provide sufficient gap for the fill material to move into the decline. No detailed failure analysis of the bolts was undertaken.

The failure occurred at the onset of rainy season in Burkina Faso but only two minor rainfall events had occurred in the preceding four weeks. The material that moved into the decline was reported as damp and there were signs of water ingress around some of the bolts in the area of the failure. However, corrosion of the bolts does not appear significant (Figure 3).

No personnel were present at the time of failure and there were no injuries as a result. However, the decline was blocked and production shut down for a month while the damage was addressed.

ROOT CAUSES

No definitive root cause has been identified for the failure of the corrugated steel tunnel. Corrosion does not appear to have played a major role in the bolt failure, but the failed bolts were not sent for further analysis.

Following the failure, other bolts in the area were inspected and anecdotally (T Kearney, 2024; pers comm), it was found that one in three bolts in proximity to the failure were loose, while further away from the failure area one in five bolts were loose.

The loose bolts may have been due to poor installation practices or potential loosening over time. The installation guidelines (Roxgold, 2024) recommended that the bolts be left loose during the installation and then torqued tight to finish which was anecdotally completed, however, it is possible that some bolts may have been missed. The loosened bolts may have allowed the tunnel plates to shift under load, applying shear forces to the bolts adjacent to the failure.

The backfill material immediately around the tunnel was designed to have engineered standards of compaction. It is unknown if this was successfully implemented to the required standards. During the excavation of the fill material following the collapse, a void was observed adjacent to the steel tunnel as shown in Figure 4. Site personnel recalled that muddy water was frequently observed running along the inside of the steel tunnel in the years preceding the failure. Therefore, there is a possibility that water flow through the backfill material has washed out the fines material over time, opening up a void that may have collapsed onto the steel tunnel although no surface expression of the failure was observed.

FIG 4 – Void adjacent to steel tunnel following initial excavation.

Alternatively, rather than a sudden collapse of material, overtime with the flow of water and fines material the backfill may have slowly worked its way between the steel plates forcing them apart, ultimately shearing the bolts and allowing some fill material to collapse into the decline which could have resulted in a self-supporting void above the failure area.

REMEDIAL ACTIONS

Immediate response

The immediate actions taken by Roxgold and the mining contractor African Underground Mining Services (AUMS) was the evacuation of all underground personnel to properly assess the situation.

A decision was then made to excavate the area around the damaged tunnel to assess the scale of the failure and plan how to remediate it. Once the tunnel was excavated, the damaged section of the tunnel was cut out and removed (Figure 5).

FIG 5 – Corrugated steel tunnel excavation and removal of damaged section prior to installation of remedial measures.

The last 15 m of the upper section of the steel tunnel still under load from the backfill material above was reinforced with 150 mm of fibrecrete to prevent further failure up the decline and any remaining loose bolts were tightened where possible.

Long-term remediation

At the time of the failure the remaining mine life of the Z55 operation was two years. Roxgold had the option to either repair the corrugated steel tunnel and re-bury it or to leave it open and remediate the boxcut area for longer term use.

The main risk to leaving the boxcut open is the potential for water ingress during rainy season as the catchment area footprint was now much larger than the decline sumps were designed for.

Roxgold decided that due to the short mine life it was not worth the cost to repair the steel tunnel and opted to leave it open instead.

MineGeoTech consultants, who were already providing geotechnical advice to the underground operations were then requested to assess the boxcut and plan the remedial actions required to ensure safe operations could continue.

In consultation with Dr Richard Moyle of GG Consultants and site personnel remedial plans included:

- Installation of sumps either side of the remaining downslope steel tunnel.

- Laying back the original fill material above the tunnel to a more stable angle.

- Installation of geofabric on the original fill material on the up-slope side of the tunnel (a34 grade or higher).

- Placement of 0.3–0.5 m fresh waste rock over the geofabric.

- Grade original and new rock fill to divert surface water away from portal as much as possible.

- Installation of run-off diversion wall around the upper boxcut bench.

- Excavation of sump on upper boxcut bench.

The aim of this remediation program was to prevent as much water as possible from entering the portal and the underground mine. Due to the increased catchment area of the boxcut this could only be reduced and not eliminated so the purpose of the geofabric and rock fill was to reduce the velocity of any surface run-off entering the boxcut to ensure the portal sumps do not get overwhelmed which could lead to flooding of the mine.

The remediation plan was successfully carried out by site personnel (Figure 6) with the addition of gabion walls either side of the upper portion of tunnel buttressing the remaining backfill material and mesh drapes on the western wall of the boxcut which site personnel deemed necessary to install. The gabions were constructed to prevent the waste material from washing out and slumping into the sumps below. The mesh drapes were installed because of the main bedding orientation on the western wall dipping into the boxcut, although it was assessed that no blocks were likely to fail, and historic cable bolts had already been installed.

FIG 6 – Z55 boxcut remediation works.

MONITORING

To ensure the long-term safety of the boxcut, it was recommended that monitoring prisms be installed to check for displacement of the gabion walls and some wedges identified on the western wall of the pit.

Six prisms in total were installed, one at the top and bottom of each gabion wall to monitor for toe deformation or toppling, and two in wedges on the west wall.

Manual survey pickups were collected weekly or more frequently if there is rainfall following a trigger action response plan (TARP) developed for site. Following rainfall events of 40 mm per 24 hrs, additional prism monitoring is required in the next 24 and 48 hrs. Following the end of rainy season in November 2024 and the consistency of the results, the monitoring was reduced to monthly.

Due to procurement delays the prisms weren't installed until February 2024 but have been successfully monitored since then and have shown no cumulative movement (Figure 7) with the scatter related to the accuracy of the survey method and atmospheric variation.

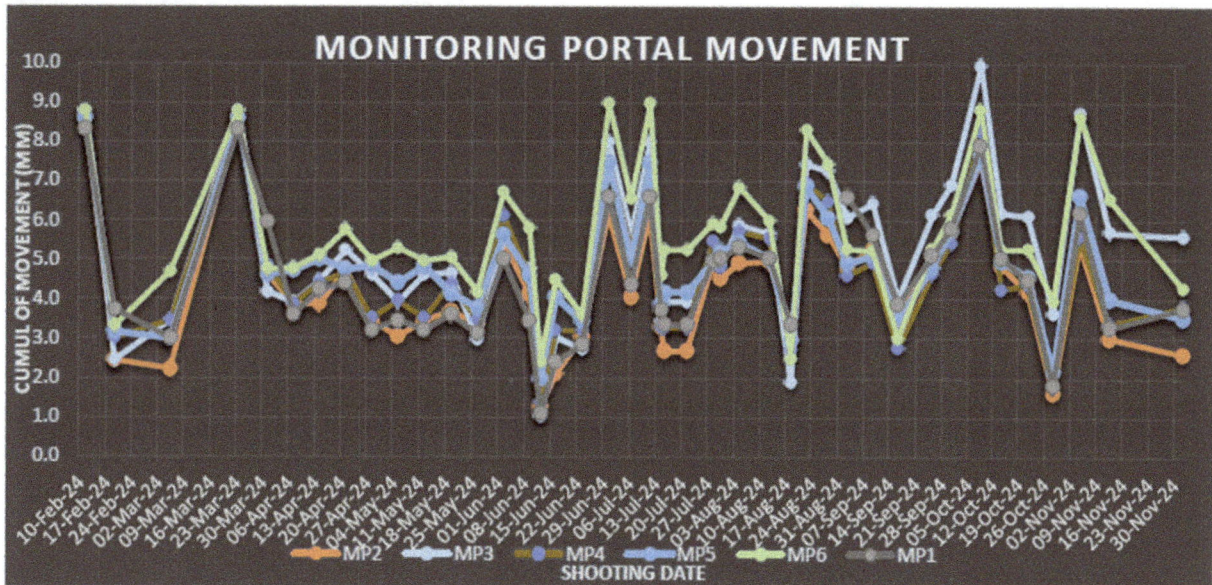

FIG 7 – Z55 boxcut prism monitoring data Feb to Dec 2024.

CONCLUSIONS

A remediation plan for the Yaramoko Z55 boxcut/portal was successfully implemented following the unexpected collapse of the installed corrugated steel tunnel structure in April 2023.

The failure was potentially the result of poor installation with numerous loose bolts reported in the vicinity of the damaged section, allowing movement of the tunnel plates producing shear forces on the bolts. Insufficient compaction of the fines material adjacent to the tunnel may also have allowed these fines to wash away, resulting in a void that could have collapsed and damaged the tunnel although this has not been definitively confirmed. Corrosion may have played a part but does not appear to have contributed significantly to the failure.

The damaged section was excavated, removed and the remediation plan successfully implemented, installing gabion walls, geofabric and waste rock, all designed to divert surface run-off from the decline or reduce the velocities of water entering the sumps.

Prism monitoring has shown the backfill material and gabion walls are stable, and the installed sumps have so far been adequate through multiple rainy seasons since April 2023.

A failure of critical infrastructure like the one suffered at Yaramoko has the potential to severely impact an operation. For Roxgold they managed to limit the impact to only one month of lost production through quick implementation of a robust remediation plan. However, the incident highlighted the need for high quality installation practices and regular inspections which could have potentially prevented the failure from occurring in the first place.

ACKNOWLEDGEMENTS

The authors would like to acknowledge Roxgold Sanu and AUMS for authorising the preparation of this case study in particular Nick Riches and Thomas Kearney.

REFERENCES

Roxgold, 2024. Corrugated steel construction information provided by Roxgold, internal documents.

The Stope Soaker 2000 – a DIY stope dust suppression system

D Tomek[1] and C Miles[2]

1. Underground Mine Supervisor, Metals Acquisition Corporation, Cobar NSW 2835.
   Email: dave.tomek@metalsacqcorp.com
2. Principal Mining Consultant, Advanced Mining Production Systems, Fremantle WA 6959.
   Email: cmiles@advancedmps.com

ABSTRACT

Bogging dry dirt is a cardinal sin, with the deadly impacts of silicosis and other dust related diseases well known. Production delays for dusting out a level from stope bogging and creating additional dust on the decline from ventilation drying out dirt in trucks should not be occurring in this era, but it's an all-too-common problem.

There are multiple solutions on the market, from cannons mounted on the back of water trucks, to portable cannons, dust curtains or a 1" hose with a nozzle. After a crew member took to Facebook asking if a better solution existed, the Stope Soaker 2000 was conceptualised.

If the Stope Soaker 2000 was to be an improvement on existing systems, it needed to address all scenarios; from watering a blind uphole stope to reaching the far wall of a long strike stope. It needed to be easy to install, work when water pressure was low and be inexpensive enough to replace if damaged.

CSA Mine in Cobar NSW is an old, deep operation with a complex raw water network that is somewhat unknown as is often the case with a legacy system. Water pressure is impacted when the demand is great, either from too many users or from large draw items, and the system can struggle to provide the necessary supply for stope watering.

Under new ownership, Metals Acquisition Corporation is encouraging the workforce to suggest improvements. The Stope Soaker 2000 is one of these suggestions.

The Stope Soaker 2000 was named as a homage to a favourite toy and is something any site can implement themselves.

This paper reviews the advantages and disadvantages of the existing stope watering systems available as used by the authors and details the Stope Soaker 2000 specifications, trials and results.

DUST 101

Hard rock mining activities generate dust during most processes, whether it be blasting large rock masses into manageable sizes for extraction, disturbed particles caused by movement of vehicles or non-rock dust from processes such as fibrecreting. Dust can also be produced by high ventilation flows disturbing particles, typically when velocities are greater than 6 m/s.

Industrial regulations focus on inhalable dust, further divided into respirable and non-respirable fractions. The respirable fraction is composed of the very fine dust which is able to reach the lower bronchioles and alveolar regions of the lung. Visible dust is generally irrespirable, meaning it is a larger sized particle which will not pass into the lungs.

Dust in stopes is initially created during the blasting process when rock is broken. Dust is further released during the bogging process, as the broken stock moves naturally and through mechanical disturbance.

Respirable Crystalline Silica (RCS) is the extremely small portion of respirable dust and is the cause of silicosis. RCS is released from minerals that contain silica including quartz, sand, sandstone, granite and slate.

REQUIREMENTS FOR DUST SUPPRESSION

Legislative requirements

Located in NSW, CSA are governed by the NSW Work Health and Safety (Mines and Petroleum Sites) Regulations 2022, with Regulation 41(1) stating:

'The operator of a mine or petroleum site must –

(a) As far as reasonably practicable, minimise the exposure of persons at the mine or petroleum site to dust and diesel particulate matter, and

(b) Ensure no person at the mine or petroleum site is exposed to 8-hour time-weighted average atmospheric concentrations of airborne dust and diesel particulate matter that is more than –

    (i) for respirable dust – 3 mg per cubic metre of air, or for a coal mine, 1.5 mg per cubic metre of air, or

    (ii) for inhalable dust – 10 mg per cubic metre of air, or

    (iii) for diesel particulate matter – 0.1 mg per cubic metre, measured as sub-micron elemental carbon.'

Most State and Territory legislation is based on the model WHS Act and Regulations developed by Safe Work Australia. As such, similar legislation clauses exist in other jurisdictions, including:

- Western Australia Work Health and Safety (Mines) Regulations 2022, clauses 49–50. Schedule 26 also outlines the requirements of the required Mine Air Quality Officer statutory position.

- South Australia Work Health and Safety Regulations 2012, Chapter 10, Part 2, Division 3, clauses 636–639.

- Queensland Mining and Quarrying Safety and Health Regulation 2017, section 136 and Schedule 5.

- Tasmania Mines Work Health and Safety (Supplementary Requirements) Regulations 2022, clause 21.

Respirable dust exposure is further limited by the crystalline silica portion, outlined in Safe Work Australia (2024a, 2024b).

Exposure limits are based on an eight-hour time-weighted average for a five-day work week, which for most roles in underground mining need to be adjusted for a 12-hour shift on a longer roster, using models such as the Quebec model or the Brief and Scala method for adjustment.

TABLE 1

Regulations on allowable dust during 8-hour time-weighted average.

	Inhalable	Respirable	Respirable crystalline silica
Safe Work Australia (2024a, 2024b)	10 mg/m^3		0.05 mg/m^3
New South Wales Resources Regulator (2022)	10 mg/m^3	3 mg/m^3	0.05 mg/m^3
WorkSafe New Zealand (2016)	10 mg/m^3	3 mg/m^3	
Resources Safety and Health Queensland (2017)	10 mg/m^3	1.5 mg/m^3	0.05 mg/m^3
South Australia Energy and Mining (2012)	10 mg/m^3	3.0 mg/m^3	0.05 mg/m^3
Mineral Resources Tasmania (2022)	10.0 mg/m^3	3.0 mg/m^3	0.05 mg/m^3
Western Australia Department of Energy, Mines, Industry Regulation and Safety (2022)	10.0 mg/m^3	3.0 mg/m^3	0.05 mg/m^3

Operational requirements

When considering stope watering operational requirements, a set of criteria was determined for the optimal solution. Current systems have not been capable of achieving all criteria, and a combination of systems has been required.

Criteria included:

- Doesn't interfere with remote guidance lasers and cameras.
- Reaches the full strike of the stope.
- Reaches the width (left and right extents) of the stope.
- Suitable for blind uphole stopes.
- Suitable for downhole stopes.
- More than a superficial cover/layer over the visible broken stock.
- Doesn't flood bogging horizon.
- Portable.
- Easy to use.
- Safe to install – does not put personnel in proximity of open voids.

An additional benefit is that a system is cost-effective to replace in case of blast or other damage.

CSA STOPES

CSA utilises varieties of longhole open stoping methods for ore extraction depending on the ground conditions and orebody geometry.

In the QTSC (Central) orebody, previously Modified Avoca methods were used, however this was replaced with reverse firing using wireless detonators for safety (Small *et al*, 2023), as shown in Figure 1. At maximum opening, stopes are approximately 30 m floor to floor and 15 m along strike. Stopes are typically less than 8 m wide and are extracted longitudinally.

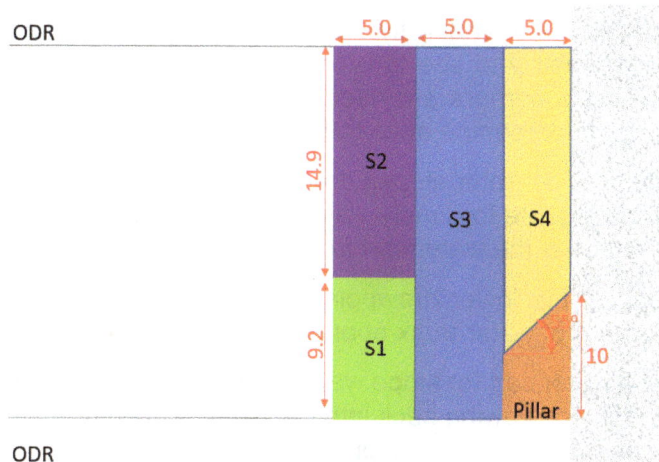

FIG 1 – Dimensions of reverse firing stope (example).

With the exception of Shot 1 (S1) which is a blind uphole firing, the remainder of the stope has top access for watering down, with the same edge location as Shot 2. After Shot 2, there is no access above any of the stope.

In the QTSN (North) orebody, stopes are usually extracted transversely with level spacing varying between 35 m and 40 m. Stopes are 20 m wide and are typically between 20–30 m along strike.

The funnel design shown in Figure 2 provides a range of challenges for stope watering. The initial three firings are blind upholes, 20 m wide with a 5 m strike. Water only reaches the broken ore directly in front of the drawpoint, with some reaching slightly to the sides. Once Shot 4 is fired, a funnel is created, promoting conventional bogging. If water added to the top of the stope does not reach far enough into the void, water will run down the funnel side and report to the drawpoint only, without reaching broken ore further into the stope.

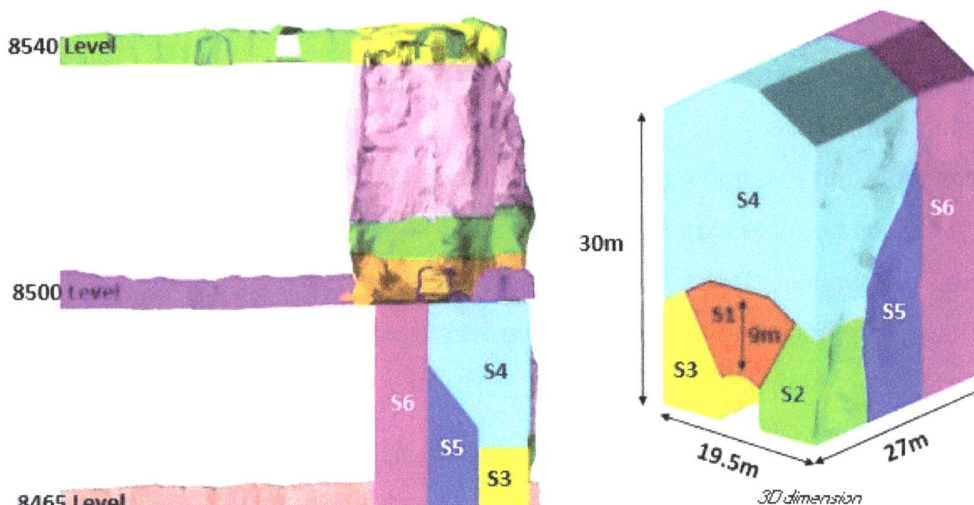

FIG 2 – Dimensions of a funnel design with extended strike (example).

EXISTING OPTIONS

Water cart cannon

CSA operate two Elphinstone water carts, the WR810 Water Cannon and the WR820 Water Tank. Their primary use is watering the extensive road network for dust suppression for the fleet of 10 haul trucks and multiple other road users.

The WR810 cannon is ideal for pushing material off the top of the stope edge without putting personnel near the brow. It is also used to clean drill sites down to bedrock. The cannon operates from an 8000 L tank with either a 27 mm or 38 mm nozzle to produce a water jet at 140 psi. As shown in Figure 3, all heaved material and fly rock has been pushed into the void, allowing for a clear visual of the void edge, ability to see tension cracks in the floor and a clean workspace for subsequent activities. Physicals barriers are used to prevent access to the void edge, and safe working distances are easier to determine once the top of the stope is cleaned off.

The water cart is currently used to water stopes down from the top and bottom accesses when the current 1" hose set-up is inadequate for timely watering of material. Whilst extremely effective and efficient for bogging, this removes the water cart from its primary task of road dust suppression.

When using the water cart at the lower drawpoint, bogging and trucking has to stop during the process to mitigate entrapment of water truck operator.

The usual application of the water cart for stope watering is when the brow of the stope is open, with water hitting the broken stock and flowing back into the extraction drive, resulting in excess water at a singular time which needs to be managed to prevent road degradation.

If the road spray whale tales are used instead of the cannon on the water carts, penetration of water into the broken stock is adequate, however is short lived. The throw of the whale tails from the bottom drawpoint do not reach the back of the stopes and are limited in the vertical reach, therefore when the stope rills, dry dirt still dusts out the level.

When the water cart is used at the top of the stope, a bund approximately 8 m back from the tipping edge is used to prevent access to the open hole. Whilst the water still reaches into the stope, the full strike length cannot be achieved with the whale tales.

FIG 3 – Top of stope cleaned off using cannon on Elphinstone WR810 Water Cannon (left: before; right: after).

The cannon, while ideal for stope watering, is not currently a long-term viable option for CSA, as the water carts need to continue with road dust suppression, and the purchase of an additional stope-only truck is cost prohibitive. The cannon, if used as part of the re-entry process after a stope firing, has the ability to clean the tipping edge off and water the entire stope in a single period. This method can water an entire stope although it requires multiple tanks of water and will occupy the truck for a few hours. In addition, on large stope firings, oxidisation of ore can be a concern at CSA with high-grade copper stopes and the process of watering the entire stope at once can accelerate this process.

Drawing large volumes of water from the raw water network, if not managed correctly, can result in development, production and raise drill downtime. Water tanks are used to assist with controlling draw at water truck fill points to manage this production concern.

Fixed location cannons

A variety of existing fixed location cannon options were considered. All of these systems struggled with fluctuations in water pressure and the majority require air assistance as well. CSA experiences low air pressure due to the complex services network and the quantity of machines requiring compressed air. When air assist is being used without a non-return valve, it is not uncommon for water to overpower the air pressure in the fixed location cannon and then back flow into the air lines, starting to flood the mine's air supply.

CSA existing cannon

CSA already had a cannon on-site for watering stopes, see Figure 4, however it was rarely used due to impracticalities and its location is often unknown. This was another in-house solution, which connected a 2" water bull hose and a 1" air line into a 1" pipe with a nozzle on the end. It was constructed on a RHS steel frame set-up for forklift slippers. If an integrated tool carrier (ITC) is not available, it requires a two person lift to get it on and off the light vehicle and still only provides a single fixed location spray. When tested with just water and no air assist, it was capable of reaching 23 m. This is suitable to reach the majority of stope strikes from the top of the void, assuming the cannon is positioned so that water clears the bund height without hitting the drive backs and shortening the throw.

FIG 4 – In-house built cannon.

DuCaR® Super Dust JET

CSA Technical Services team procured a DuCaR® Super Dust JET agricultural spray to also trial stope dust suppression. This spray is designed for dust control and suppression wash irrigation, with a minimum jet length of 38 m at 44 psi when 733 L/min is provided.

A disused drive was located for initial testing and the Super Dust JET was connected to a 1" mine hose. It was discovered that the nozzle on the spray head was too large for the available water supply, with the initial reach 6–7 m. The nozzle was modified to reduce the output diameter resulting in approximately 14 m of reach, Figure 5.

FIG 5 – DuCaR Super Dust JET.

The spay was then connected to a 2" bull hose to increase water supply to the spray. This resulted in approximately 40 m of spray reach after the fixed frame was tilted to lower the spray height and prevent the spray from hitting the backs. The unit is capable of further reach if not limited by back height and could extend further into the stope if set-up at the lower brow of the stope.

This product has potential for being used at the base of stopes, however in remote loader operating areas, the services are only available in the shoulder, not at ground level, making connecting hoses impractical without working at heights equipment. Due to the size and placement required of the unit, it would not be possible to use the Super Dust JET and bog the stope at the same time. At the base of the stope, the water spray would still have the issue of hitting the backs, when placed behind the standard safety bund, thus restricting achievable strike again. When not in use on the extraction level, the unit needs to be safely stored in close proximity to the stope.

If this unit is placed unattended at the top of the stope to water whilst bogging, there will be no way of monitoring water flow, resulting in the potential use of too much water, taking away from other operational tasks. A person would need to leave their current job site to turn the spray unit on or off, which is problematic with the lack of available light vehicles. If the loader operator was to manage the spray, bunds would need to be placed every time they went to monitor or adjust the Super Dust JET.

The 12 kg spray is fitted with wheels to assist with movement, however still proved cumbersome to lift on and off the back of the light vehicle to move between stopes.

Magnum Australia water cannons

Magnum Australia's water cannon is also used within the industry. When connected to 1½" BSP inlet, the MM40 (Figure 6) throws 30 m with a 7 bar (101.35 psi) input. This has been used at a different operation and proved very effective with a long throw when set-up at the top of 30 mW × 30 mL × 90 mH stopes, however the 16 kg cannon also required a frame to be made to hold the cannon in place. With correct design, the frame enabled easy movement between stopes by mechanical means, when an ITC was available. An optional fan spray control is available however this site was using a singular nozzle.

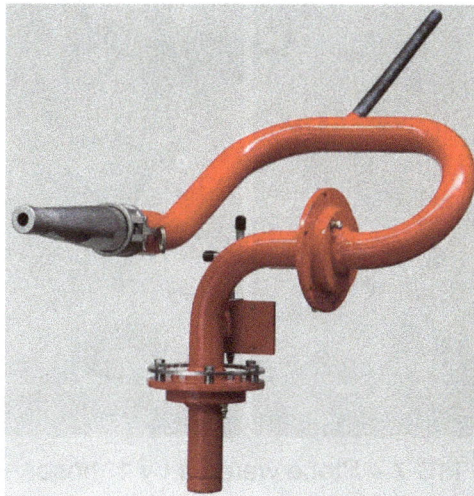

FIG 6 – Magnum Australia MM40 Water Cannon.

The next model up is the MM65, throwing 60 m of water at 7 bar (101.35 psi) input. At 36 kg plus frame weight, this cannot be moved without an ITC. Depending on the model, the MM65 comes as a fixed nozzle, or can be controlled with an electric or hydraulic motor.

Dust curtains

Dust curtains are a line or arc of sprays set-up on a poly line installed around the drawpoint. The dust curtain can be set-up at either upper or lower drawpoint, depending on the ventilation direction and dust movements. This only provides superficial dust suppression and whilst it may control surface dust that interferes with remote bogging guidance lasers or cameras, it does not wet the majority of material, requiring further dust suppression in the stockpile, prior to loading trucks.

Spray lines in backs

One practice used at sites is setting up a 25 mm poly line in the back, with holes cut in to use as a spray line post firing the breakthrough firing of a stope. This method allows water to reach the entire strike of the stope, which can provide good dust suppression from the first bucket to the last.

The downfall of this system is it requires the spray line to be installed prior to stope charging, sometimes before stope drilling to preserve hole collars. The line is at risk of damage from the drill rig if installed too early, or from fly rock during blasting. If holes are too small, too big, or the line is damaged, there is no way of modifying, repairing or replacing the line once the void is created.

This method is also only appropriate for breakthrough stopes and has limitations on water spread in wider stopes.

Hose at drawpoint

The most common method of stope watering at CSA is a 1" hose with a 0.5" Minsup nozzle fitted as shown in Figure 7. When placed at the bogging horizon, the hose has a limited reach and due to the height of the spray, can interfere with remote bogging guidance lasers and cameras. To minimise spray interference with bogging, the hose location is relatively fixed, only wetting one part of the broken stock.

FIG 7 – Stope watered by 1" hose.

When placed at the top of the stope the loader operator, supervisor or other person needs to continually go to the upper level and reposition the hose to better wet the broken stock. At CSA, remote bogging is conducted from three fixed hut locations, sometimes quite a distance from the stope location, requiring a light vehicle to reach the stope.

The hose with a nozzle is usually inadequate in managing dust levels to an acceptable level and is required to be used in conjunction with the water cart to provide suitable coverage.

STOPE SOAKER 2000

The concept

After years of operator frustration, when a crew member took to Facebook asking for suggestions on what other sites were using for watering down stopes, an idea formed. There were lawn sprinklers already in the garden shed, that were known to have a good reach, coverage and pressure off a small water input. The question was asked, 'Could a simple, cheap, replaceable product be suitable for this application?'.

The concept was four garden sprinklers, two at the top of the void and two at the bottom, with one secured to each wall to provide maximum coverage. A short length of 1" hose could be connected between the water dropper and a male 1" Minsup fitting screwed into the inlet of a 25 mm sprinkler based on the geotechnical lessons learnt from prior stoping practices and to improve the safety of underground personnel via safe and efficient blasting techniques.

Design optimisation

T-drifts for slots

Wing slashes (taken as a development slash cut and supported and then a side winder is left unsupported) and longwall slashes (blasted into the wing slashes and left unsupported) were common development practices at the Musselwhite Mine to generate enough void for the initial toe blasts of the PQD transverse stopes in 2020 and 2021. However, the unsupported large span of these slashes combined with entirely undercutting the stopes within strongly-folded geological formations have resulted in significant ground failures in 2021 and 2022. The cross-cuts were experiencing ground issues while they cross the 4F mafic contact impacting the stand-up time.

To mitigate the unsupported large span which was posing safety risks and impacting the stoping cycle due to oversize material at the drawpoints, the development layouts of transverse cross-cuts have been revised to accommodate T-drifts, which are mined perpendicular to the cross-cuts for slot raises in 2022 (Figure 8). As the T-drifts were providing enough void for blasting and removing the slashing requirements, they became more favourable for the ground conditions, and also production drilling was utilised to get the full recovery of stope walls instead of using jumbos for the slashes. Introducing the T-drifts has reduced the development requirements by eliminating the wing slashes at the PQD, improved stoping cycle time by reducing time spent on the same headings and allowed allocating mining resources accordingly to other mining areas as they needed.

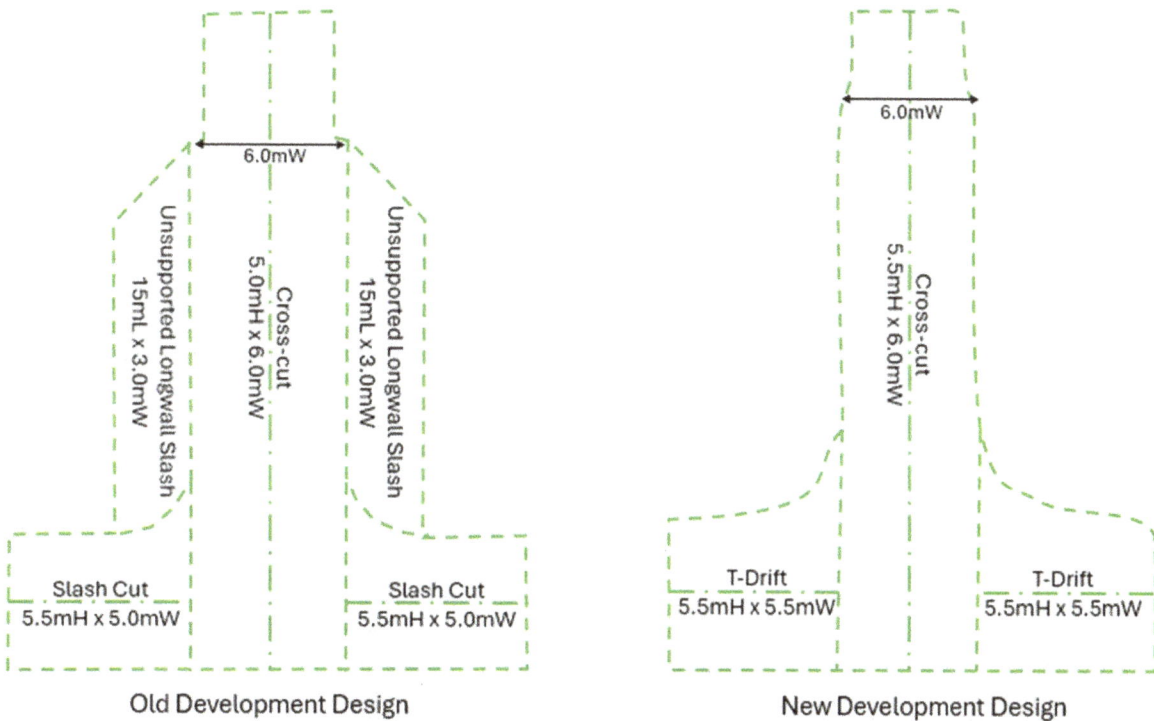

FIG 8 – Plan view of development designs showing wing slashes and unsupported longwall slashes on the left and new development layout including T-drifts on the right.

Ground control measures

Musselwhite Mine has been introduced to new ground support standards to mitigate geotechnical challenges and to improve the stability of stopes. Some of the updates in the past three years included introduction of adequate cable bolting (Figure 9), drawpoint support with modified shotcrete arches and tight-filling of voids using mobile CRF plant and rammer jammer. Increased cable length and density based on the back hydraulic radius from Hutchinson and Diederichs (1996) cablebolt design charts along with dumped cable bolts (Figure 10) to cross the rock foliation (typically 55° to 80°) have helped achieving stable stopes. Besides these, plated and tensioned cablebolts in the backs and brow have assisted obtaining better load transfer between the surface support and deep embedment.

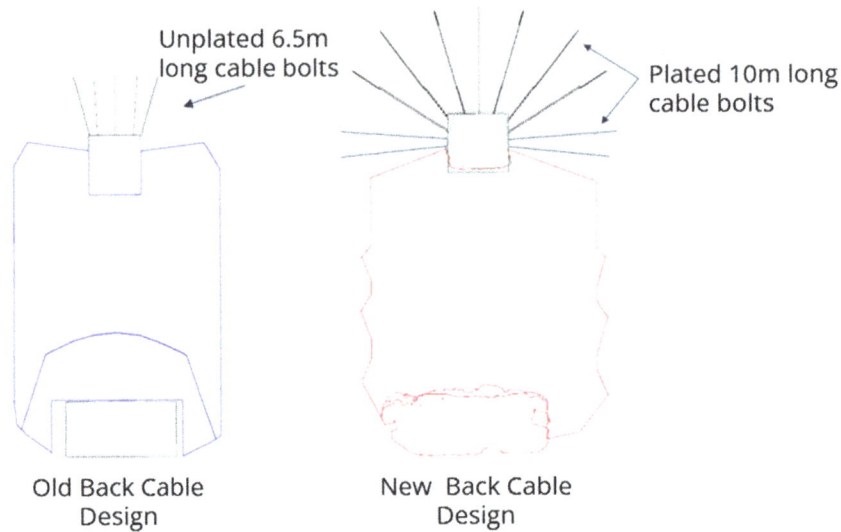

FIG 9 – Cablebolt designs at the PQD stopes (cross-section view looking west).

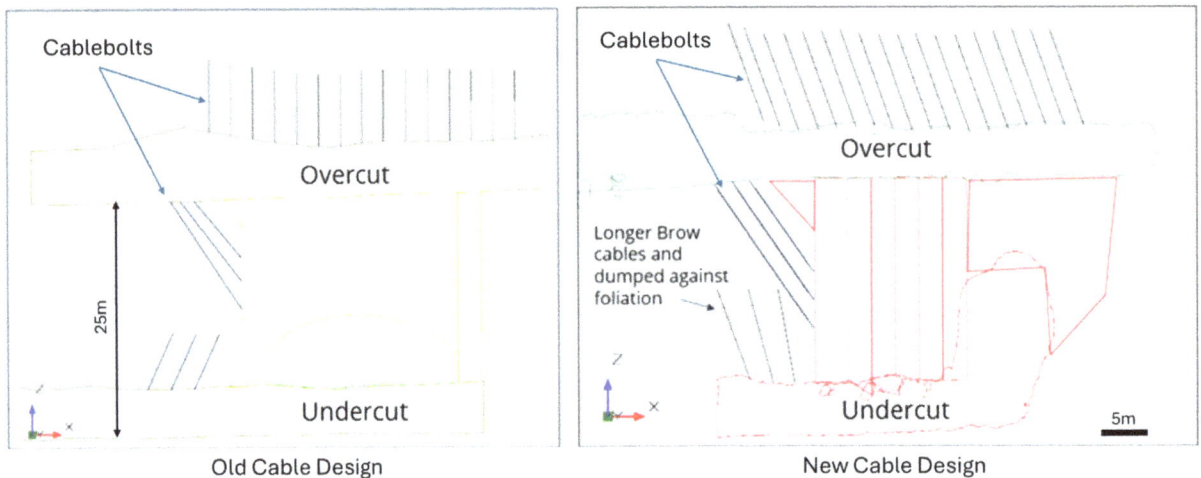

FIG 10 – Old cablebolt design with 6.5 m cables installed vertical in the back (on the left) versus New cablebolt design with 10 m cables dumped against the dip of foliation (on the right) at the PQD stopes – Long section view looking north.

A fully mechanised Epiroc Cabletec rig has been used to install the cables in the stope crown, walls of the overcut and the brow, and manual cable installations have been conducted in the floor dipped to the HW. Besides these, the MWM has replaced MacLean roof bolters with Sandvik DD422i jumbo drill rigs and has conducted trials of Mechanical Dynamic (MD) bolts at the secondary transverse stopes to explore the possibility of replacing resin rebar with the MD bolts. The performance of the MD bolts was exceptionally well without any indication of damage. The change management process of implementing new ground support was presented at the SME Annual Conference in 2024.

As part of the new design criteria, unsupported versus supported back hydraulic radius calculation has been utilised during stope design process which helped eliminating failures associated with large unsupported back spans. Besides these changes, pre-conditioning and buffer rings have been employed to manage the high-stress and ensure the HW stability which yielded better results than the conventional drill and blast pattern which is discussed under 'Drill and Blast Optimisation' section of this paper.

Drill and blast optimisation

Dropped shoulders

To mitigate some of the earlier stope failures attributed to lack of shoulder confinement due to the rectangular shapes with upward production holes, the shoulders of stopes are dropped by steepening the production holes to 35–38° from horizontal which assisted forming more stable shapes during excavation and post-production activities. This design revision also required excluding loading of the flat and upward production holes for already drilled off stopes and the new stopes are designed without the flat and upward production drill holes to provide a more stable shape. Also, the previous flat-back stope designs included toe shots which were completely undercutting the entire stope. For the revised dropped-shoulder designs with the help of arched-shaped toe-shots, the span of toe-shots are reduced to ensure adequate pillar between the toe-shot and top-cut level prior to the mass blasts.

Two sets of angles have been employed for the dropped shoulders: 35° shoulders are used for primary stopes, and 38° for secondary stopes (Figure 11). The material left around the dropped shoulders are mostly recovered with the stopes above by extending and blasting the production drill holes. The planned recovery assumptions have been updated to incorporate the dropped shoulders for Mineral Resources and Mineral Reserves (MRMR) estimations and production forecasts.

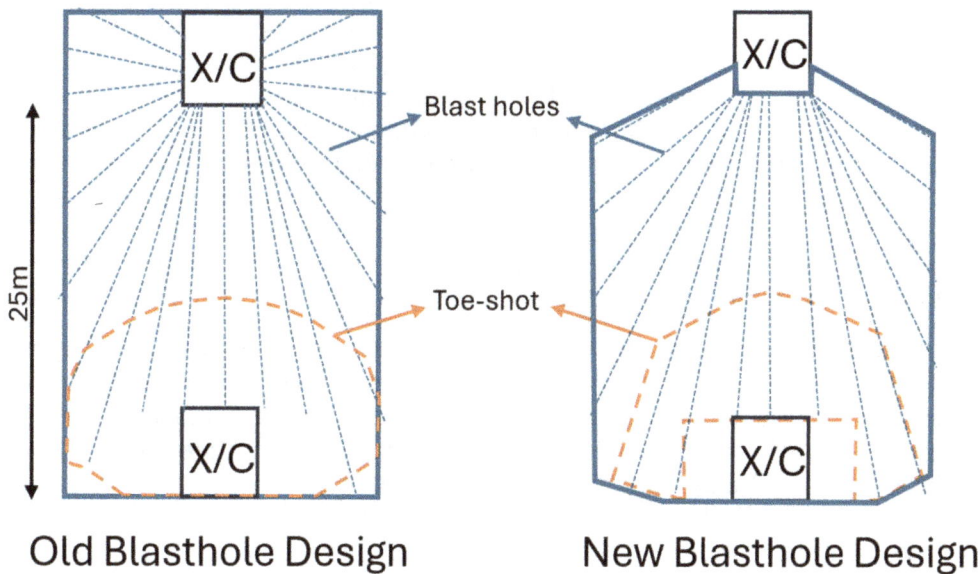

FIG 11 – Previous flat-back stope design with upward holes (on the left) versus New stope design with dropped shoulders providing more stable shapes (on the right) – Cross-section view looking west.

Blasting

Along with disciplined sequencing based on the geotechnical recommendations, the Musselwhite Mine has revised the undercutting of transverse stopes by replacing the horizontal void blast designs with a vertical slot design (Figure 12). The new blasting sequence has eliminated the wide unsupported back span after the large void blasts and increased the height of slot raise cap from 10 m to 12–13 m to maintain a stable pillar for re-entry at the overcut level. Successful implementation of the revised blasting sequence has allowed implementation of the double-lift stoping at the PQD to facilitate stoping cycle time with higher production rates that attained better stope recovery.

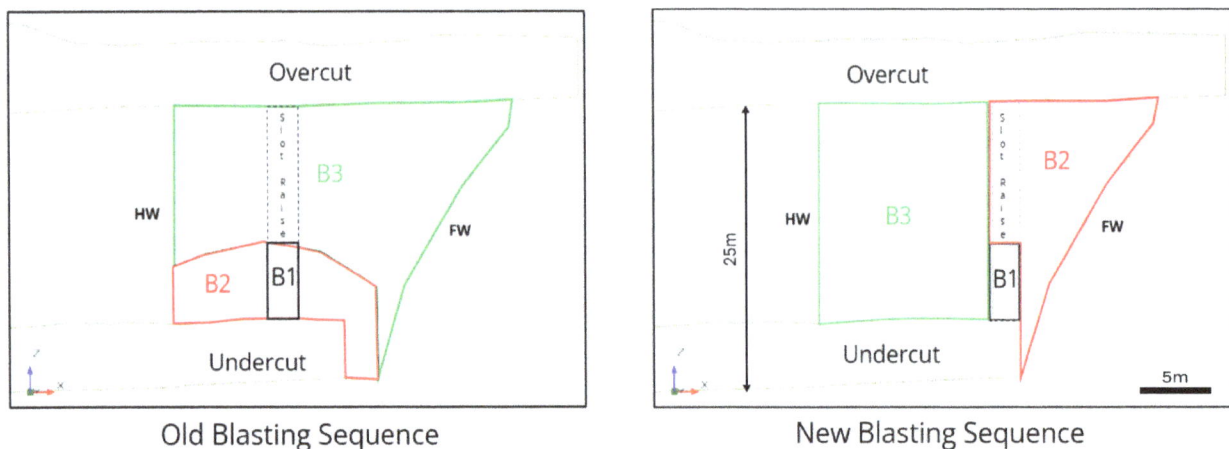

FIG 12 – Blasting sequence of the PQD transverse stopes (long section view looking north).

Transverse stopes are charged (loaded) with emulsion and detonated with electronic detonators. The HW of stopes are now designed vertically to account for lithology and to minimise undercutting of the top level.

One of the biggest changes to blasting was moving away from a chevron blasting method (V-shaped blasts toward the initial void) to a sliced bread method (vertical blasts of each ring) using a slot raise as the stope designs were not optimised for the rock types which were affecting the fragmentation and higher costs associated with oversize material. Another major revision was to timing of the blasts as it was leading to poorer blast outcomes and overbreak on the HW. When blasting transverse stopes, detonator delays have been decreased and the longer delays on the stope boundary have been increased to prevent further overbreak. In the new method of transverse timing, the delays have been done relative to each other, with a specific delay to start the next ring back to ensure the rings are offset. This offset depends on the burden as it is related to amount of time that takes the previous ring to move out of the way and the typical rule of thumb is 20–30 m/s per metre of burden.

At the Musselwhite Mine, double priming has been used as a security measure in case one primer gets destroyed. There has been a significant focus on the primer location whilst considering the lithology for optimal results and providing specific instructions to the blasting crew to eliminate sympathetic primer interactions in the adjacent holes. Two primers on the same delay have been used for interior holes to increase fragmentation and offset primer delays have been employed for both the shoulder holes and the HW holes including the intermediary HW between individual blasts. The primers have been staggered and the boosters have been spaced at a minimum of 2 m from each other in a hole. Improvement on the primer location has significantly changed the fragmentation profile by allowing further fragmentation of material in the air. The blasting crew is now provided with the designed booster locations from the toe on the charging instructions. Some of the rules of thumb for transverse blast timing are listed below:

- Hole-to-hole delay is 20 ms.

- Hole delay on the second to the last shoulder hole is 40 ms.

- Hole delay on the last shoulder hole is 70 ms.

- Hole-to-hole delay on the last HW ring is 40 ms to prevent damage.

- Inter ring delay is 100 ms.

- Increase the delay by 50 ms (100 to 150 ms) after five rings have been blasted.

- Double the delay (100 to 200 ms) for the HW ring.

- All primers are timed on the same delay except the shoulder and HW holes which have a 10 ms offset with the bottom primer going first.

- Inner hole primer spacing to be greater than 2 m ensuring the top primer is buried in larger than 2 m of emulsion and hole to hole staggering to be larger than 1 m.

- Primer placement at toe and collar is staggered to keep a minimum 2 m separation to avoid sympathetic detonation and to increase fragmentation.

The burden and spacing was 25D * 102 mm hole diameter (2.5 m burden and spacing) in the past. However, that was exceedingly small and too conservative considering the rock mass properties. In 2023, vibration monitoring was used for some transverse stope blasts and the results for the 1345T000PQD third blast showed that 10 per cent of the drill holes never went off. That was likely caused by the drill holes being too close together and destroying each other before they could detonate. For this reason, 27D * 102 mm hole diameter (2.7 m burden) has been recently used. Some of the rules of thumb for transverse blast charging are listed below:

- Stem all drill holes with 1.5 m of stemming using 19 mm crushed screened bagged material, except the HW holes and shoulder holes.

- For the charge column greater than 10 m, place an airbag on top of the emulsion before pouring the stemming down the hole to avoid the stemming getting buried in the emulsion.

- Charge collar lengths to consider lithology (location of the Unit 4F contact for the PQD stopes).

- Stagger all drill holes (for all rings other than the raise ring and the two rings around the raise).

- Shoulder holes to have a long charge collar and short loading column.

Void ratio used was complex and inadequate since it was overestimating the void required by assuming that the void expands and negates the inactive void resulting in choked firings and bridges. The void ratio calculation has been adjusted with a more accurate method which utilises 30 per cent active void (regardless of the stope dimensions) with consideration of swollen ground and landed material based on timing of the blast. The revised calculation takes into account the void available at the undercut drive and the raise, the blast volume, the swollen volume of the raise, and the active void volume taking into account the rill angle of broken material. Using the revised calculation and a void ratio of 30 per cent (instead of previous 50 per cent) have decreased the duration between blasting and mucking.

To ensure the accuracy of drilling, drill holes are surveyed by a Carlson Boretrak2 tool and gyro-based data is captured for three to four rings (all accessible holes in each ring). The mine planners analyse the data, and if the hole deployment data meets the tolerance, no further Boretrak data is required for the upcoming blast. If the hole deployment data does not meet the tolerance, more Boretrak data may be required, and the rings are evaluated case by case to verify re-drilling requirements.

The raise pattern at the MWM is called a shielded cut raise, as the cut blastholes (the first four holes) are hidden in the reamers to shield them from each other's shock energy whilst avoiding radial cracking. This cut is designed to breakup to 25 m in a single blast. The MWM has stopped pulling the raises in 3 m intervals (Özen, 2022) and instead commenced taking 8–11 m long shots depending on the raise length. The spiral blasting sequence in the raise has provided good fragmentation. The same standard timing on all raises has been used irrespective of the length of raise. In the first four holes of the raise, the bottom primer is fired 10 ms prior to the top primer to crater out the bottom of the raise whilst eliminating too much material being flushed out of the reamers at the same time. All other holes in the raise have primers on the same delay to utilise high energy which yielded better fragmentation. The drill holes in a raise are remarkably close together, evidencing that staggering of the primers has been important. The first four primers must climb up the holes to avoid being destroyed by each other as the blasting front goes up the raise.

Stemming was used for all the holes except the reamers. That situation has led the reamers to become a large vent for the gas energy, causing heaving and cratering at the top level. To minimise rehabilitation requirements after the blasts, stemming has been added at the top 2 m of all reamers to confine the gas energy, which has yielded better fragmentation and less cratering and riffling at the backs. Some of the rules of thumb for transverse downhole raise timing are listed below:

- Spiral sequence of blasting with 400–300–200–100 ms timing.

- Offset primer delays on the first four holes of the raise.

- Stem all holes in a raise pull using 2 m of stemming.

- Stem reamer holes at the top 2 m to confine gas energy and avoid heaving/cratering.

- Staggered primers on the first four holes (climb up the raise 1–2-3–4 m from toe).

Recovery and dilution

The short-term stope designs have a calculated dilution figure using the expected ELOS from the back and HW provided by Rock Mechanics. The mine planners measure the surface area of the HW to estimate the volume of dilution expected from the HW and for the sake of simplicity, the back span of the stope is measured by drawing a flat shape over the footprint in the plan view (35/38 degree shoulders are negligible in this method). The dilution volume relative to the total blasted volume of the stope is the percent dilution and is then applied to all the blasts in the sequence. The target recovery of a short-term design stope is 95 per cent, and different dilution figures are used for primary versus secondary stopes. When the short-term design is not available, the long-term stopes (reserve shapes) and updated dilution estimates are used in the schedule. The estimated dilution for primary stopes is 11 per cent and for secondary stopes is 17 per cent. Both dilution and recovery assumptions are re-assessed every six months as more data points are collected and reviewed for the drill and blast results (Table 1).

TABLE 1

PQD stope recovery and dilution figures in 2023 (planned versus actuals).

Transverse stoping type	Planned dilution (%)	Actual dilution (%)	Planned recovery (%)	Actual recovery (%)
Primary	11	17	94	95
Secondary	17	8	94	89

The target recovery of long-term design follows the assumptions outlined in the cut-off grade instructions and these long-term shapes incorporate the dropped shoulders in the MRMR estimates which ensures the minimal polygon deviation between the long-term and short-term design assumptions (better scheduling compliance). Regarding the dilution and recovery of the long-term solids as part of the MRMR process, the estimated dilution for primary stopes is 11 per cent and for the secondary stopes 17 per cent with the target recovery of 93 per cent for both primary and secondary stopes.

Buffer rings and pre-conditioning rings

Access and infrastructure at the PQD are located on the HW side of the orebody. Historically, there has been up to a maximum of 5 m overbreak on the HW with an average HW ELOS of 1.0–1.5 m. Although there have been significant improvements on the cable bolting program, including plating and tensioning all cables and dumping them against the foliation to prevent rock mass failure and unravelling, implementing buffer rings in the HW has shown remarkable success since its inception in 2023.

The buffer rings are consisting of a ring of production holes drilled as the same drill pattern of regular transverse stope blastholes and offset by 0.5 m from the leading edge of the blast being taken towards the HW. The buffer rings are not loaded or stemmed when the final blast is fired. The purpose of the buffer ring is to reduce the blast damage to the HW by venting gas energy from the last ring fired and to provide a plane of weakness or perforation to dissipate the radial cracking by eliminating propagation of the blast energy into the HW (Nichols and Sinaga, 2023). The buffer rings have been successfully implemented with minimal to no damage on the HW of transverse stopes at the PQD and there was less oversize at the drawpoint contributing to higher production rates.

Besides the implementation of the buffer rings, the MWM has conducted trials of pre-conditioning rings which are densely drilled holes in the FW of the transverse stopes to create a stress shadow prior to production drilling. In order to minimise the stress induced ground failures in the secondary stopes, drilling along the FW with a series of sacrificial holes (30–60 cm apart) creates an open crack through the drill holes which will shed the stress away from the secondary stope and assist the stability of the stopes. As opposed to the buffer rings, the pre-conditioning rings are loaded and blasted as the first step in stoping cycle and then the rest of production drilling and stope blasts commence. The goal is to fragment the rock mass in the FW creating a stress curtain which prevents the horizontal stress flow into the core of the stope and to reduce the potential hole squeezing of production holes. Due to the high-stress ground conditions, there had been up to 15 per cent of hole cleaning + redrills (approximately 950 m per stope) at the PQD.

The trials of pre-conditioning and buffer rings at the 1370T090 stope (a secondary transverse stope in high-stress area, Figure 13) were successful. The second blast showed the half barrels in the intermediary HW indicating that the blasting did not damage the rock mass to the point of marking the holes in the rock which had never been seen at the Musselwhite Mine before (Figure 14).

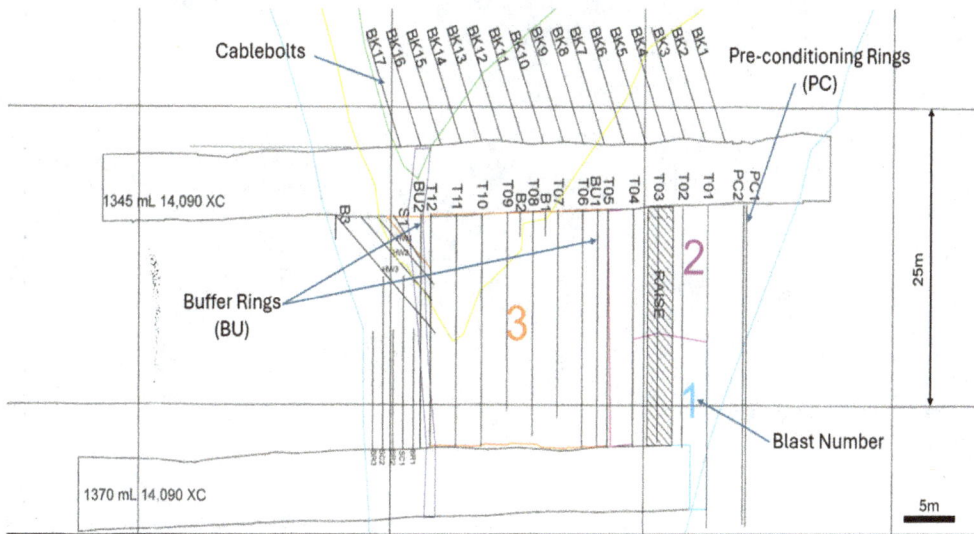

FIG 13 – The 1370T090PQD stope design with the buffer rings and pre-conditioning rings (long section view looking north).

FIG 14 – Isometric view of the CMS results for the buffer rings at the 1370T090PQD with the half barrels.

Trials of pre-conditioning rings were also conducted at the 1295T910 (a secondary transverse stope) and the 1345T090 (a secondary transverse stope, Figure 15) which had single-string of 40 mm Powersplit packaged explosives and 3 m of stemming in every hole. The results of pre-conditioning blasts were successful with no signs of hole fracturing. The stemming height and dropped explosive column height has reduced the blast damage at the backs and cratering around the collars was minimal compared to the previous pre-conditioning rings at the 1370T090 which had emulsion and an uncharged second pre-conditioning ring. In the hole#16 of the south pre-conditioning ring at the stope shoulder, there was no evidence of fracturing until 6.6 m depth where it started showing some minor cracks before the hole was clogged (Figure 16). It was unusual that no stemming was left in the hole, whereas some of the other holes with intact collar pipe had stemming still packed in the collars. Some uncertainty was remaining on how well the packaged explosives fractured the ground in this stope configuration. Overall, the 1295T910 and the 1345T090 stopes had successful results of the pre-conditioning and buffer rings to mitigate the stress-related issues.

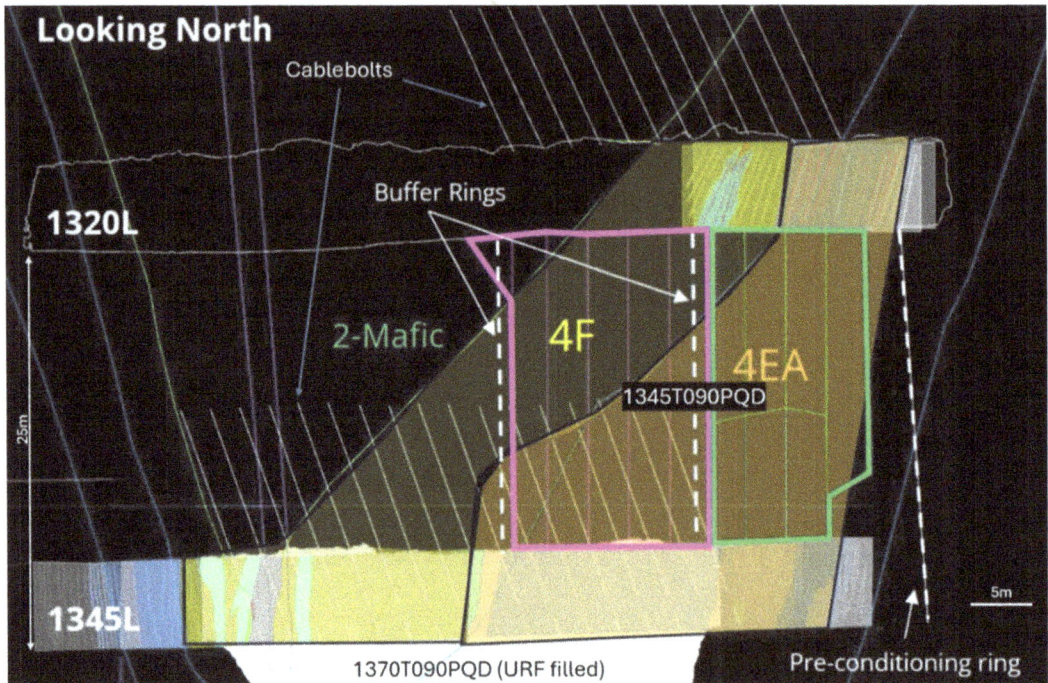

FIG 15 – Long section view of the 1345T090PQD stope showing buffer rings, pre-conditioning ring and different lithological units (4F, 4EA, 2-Mafic).

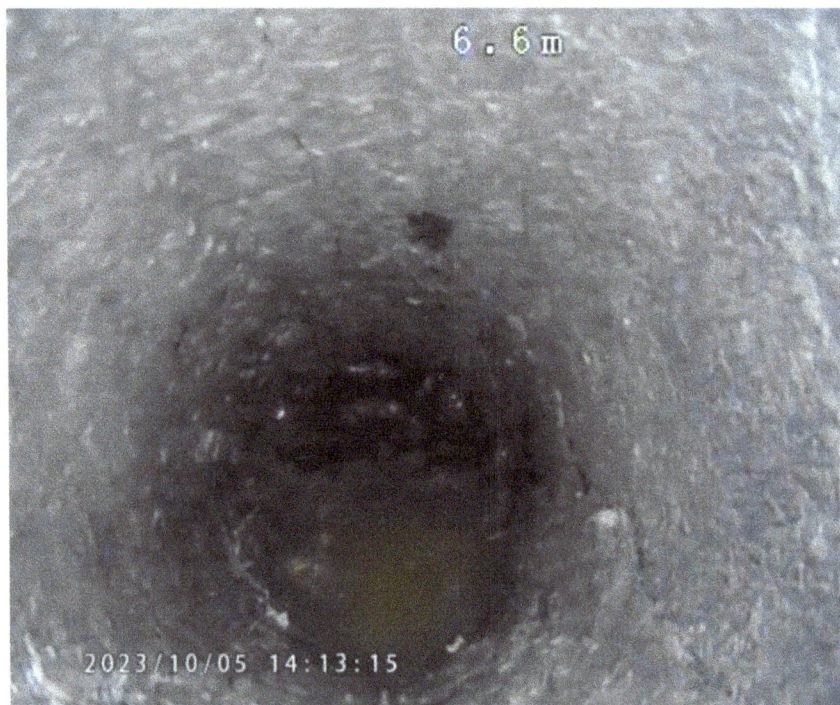

FIG 16 – Photo of the pre-conditioning ring hole#16 at the 1295T910 stope with no signs of fracturing and some minor cracking below 6.6 m.

Although the buffer rings and pre-conditioning rings have increased the drilling and blasting costs, it was negligible as a result of minimised requirements for hole cleaning and re-drills (less than 5 per cent), stable interim HW providing safer work platforms to underground operators and equipment and higher production rates (20–30 per cent) from the stopes that offset the time spent for these additional tasks.

Double-lift stoping

The first double-lift transverse stope, the 1395T111PQD, is situated nearly 1.4 km below surface and was produced in late 2022. This stope was originally designed as two individual stopes with high exposure to the HW undercutting but later the double-lift option was proposed to have adequate ground control measures whilst utilising the maximum void available to decrease number of blasts.

A number of field risk assessments were held during the planning phase to ensure controls are in place to mitigate any potential HW and back failures. Some of the mitigation measures included increased cablebolt density, surveying all production holes, implementing smooth wall blasting, reducing drill hole deviation, and installing shotcrete arches.

For development layouts, drift dimensions of 5.0 m high × 5.0 m wide was used at 1345 RL 14 111 XC which hosts the drilling and blasting at the top level. At 1370 RL 14 111 XC, the mid-level of the double-lift stope, a drift with dimensions of 5.0 m high × 5.0 m wide was developed. At 1395 RL 14 111 XC, at the bottom sublevel access drift was mined at 5.5 m high × 6.0 m wide to serve as the drawpoint and T-drifts with dimensions of 5.0 m high × 5.0 m wide for slots were mined. The T-drifts were developed to help increase the initial void ratio as discussed in the earlier section of the paper.

Drilling of the 1395T111 double-lift stope required thorough controls over which mining activities were going to take place at each sublevel. A slot raise of 6.0 m × 6.0 m was drilled at 1345 RL and reamed at 203 mm diameter with Boart Stopemaster. The remainder of 111 mm diameter blastholes were drilled with a Sandvik DL432i. The burden of 2.5 m and spacing of 2.5–3.1 m were used. At 1370T111 (mid-level), the same slot raise was used and production holes at a diameter of 102 mm was completed by the Boart Stopemaster. The burden of 2.5 m and spacing of 2.5–3.1 m was used. To ensure the QA/QC of drilling, all drill holes were gyro surveyed utilising the Carlson Boretrak2 survey tool, and the results were within 10 per cent tolerance for all blast panels. During the blasting of the toe shots, approximately 2 t of Orica Subtek V explosives with a powder factor (PF) of 0.6 kg/t and void ratio of 47 per cent was blasted to get planned tonnes and grade (T&G) of 3.7 kt at 7.1 g/t Au. At the second cap shot, nearly 13 t of explosives with a 0.7 kg/t PF at void ratio of 33 per cent was blasted to mine a planned T&G of 20 kt at 6.5 g/t Au. The entire mass third blast had a planned T&G of 38 kt at 7.7 g/t Au via blasting 25 t of explosives with a PF of 0.7 kg/t and void ratio of 43 per cent.

With successful implementation of the first double-lift primary stope (Figure 17), some benefits realised were having flexibility in operational schedule such as alternating drilling and cabling activities in different sublevels, ensuring uninterrupted production cycle by utilising higher void ratios during charging and blasting, increasing daily production rates up to 2000 t/d in comparison to average of stoping rate of 650–750 t/d, deferring the CRF backfill activities of the first-lift whilst bringing ounces forward. Learnings from the first double-lift stope were employed for other stopes and more double-lift stopes were included in the production forecasts to achieve higher ounce profile.

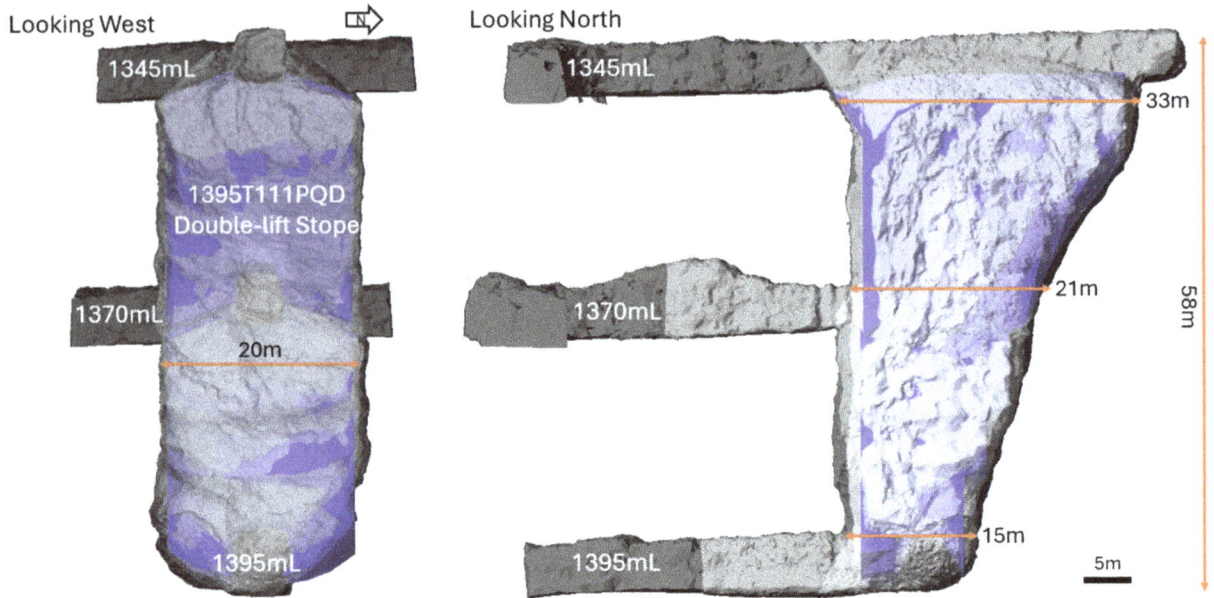

FIG 17 – The 1395T111 double-lift stope was successfully mined (design in blue versus the final CMS in grey).

CONCLUSIONS

The Musselwhite Mine has made significant improvements on transverse longhole stoping practices at the PQD orebody in the past two years. With adoption of optimised drilling and blasting practices, incorporating geotechnical and geological input into the design parameters and establishing adequate ground control measures, the result of stoping improvements is evident that the recent transverse longhole stopes at the PQD have yielded a 5–7 per cent dilution on a target planned dilution of 17 per cent and a 90–94 per cent recovery for both primary and secondary stopes.

It is important to note that the recovery of transverse stopes at the PQD have been increased from low 70 per cent in 2021 to high 90 per cent in 2023 with the highest being 99 per cent for the first double-lift primary stope and the dilution figures went down from high 30 per cent to as low as 5 per cent as shown in Table 2.

TABLE 2

PQD Stope Recovery and Dilution after Optimisation.

Stope	Dilution (%)	Recovery (%)	Design comments
1320T954PQD (secondary)	5%	90%	Buffer rings implemented, no preconditioning, square corners difficult to break and muck on first lift, new priming and timing standards used, finely broken muck.
1370T090PQD (secondary)	6%	93%	Buffer rings implemented, preconditioning implemented, new timing and priming standards used, finely broken muck.
1345T000PQD (secondary)	7%	94%	Buffer rings implemented, no preconditioning, old timing, and priming standard.
1395T201PQD and 1420T201PQD (double-lift primary)	14%	99%	Buffer rings only implemented on top lift, shotcrete arches not installed in intermediary level, old timing and priming standard used.
1345T111PQD (primary)	19%	95%	No buffer rings used, old timing and priming standard used. This primary stope is mined after 1395–1370T111 double-lift primary stope is filled.

Application of the buffer rings to minimise the HW damage and introduction of pre-conditioning rings on the FW to reduce the hole squeezing and re-drilling requirements have resulted in considerable progress on the daily production rates from the stopes within the high-stress ground conditions. Although the buffer rings and pre-conditioning rings have increased drilling and blasting costs, it was negligible compared to the reduced amount of hole cleaning and re-drilling while providing stable interim HW for safer work platforms to underground operators and equipment. Along with the optimised drilling and blasting activities, the implementation of the buffer rings and pre-conditioning rings helped achieve up to 30 per cent higher production rates with minimal oversize material at the drawpoints which offset the time spent on additional drilling and blasting tasks.

Extracting and backfilling the primary stopes as quickly as possible assisted mitigating failure of the HW and FW thus minimising dilution in the secondary stopes. Besides these, the double-lift stoping has been successfully employed and resulted in the highest daily stoping rates over 2000 t/d.

As the next steps, the Musselwhite Mine will be evaluating the results of the pre-conditioning rings for the secondary stopes and will continue optimising fragmentation by revising the burden and spacing based on the data collected from vibration monitoring and particle size distribution.

ACKNOWLEDGEMENTS

The author would like to acknowledge Newmont and Musselwhite Management for granting permission to publish learnings from transverse longhole stoping practices at the Musselwhite Mine and wishes to recognise the collaborative efforts by Technical Services and Operations teams for implementing improvement ideas in a safe and timely manner. Additionally, the author would like to extend thanks to his family and friends for their continued support whilst working at remote mines.

REFERENCES

Hutchinson, J and Diederichs, M, 1996. *Cablebolting in Underground Mines*, BiTech Publishers Ltd.

Mkadmi, N, 2022. Newmont Musselwhite Mine, Canada, 1320T887 and 1345T977 Stopes Back Analysis [unpublished report].

Nichols, L and Sinaga, F, 2023. Newmont Musselwhite Mine, Canada, MEMO – Buffer Ring Design Requirements [unpublished report].

Oswald, W, 2018. Geology of the banded iron formation-hosted Musselwhite gold deposit, Superior Province, Ontario, Canada/Géologie du gisement aurifère encaissé dans des formations de fer Musselwhite, Province du Supérieur, Ontario, Canada, Doctoral dissertation, Université du Québec, Institut National de la Recherche Scientifique, Centre Eau Terre Environnement.

Oswald, W, Malo, M, Castonguay, S, Dubé, B, Mercier-Langevin, P, McNicoll, V and Biczok, J, 2015. Geological Setting of the World-Class Musselwhite BIF-hosted Gold Deposit, Ontario, Canada, 13th SGA Biennial Meeting on Mineral Resources in a Sustainable World.

Özen, I, 2022. Newmont Musselwhite Mine, Canada, MEMO – Blasting Sequence of Ventilation Raise [unpublished report].

Özen, I, Earle, N, Small, R, Bzdok, M, Mauro, V, Thibodeau, D and Weirmeir, D, 2023. Newmont Musselwhite Mine, Canada, Qualified Persons Report [unpublished report].

Perry, A, Hashemi, A, Özen, İ Ö and Anongos, V, 2024. Optimizing stope design in a deeper mining zone at the Musselwhite Mine, *CIM Journal*, pp 1–12. https://doi.org/10.1080/19236026.2024.2409056

Technology and innovation

Establishing a comprehensive underground geotechnical convergence monitoring program using the latest SLAM technology

M Burns[1], K H B Chu[2], G Davis[3] and E Jones[4]

1. Superintendent Geotechnical Engineer, MMG, Rosebery Tas 7470.
   Email: michael.burns@mmg.com
2. Rock Mechanics Engineer, Engenex, Hobart Tas 7000. Email: bernard.chu@engenex.com
3. Senior Product Manager, Emesent, Brisbane Qld 4064. Email: greg.davis@emesent.io
4. Principal Rock Mechanics Engineer, Engenex, Sydney NSW 2065.
   Email: evan.jones@engenex.com

ABSTRACT

Convergence monitoring in underground mining is critical for enhancing safety, operational efficiency, and ground stability management. LiDAR-based monitoring, particularly using SLAM (Simultaneous Localization and Mapping) technology, offers real-time, high-resolution data to track rock mass deformation and identify stability risks. Traditional methods, such as extensometers and total stations, are limited by coverage, accuracy, and data density. The integration of Emesent's Hovermap SLAM hardware and their Aura Convergence Monitoring processing module enables rapid, repeatable convergence monitoring over large mine areas, improving early detection of convergence zones, facilitating timely intervention.

This paper presents the implementation of these technologies at the MMG's geotechnically challenging Rosebery mine in Tasmania, detailing the innovative alignment and processing workflows in Aura that reduce user subjectivity and optimise data processing. It outlines the TARP established to incorporate the new technology while attempting to align with processes already implemented on-site. Challenges with adopting the technology and software are discussed providing the readers with a firsthand account of the advantages and disadvantages of the workflow.

ROSEBERY MINE INTRODUCTION

The Rosebery Mine is an underground base metal operation located on the west coast of Tasmania, Australia. It primarily produces zinc, lead, and copper concentrates, along with gold and silver as a by-product. The mine has been in operation since 1936 and was acquired by Minerals and Metals Group Limited (MMG) in 2009. The active mine area is approximately 4 km along strike with the deepest mining conducted at depths of over 1800 metres below the surface, Figure 1.

The *in situ* rock mass conditions at Rosebery are generally consistent across the mine with good rock quality designation (RQD) values between 60 and 100 per cent in most lithological units.

The Rosebery rock mass is principally characterised by a prominent east dipping foliation and a dominant west dipping joint set. Chlorite/sericite alteration and the intensity of foliation is variable, typically increasing in the immediate vicinity of the ore lenses and along strike towards the north.

The major and minor principal stresses (Sigma 1 and 3) are oriented perpendicular to and parallel (down-dip) to the prominent foliation respectively. The intermediate principal stress (Sigma 2) is essentially horizontal and parallel to the strike of the ore lenses (and foliation).

The rock mass strengths are in the range of 130–160 MPa. The principal stress versus rock mass strengths at Rosebery imply challenging mining conditions at depth. Figure 2 shows the relationship of the dominant structures and stress orientations to a north–south, (ore strike parallel) drive. At an RL of 1500 m the principle, intermediate and minor stress magnitudes are approximately 90 MPa, 60 MPa and 40 MPa respectively.

Changes in the stress regime in strike parallel drives ahead of the stoping front presents as shearing along the dominant east dipping foliation. Convergence is most extensive in the hanging wall shoulder and lower footwall side walls of ore strike parallel drives. East–west drives, being more favourably oriented with regards to structure and principal stress direction, experience only nominal deformation.

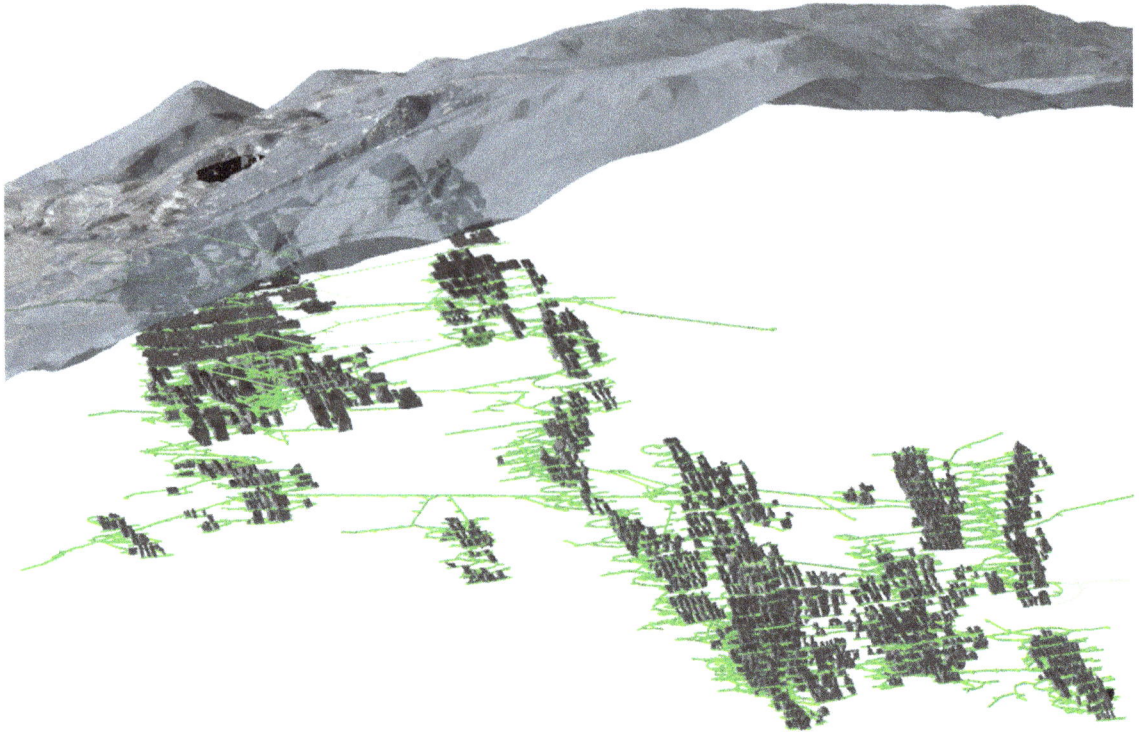

FIG 1 – Perspective view of the Rosebery Mine below Mount Black. The deepest mine levels are approximately 1800 m below surface.

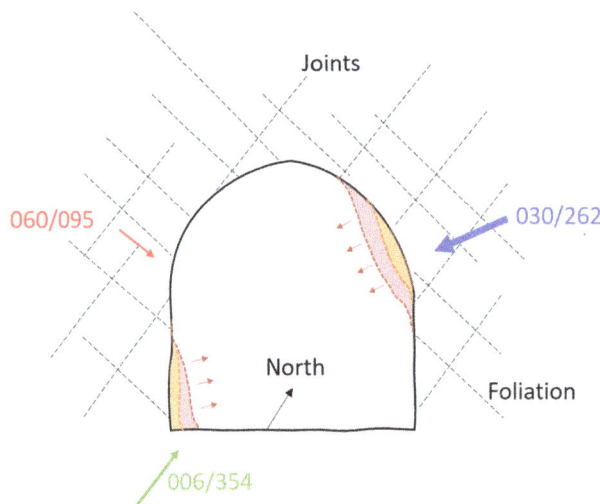

FIG 2 – Dominant structures and stress orientations related to a north–south development drive.

Over the course of extraction an empirical approach has been taken to optimise development layout and adjust the application of ground support to mitigate drive convergence issues.

Depending on depth, distance ahead of the stope front, age of excavation and the variability of foliation and alteration intensity, converge has been observed between 0.1 and >1.0 m.

MINING METHOD

In mining areas expected to experience high rates of convergence, a centre out mining method has been adopted. Access to the orebody is via parallel footwall drives and perpendicular cross-cuts.

A just in time (JIT) development strategy was implemented to reduce excavation vulnerability and exposure time by limiting the development ahead of the stoping fronts. A bottom-up mining method results in slow lateral advance as stopes are mined up in columns over multiple levels. Figure 3 shows a plan view for a production level at approximately 1500 m RL with the relative timing and direction of development and stoping.

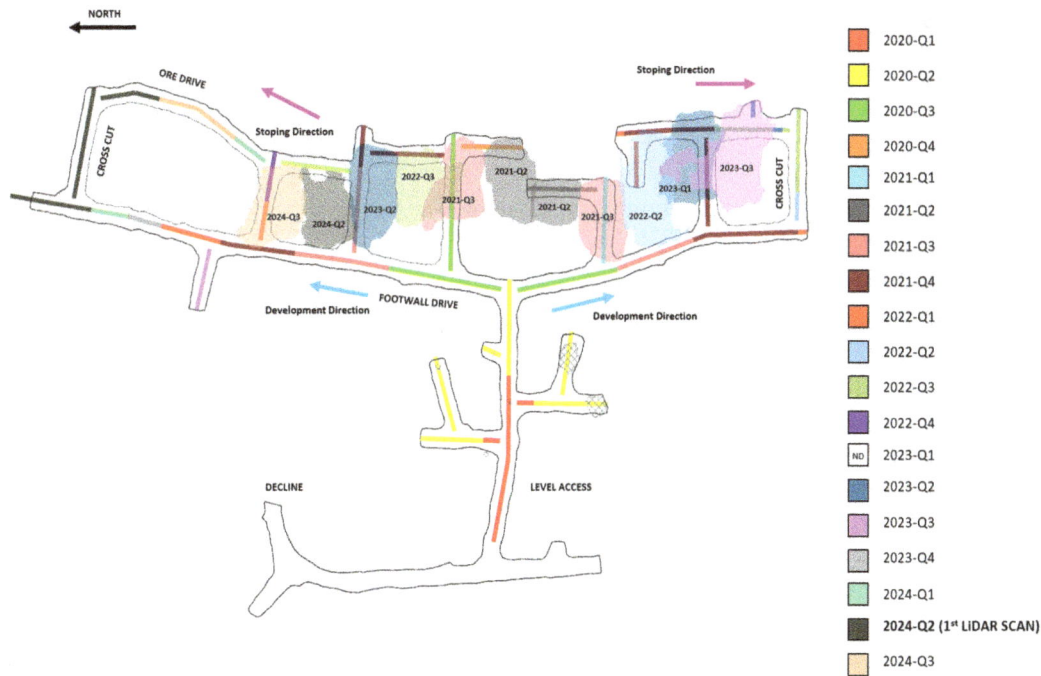

FIG 3 – Plan view of a production level at Rosebery mine set-up for a centre-out mining.

GROUND SUPPORT SYSTEMS

Different ground support systems and strategies have been used to manage the converging rock mass. Figure 4 illustrates the different support systems and where they were installed during development. All primary rock bolts in the following are 2.4 m Kinlocs on a 1.20 m ring × 1.00 m row pattern.

1 — Weld Mesh FL-FL, 2.4m Kinlocs 1.2m x 1.0m, 6.0m CBs 1.2m x 1.7m GL-GL, Fibrecrete FL to 2m.

2 — Full profile 75mm Fibrecrete, Weld Mesh GL-GL, 2.4m Kinlocs 1.2m x 1.0m, 6.0m CBs 1.2m x 1.7m GL-GL.

3 — Weld Mesh GL-GL, 2.4m Kinlocs 1.2m x 1.0m.

4 — Weld Mesh GL-GL, 2.4m Kinlocs 1.2m x 1.0m, 6.0m CBs 2.4m x 1.7m GL-GL, Fibrecrete FL to 2m.

5 — Weld Mesh GL-GL, 2.4m Kinlocs 1.2m x 1.0m, 6.0m CBs 2.4m x 1.7m GL-GL.

FIG 4 – Ground support system application.

Support system 1

Floor-to-floor welded wire mesh with a dense bolt pattern consisting of Kinlocs, plus 6.0 m plain strand cables in a 1.5 m ring × 1.7 m row pattern. A fibrecrete layer below grade line followed the primary support.

Support System 2

75 mm floor to floor fibrecrete with welded wire mesh grade line to grade line and the same dense bolt pattern as System 1.

Support System 3

Welded wire mesh with Kinlocs grade line to grade line.

Support System 4

Welded wire mesh grade line to grade line with moderate bolt density of Kinlocs and 6.0 m plain strand cables in a 2.4 m ring × 1.7 m row pattern. A fibrecrete layer below grade line followed the primary support.

Support System 5

Same as 'Support System 4', with cables installed as support upgrades with approaching stope front.

Support Systems 1 and 2 were found to be effective at the onset of convergence along the FWD due to the high bolt density and the confinement of the fibrecrete and its ability to distribute load to mesh and bolts (System 2).

The cross-cuts, supported by a lower capacity system (Support System 3) proved adequate due to the favourable orientation of the drives being perpendicular to the dominant foliation and parallel with the principal stress.

Support System 4 installed on the ore drives proved to be successful in enabling stope extraction without the need for significant rehabilitation or upgrades.

In the northern end of the level, Support System 5 was trialled to facilitate improved development rates. Support upgrades (addition of cables) were scheduled and installed as the stope front advanced. It was found that the onset of rock mass damage, surface support deformation and Kinloc bolt failure occurred much earlier than that observed for Systems 1 and 2 and the life span of the primary support was much reduced before rehabilitation was required.

Each of the above systems and strategies of ground support application are in line with accepted industry practices for the management of converging ground. As will be seen in the following sections a failure to implement a robust strategy for the monitoring and maintenance of the ground support systems resulted in unsatisfactory long-term performance in some areas of the production level.

HISTORICAL MONITORING AND REHABILITATION

Historically, ground convergence has been monitored principally through visual inspections and distometer measurements. While smart cables were trialled as a monitoring solution, their effectiveness was limited by inadequate area coverage and vulnerability to machine damage.

Consequently, the propagation of stress-related rock mass damage and convergence often went undetected due to the ad hoc nature of monitoring and damage mapping. This led to the need for extensive rehabilitation efforts when excessive loading compromised the capacity of ground support rendering it inadequate for sustaining safe active headings. In severe cases the movement of mining equipment was obstructed due to the extent of drive convergence. Where this occurred, the degree of deformation and lateral extent of damage was determined using a Continuous Measurement System (CMS) tool, and the results were compared to the initial development survey. Figure 5 illustrates a notable example of significant deformation in the sidewalls and backs of a footwall drive at approximately 1500 m RL.

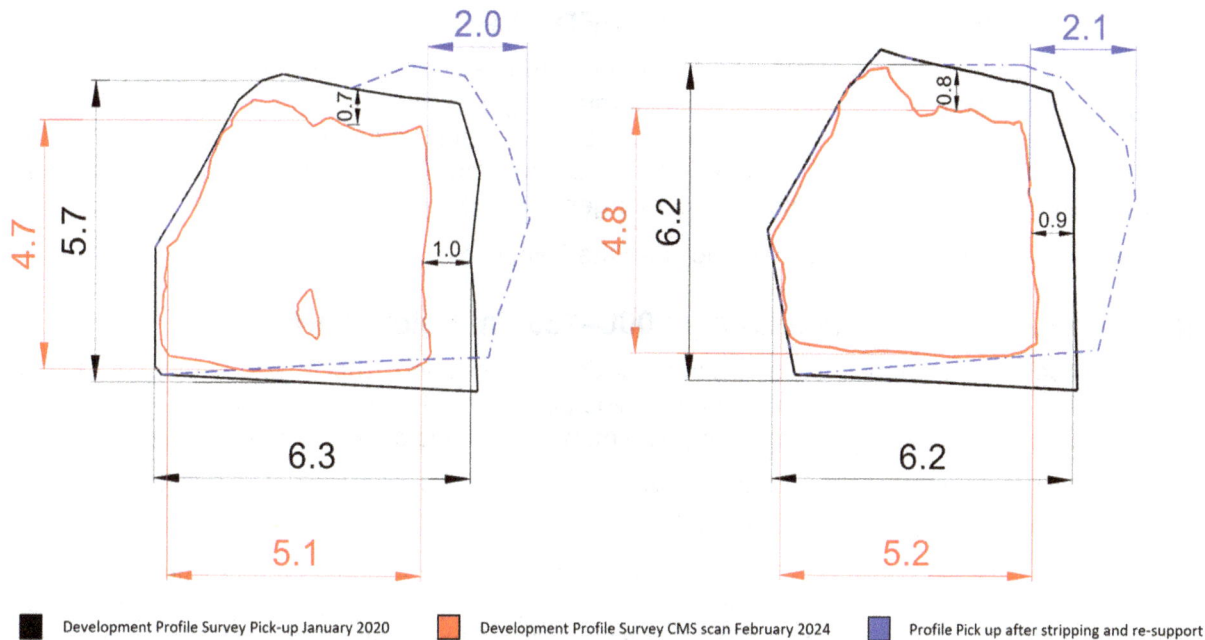

Development Profile Survey Pick-up January 2020   Development Profile Survey CMS scan February 2024   Profile Pick up after stripping and re-support

FIG 5 – Example of significant drive convergence in a footwall drive and the profile post stripping and rehabilitation.

Figure 6 shows the rehabilitation areas along the case study level. In addition to formal rehabilitation plans, numerous shift boss instructions were issued along the footwall drives to repair heavy vehicle damage where drives were becoming too narrow. Ineffective monitoring resulted in reactive, unscheduled rehab plans and support upgrades often resulting in delays to production. In severe case as seen in Figure 5, significant stripping and re-supporting pause production on the level for over a month. Such practices are not sustainable, a more proactive approach to convergence monitoring is needed.

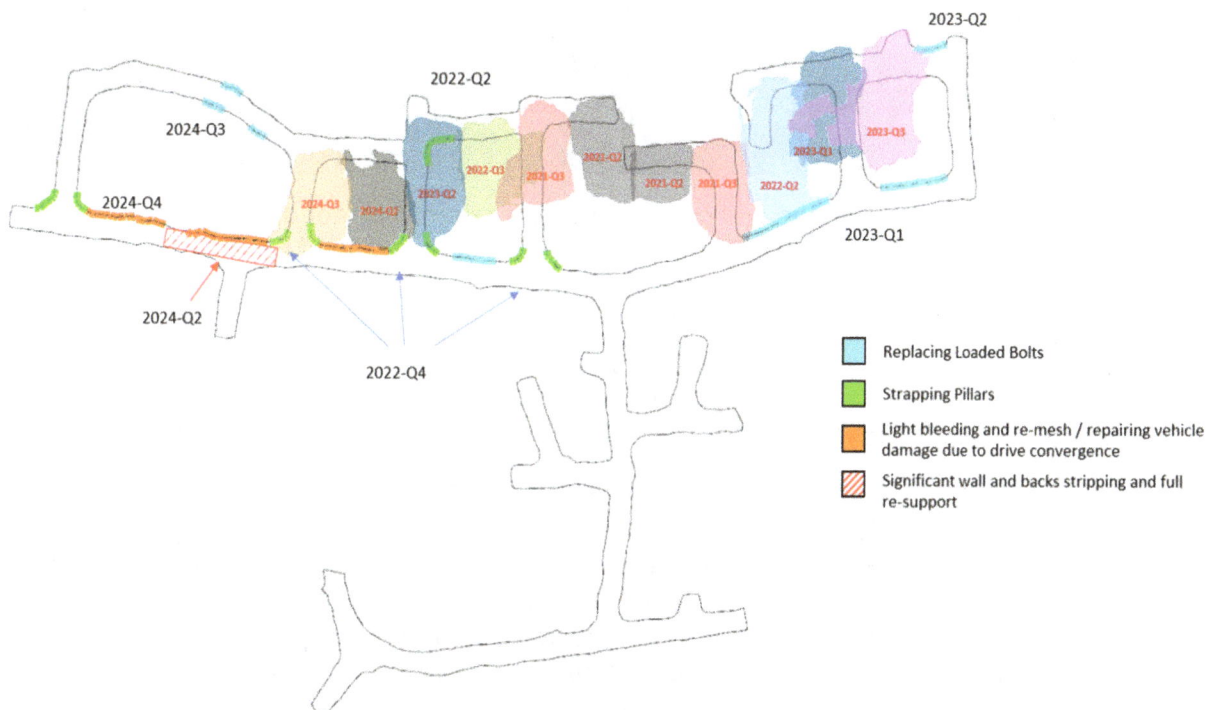

Replacing Loaded Bolts

Strapping Pillars

Light bleeding and re-mesh / repairing vehicle damage due to drive convergence

Significant wall and backs stripping and full re-support

FIG 6 – Rehabilitation Areas with date against the progression of stoping.

EMPIRICAL OBSERVATIONS DEVELOPED HYPOTHESIS

In early 2024, the Geotechnical Department implemented a proactive approach to improve convergence management. Due to a limited understanding of convergence rates and extents across the mine and at various stages of production, a damage mapping campaign was initiated to help focus the monitoring program and generate first pass hypothesis on why certain areas of the production levels experienced problematic convergence while others did not.

Key findings from the damage mapping included the following.

Intermediate levels, approximately 1000–1350 m below surface

- Minor rock mass damage observed in older drives (>2 years), characterised by minor loading of light ground support, and yielding of Kinloc bolts in the eastern shoulders of strike-parallel drives. Convergence of approximately 100 m, with no signs of failure of surface support.

- New strike-parallel developments experienced convergence within weeks of excavation, resulting in yielding of some Kinloc bolts.

These observations suggest a rapid period of stress relaxation immediately after excavation, followed by minimal further loading during the subsequent stope production period. No conclusive evidence from the damage mapping was found to suggest variability in the rock mass behaviour associated with geological structures or position relative to the stoping front. Figure 7 shows a typical examples of deformation in strike parallel drives at intermediate depths.

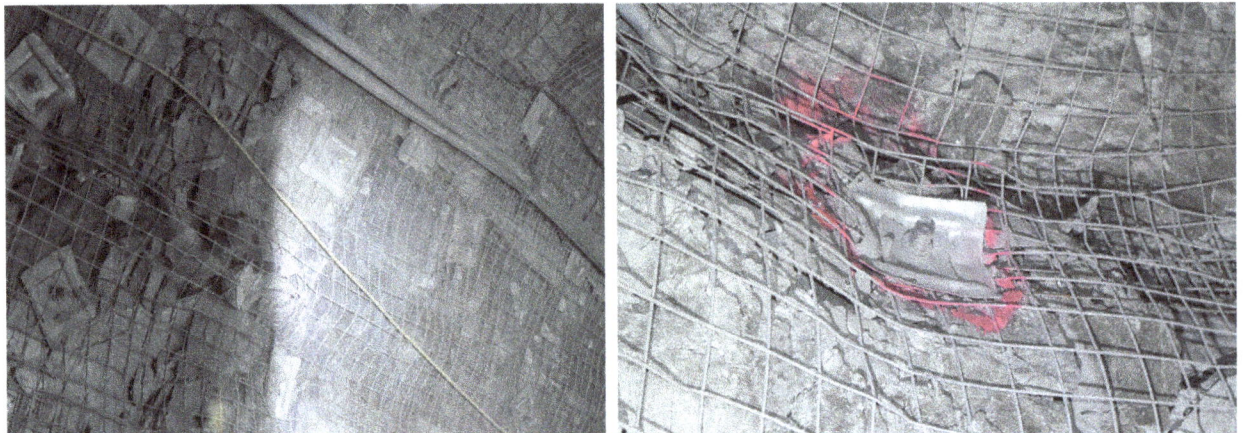

FIG 7 – Light rock mass damage and convergence in eastern shoulder of intermediate depth levels. Light loading of bolt plates with occasion yielding of Kinlocs.

Deep levels >1350 m below surface

- Light loading of ground support was observed in the FWD of deep levels in the oldest development sections, suggesting that the centre-out mining method causes a stress-shadowing effect on the FWD behind the advancing fronts once they move north and south.

- Minimal rock mass damage was noted in east–west cross-cuts.

- Increased rock mass damage and loss of the drive profile observed approximately 50–100 m north and south of the access. Historical support upgrades and rehabilitation had been required. The rehabilitated ground support showed little signs of loading. This supported the above observation that loading occurs as stope fronts approach, once past, the FWD becomes stress shadowed and does not experience further loading.

- Significant loading of ground support and notable profile convergence was evident 0–40 m ahead of the stope front in the northern FWD in the primary development ground support.

- Light deformation and loading of primary ground support was observed 40–60 m ahead of the stope front on FWD, with no evidence of rock mass damage or support loading beyond 60 m.

- Light loading of rock mass and ground support 0–30 m ahead of stope front on OD with little to no damage beyond 30 m.

- The highest areas of convergence over long strikes were associated with regions of intense foliation, typically found further north.

- Points of high convergence and heavy ground support upgrades coincided with fault intersections and the locations of historical cement rock fill (CRF) mixing bowls.

- Areas of high convergence typically initiated at intersections where fillets were not strapped and between closely spaced east–west cross-cuts or where offset three-way intersections existed.

- Support systems without primary fibrecrete application before meshing showed significant rock mass damage, loading of bolt plates and deformation of the surface support.

Figure 8 shows examples of deformation in strike parallel drives for deep levels.

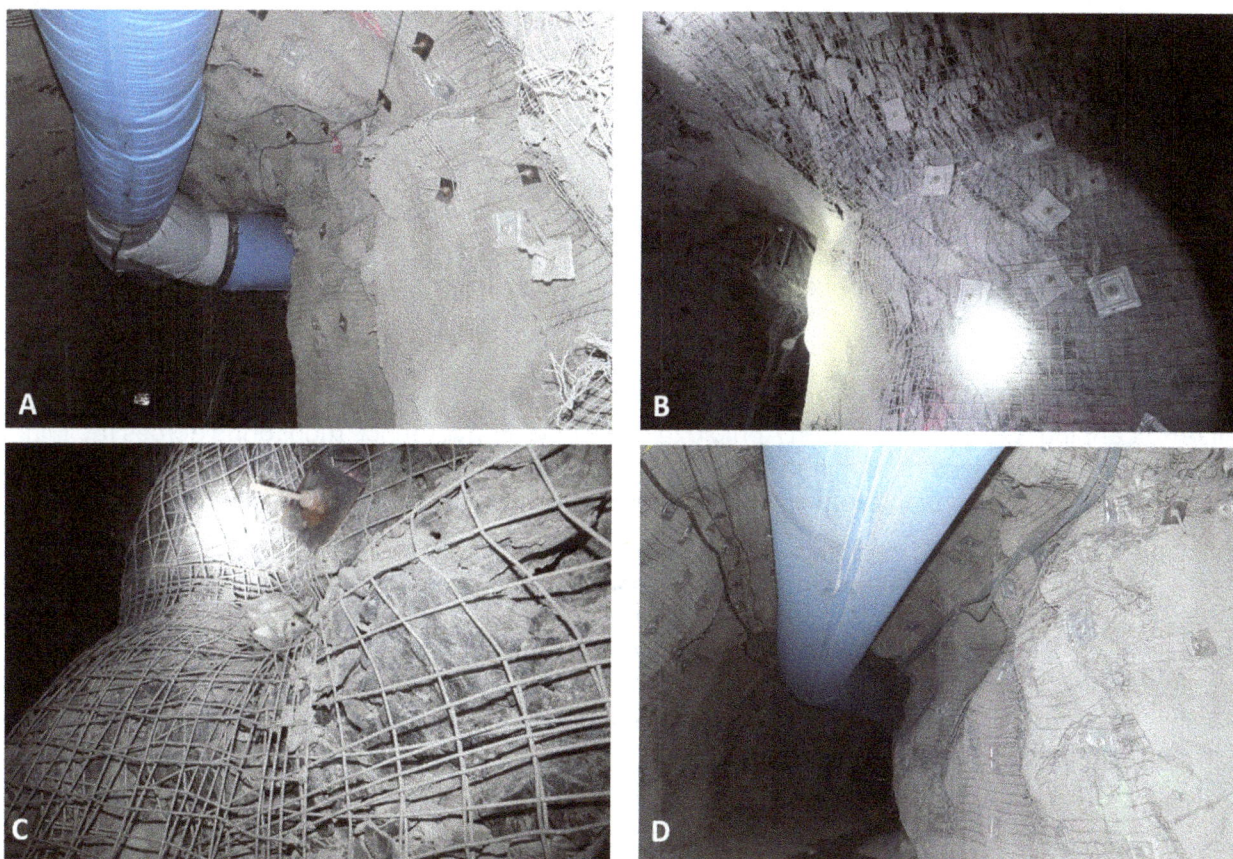

FIG 8 – (a) Convergence of primary support in eastern wall of FWD, Osro strap support upgrade with cables. (b) Intense foliation showing shear movement in eastern shoulder of FWD (viewed from inside east–west cross-cut), rehabilitation ongoing in the FWD. (c) Extensive rock mass damage and loading of primary bolts and mesh in support system with no first pass fibrecrete. (d) Rehabilitated FWD after stripping and supporting right wall. Note good performance of the left wall compared to right, related to orientation of foliation.

Figure 9 shows an overlay of geological faults, rock mass foliation intensity and design features on the case study level. The relationship between these features and the relative severity of rehabilitation provides an indication of how to best set-up the targeted LiDAR monitoring and ground support management program.

CRF mixing bowls had not previously been identified as a causal factor of FWD instability. Amending the level design process to locate mixing bowls in old cross-cuts was a easy and simple solution to improve the expected performance of the drives.

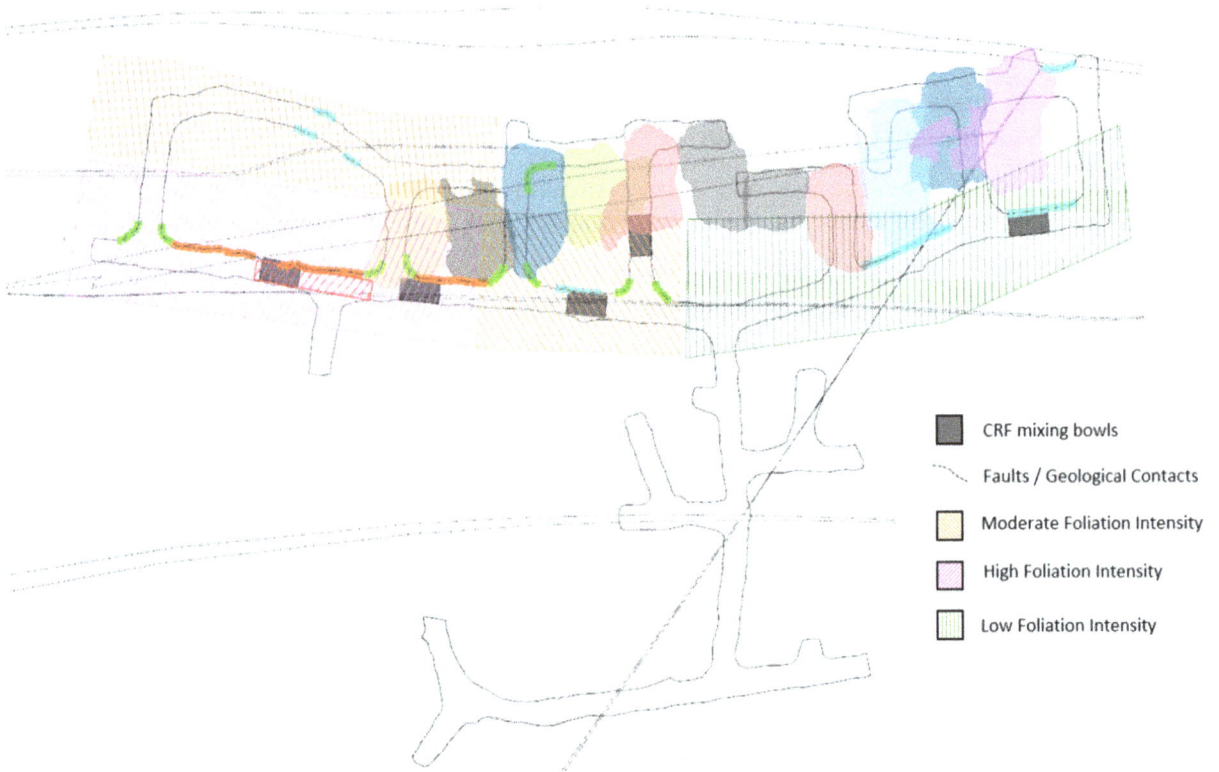

FIG 9 – Overlay of rock mass observations, geological structures and design features relative to rehabilitation.

IMPLEMENTATION OF LiDAR MONITORING

LiDAR scanning was implemented in April 2024 to improve convergence monitoring across the mine. Scanning frequency was determined based on historical deformation patterns, with monthly scans conducted for deep production levels and quarterly scans for intermediate levels.

Initial baseline scans were captured for new developments to establish reference data, which would serve as a benchmark for subsequent measurements. It is expected that these scans will enable the assessment of rate and cumulative convergence and provide a means to estimate the residual capacity of ground support and when to implement maintenance, upgrades, or rehabilitation. Additionally, it is hoped that the regular monitoring will provide good data to develop a deformation versus time graph for Rosebery at different depths, as presented by Hadjigeorgiou and Potvin (2023), Figure 10.

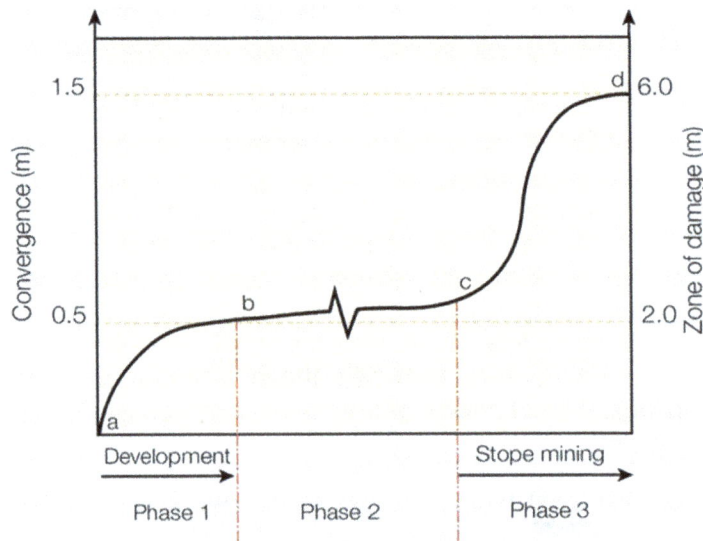

FIG 10 – Phases of deformation after Hadjigeorgiou and Potvin (2023).

EARLY RESULTS

To date, 73 LiDAR scans have been conducted across 19 levels of interest. In areas where convergence had already occurred and support upgrades were implemented, the monitoring program has provided valuable data on whether convergence ceases or continues at a slow rate after the stoping fronts pass.

In newly developed areas, the monitoring has offered insights into deformation rates and ground support behaviour. For example, at measured deformations of 100–150 mm, the internal bars of Kinloc bolts were observed to yield.

Monitoring of open development on production levels ahead of the stope front has provided highly valuable data that would be difficult to obtain through other means. Figures 11 and 12 illustrate the relative change over time in a section of FWD ahead of the stope front on a production level located 1560 m below surface. These convergence versus time plots assist in the understanding of when to plan for support upgrades to minimise disruption to production activities.

Repeated monitoring is providing the site team with real-time data on the rates and extent of convergence, facilitating the re-evaluation of mining and ground support strategies at depth at Rosebery.

The ease of visualising convergence using LiDAR scanning has proven instrumental in engaging the Mine Management and Operations teams, underscoring the importance of allocating resources for the periodic upgrading and maintenance of ground support. To further formalise the actions, a Trigger Action Response Plan was developed and is presented in the next section.

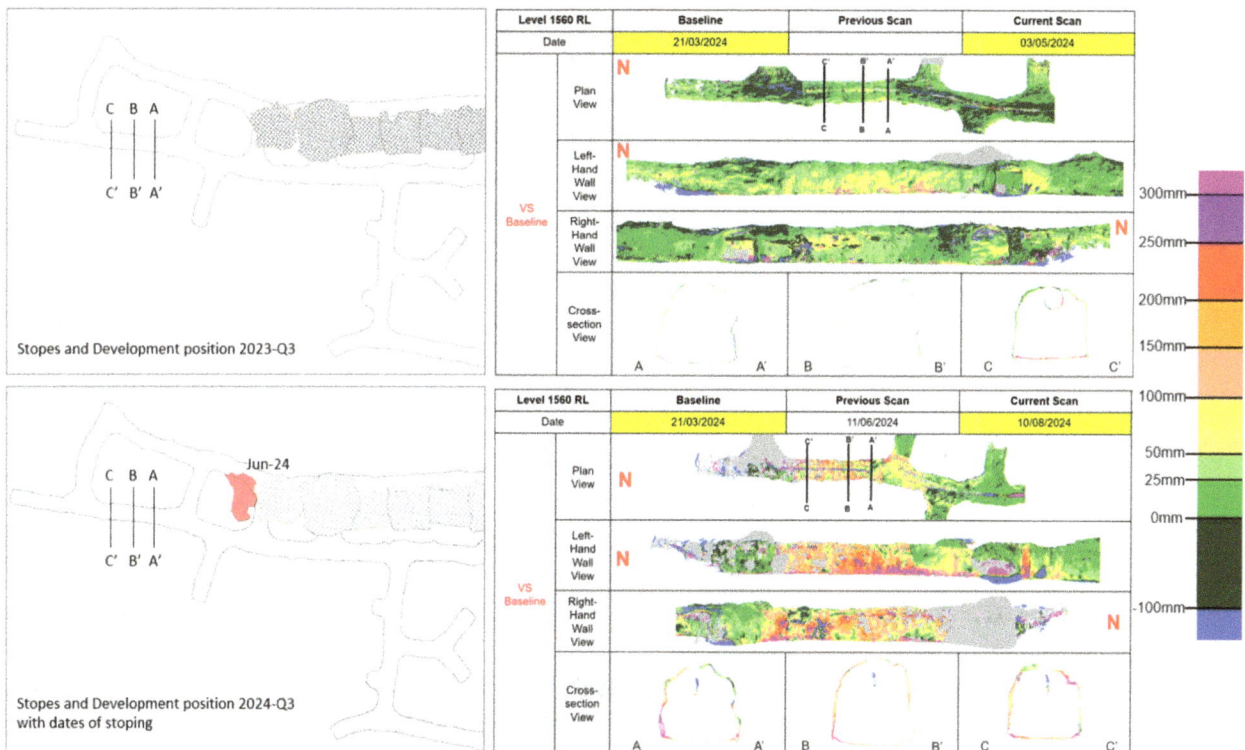

FIG 11 – Measured deformation during a period between stoping (top). Measured deformation after stoping compared to baseline (bottom).

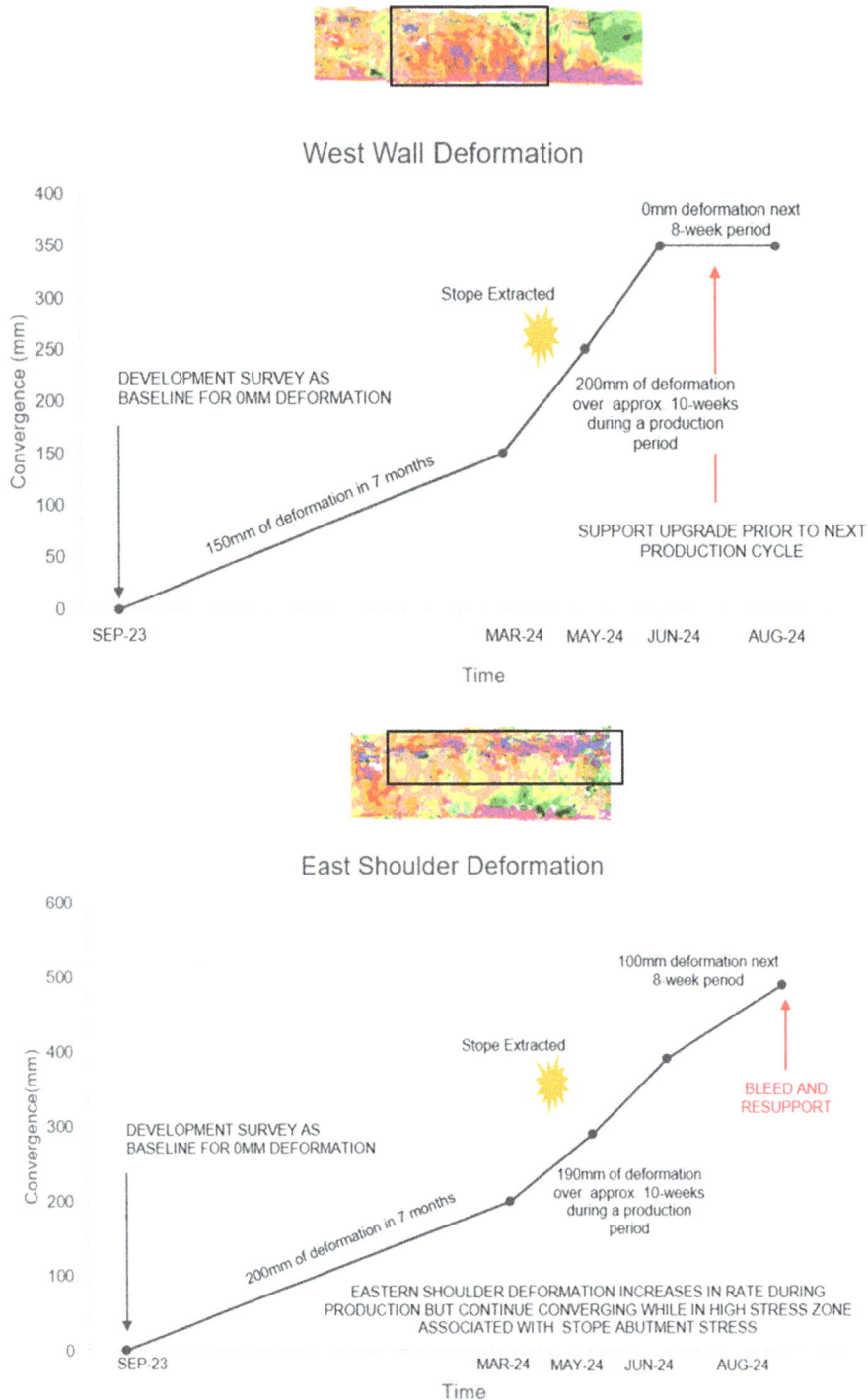

FIG 12 – Cloud Compare M3C2 analysis of areas with high rates of deformation allows time versus deformation graphs to be generated.

PROACTIVE SUPPORT MAINTENANCE FROM TRIGGER ACTION RESPONSE

MMG Geotechnical engineers developed a Trigger Action Response Plan (TARP) to systematically apply an appropriate action for key convergence thresholds. The objective was to ensure all personnel (engineers, supervisors and operators) were clear and in agreeance on the requirements at each stage. The proposed TARP (Table 1) was developed based on anticipated deformation values and the expected ground support performance from empirical observations made in the damage mapping exercise for each deformation level.

TABLE 1

The initial ground deformation TARP being applied.

TARP level deformation	Rock mass and ground support observations	Geotechnical engineer action and response
Level 1 Minor Squeezing <100 mm	**Rock Mass** Shallow depth of damage, minor foliation dilation, no slabbing **Bolts/Plates** Minor loading occasionally observed **Surface Support** Cracking and minor bulking in FRS Slight mesh deformation	Increase scanning frequency to ~monthly
Level 2 Moderate squeezing 100–200 mm	**Rock Mass** Increasing depth of damage, localised slabbing below grade line **Bolts/Plates** Rebar being loaded and stretched, Point anchored bolts show initial slippage. Kinloc bolts starting to fail, plates visibly deforming. Multiple split set plates yielding 3–4 primary bolts failed within 10 m² (~ size of mesh sheet) **Surface Support** Existing FRS cracks widening and extending. Mesh loading but not yielding	Deformation trend reported to senior Geotech. Short-term/Low Exposure excavations Increase visual inspections. Scan frequency increased to every 2 weeks. Long-term/High Exposure excavations Spot bolt as required to re-support to original ground support standard. Consider infilling bolts to increase bolt density. Consider scheduling cable bolts, if not already installed. Spray FRS below grade line (If not already there).
Level 3 Major Squeezing Limit State reached. 200–300 mm	**Rock Mass** Moderate depth of damage, bulking to shoulders/walls **Bolts/Plates** Occasional rebar failing, multiple primary bolt plates ejected. Point anchored bolts failing and pulling out Cable bolt tails unravelling, plates visibly loading 4–6 failed primary bolts, or 1–2 cables, within failed within 10 m² (~ size of mesh sheet) Mesh overlap compromised (1 failed bolt on overlap) **Surface Support** Shotcrete slabs detaching behind mesh Isolated mesh strands snapping	Increase scanning frequency to weekly, monitor for damage spreading. Short-term/Low Exposure excavations Plan and schedule rehabilitation in areas exceeding 250+mm. Replace failed bolts. Increase support capacity with cable bolts Long-term/High Exposure excavations Barricade and limit access to drive Plan and schedule rehabilitation in areas exceeding 250+mm deformation Areas within 50 m of stope to be completed prior to firing. Short-term access may involve increase total bolt density only.
Level 4 Extreme squeezing Serviceability limit exceeded 300–400 mm	**Rock Mass** Widespread, excessive damage. Heavy bulking in shoulders/walls **Bolts/Plates** 6+ failed primary bolts, or 2+ cablebolts, within 10 m² (~ size of mesh sheet) All point load bolt capacity lost, or assumed lost. Kinlocs shearing **Surface Support** Shotcrete integrity and adhesion lost Multiple mesh strands failing	Short-term/Low Exposure excavations Rehabilitation scheduled. If no near field seismic risk, inspection by shiftboss or geotechnical engineer to enable personnel undertake short-term working. Long-term/High Exposure excavations Barricade/no-entry area and schedule rehabilitation.

TARP level deformation	Rock mass and ground support observations	Geotechnical engineer action and response
Level 5 Ultimate Limit 400+ mm	Rock Mass	Barricade and restrict all personnel access
	Severe wall movement. Failure imminent	
	Bolts/Plates	
	No dynamic capacity left in ground support system	
	Cablebolts or primary bolts sucked into walls/shoulders.	

SCAN PROCESSING, ALIGNMENT, AND INTERPRETATION

Rosebery has been using Emesent's Hovermap LiDAR scanner for undertaking all mobile LiDAR scanning and processing the raw data to a usable.laz format. The alignment and distance comparison methodology loosely follows the workflow outlined by Jones, Ghabraie and Beck (2018). The workflow described is a manually intensive workflow. Although this workflow yields a usable solution, multiple restrictions exist when applying on-site:

- Requires manual segmentation and alignment steps tying up valuable personnel resources.

- Segmentation and alignment is time intensive.

- The segmentation and alignment steps are subjective, leading to repeatability issues.

In April 2024, Emesent released a trial for selected mines to undertake beta-testing of their Deformation and Convergence Module within their Aura software. Rosebery has since been using this model, learning how best to implement it, providing industry feedback for future updates.

AURA SOFTWARE DEFORMATION AND CONVERGENCE MODULE

To evaluate the benefits and limitations of this new method it was compared against the Cloud Compare's M3C2 (Multiscale Model to Model Cloud Comparison) workflow which has traditionally been used for comparison of LiDAR scans.

The Cloud Compare MC32 process is based on a rigid transformation to align the reference and new scans to observe differences between scans. This process involves moving and rotating the new scan to fit the reference scan. In rigid alignment, both reference and new scans remain unchanged, with only their location and orientation (transformation) adjusted. This approach has been observed to work well in ideal conditions where scans are highly accurate, and the only differences are environmental changes or where larger data sets are segmented into smaller sections to eliminate drift. This approach has shown to be less suitable for large-scale environments such as long tunnels. In these cases, inherent inaccuracies and potential drifts in the scans lead to misalignments, resulting in errors in detecting true deformations and convergence.

In contrast the Emesent Aura Convergence monitoring workflow uses a SLAM algorithm to apply a non-rigid (elastic) alignment method. In this approach, the first scan is used as a reference and remains unchanged, while the second scan undergoes elastic modifications to align locally with the reference scan through non-rigid transformation. This process is based on the identification and matching of 3D features in both scans, accounting for drifts and errors. This local alignment has shown to accurately detect changes when monitoring convergence. This method has also shown to deliver precise alignment of extensive scans, up to kilometres of underground tunnels, eliminating the need for the data segmentation process typically used to adjust for scan drift. Figure 13 shows an example of true tunnel deformation that can be masked due to drift in long scans when using rigid alignment. Whilst non-rigid alignment accounts for drift and highlights the true tunnel deformation as the difference between Scan 1 and Scan 2.

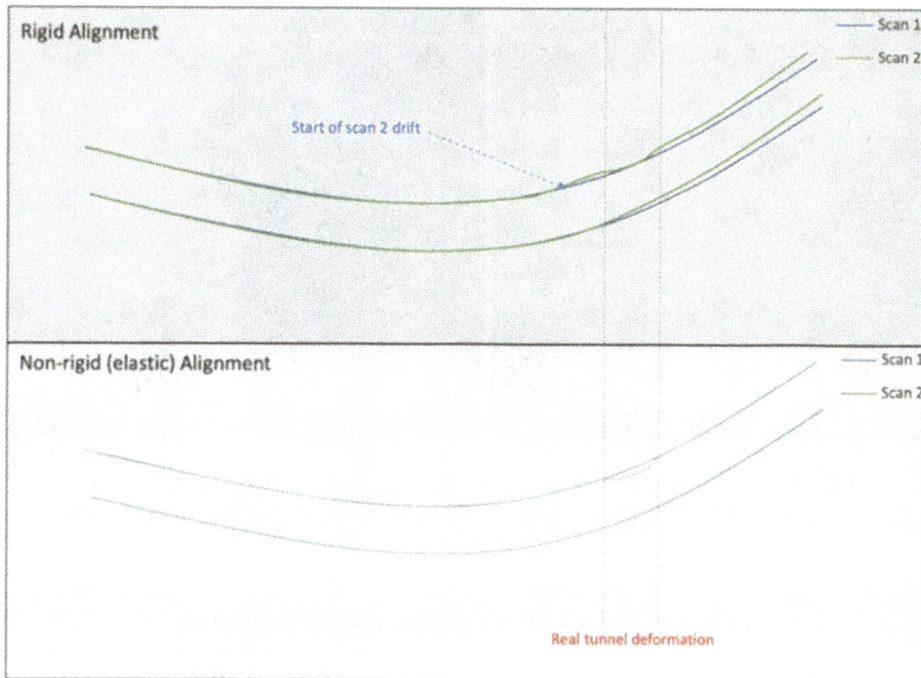

FIG 13 – Example showing the difference between a Rigid Alignment versus Non-rigid Alignment.

After alignment of scans, the Emesent Aura software meshes the reference scan and measures the distance from the points in the second scan to the mesh as shown in Figure 14.

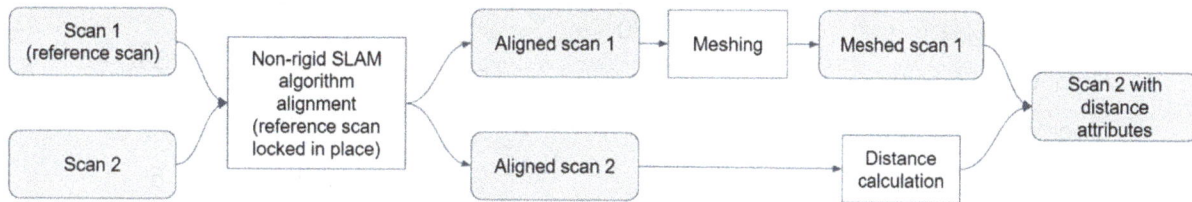

FIG 14 – Emesent Aura Convergence monitoring workflow.

The output is a point cloud with point to mesh distance attributes. While the output accurately showed changes between scans, there is thickness to the point cloud meaning convergence is viewed from inside the tunnel rather than being visible from outside. In contrast, Cloud Compare's MC32 takes an average of all the points and creates a single plane which can be viewed outside of the drive which the site engineers find useful for easy identification of areas of interest or concern. The Aura output is not georeferenced, so this process needed to be done in Cloud Compare.

Like Cloud Compare the Aura module displays convergence as a coloured 'heat map'. Aura's colour scale is autogenerated based on the deformation range within the comparison. The scale is adjustable but unlike Cloud Compare cannot be set to pre-defined deformation categories. This prohibits quick interpretation of the measured convergence against pre-defined limit states in the TARP. Figure 15 shows an example of the convergence output from the Aura model next the scanned drive.

A significant advantage of the Aura Convergence Module over Cloud Compare is the faster rate of processing.

A time comparison was undertaken of each workflow, including the automated steps, where the operator is not required and manual steps, where the operator needs to provide input. It was observed that the Aura workflow saved 1 hr and 50 mins of manual processing time per scan as detailed in Table 2.

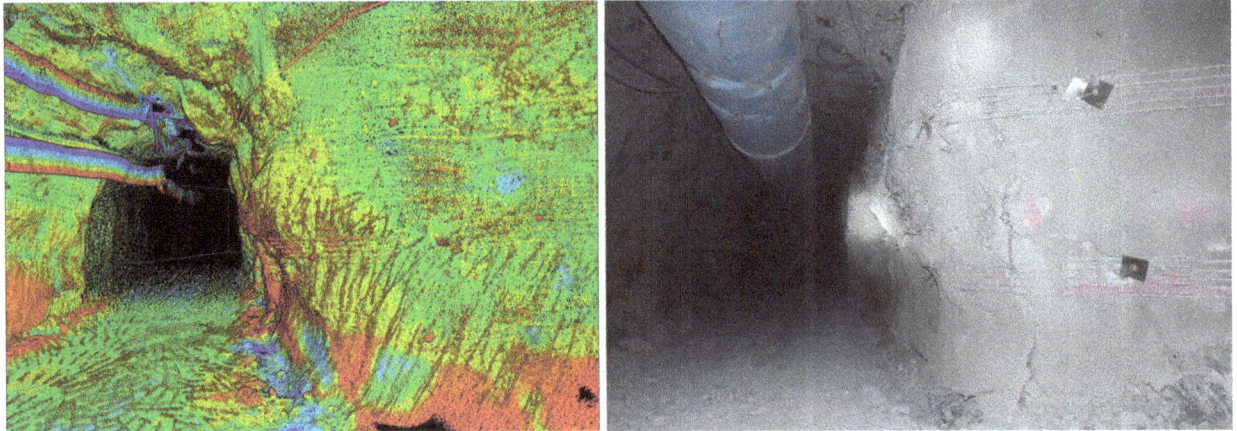

FIG 15 – Convergence comparison output from Aura Module.

TABLE 2

Breakdown showing each step and the time required for a Cloud Compare workflow using M3C2 and beta-version of Aura Convergence Module.

	Time (min)			
	Cloud Compare workflow		Aura workflow	
Processing prior to comparison	**Automatic task**	**Manual task**	**Automatic task**	**Manual task**
Raw file to.laz format	20		20	
Point cloud cleaning		30		5
Geo-reference to baseline		15		15
Align secondary scan to baseline scan		90	30	5
Time	155		75	
Deformation comparison and calculation				
M3C2	30	5	N/A	N/A
Aura convergence module	N/A	N/A	30	5
Time	35		35	
Visualisation				
Assign colourisation, legend etc		5		5
Time	5		5	
Total time	200		115	

CONCLUSION

Managing convergence has been and continues to be a challenge at Rosebery. Efforts to manage converging excavations historically started with qualitative inspections, distometer measurements and escalated to CMS surveys in significant cases. These strategies were largely reactive and caused inefficiencies to production and development cycles.

The introduction of a LiDAR based monitoring program has given Rosebery the ability to determine where, when and how much convergence has occurred. This has enhanced the decision-making processes for site engineers in conjunction with the development of a TARP, helping to mitigate the risks associated with deteriorating ground support.

The application of the Aura Convergence Module has significantly improved the workflow speed making the LiDAR monitoring program more practical to implement and sustain.

This paper is not intended to serve as a direct strategy for other mines but rather to present the underlying issues contributing to the problem of drive convergence at Rosebery, how they have been investigated and how they are now being managed.

The causes and extent of drive convergence will vary from mine site to mine site, it is hoped that this paper will find use as a framework of ideas for other practitioners who experience drive closure in their operations.

RECOMMENDATIONS AND FUTURE WORKS

While the initial implementation of LiDAR scanning and use of the Aura convergence workflow has yielded valuable insights, further work is needed to fully integrate the system into the mine's long-term ground support and mine development strategy.

The continued analysis and interpretation of the data in relation to the variable rock mass conditions will be used to optimise the monitoring program. Areas of particular focus for future work will include relating foliation intensity to rate and extent of convergence, further understanding of the relationship between support capacity and convergence, how and where surface support and rock mass reinforcement is best applied and how proactive support maintenance in combination with just in time development can be configured to minimise the occurrence of extensive, costly and time-consuming rehabilitation.

ACKNOWLEDGEMENTS

MMG Rosebery for providing access to the case study and site data. Emesent for provision of Aura Convergence Module.

REFERENCES

Hadjigeorgiou, J and Potvin, Y, 2023. Ground support guidelines for squeezing ground conditions, *Journal of the Southern African Institute of Mining and Metallurgy*, 123(7):371–380.

Jones, E, Ghabraie, B and Beck, D, 2018. A Method for Determining the Field Accuracy of Mobile Scanning Devices for Geomechanics Applications, *Proceedings of 10th Asian Rock Mechanics Symposium*, ISRM, Singapore.

Subtek™ 4D™ – Optimised underground blasting performance through improved energy distribution and control

S Evans[1], B Taylor[2] and M Brayshaw[3]

1. Lead Engineer – New Underground Technology, Orica, Perth WA 6007.
   Email: sam.evans@orica.com
2. Senior Manager – Commercialisation, Orica, Kurri Kurri NSW 2327.
   Email: ben.taylor@orica.com
3. Implementation Technician – Technologies, Orica, Kurri Kurri NSW 2327.
   Email: mick.brayshaw@orica.com

ABSTRACT

Bulk explosive technology for underground mining has not seen any significant technological advances since the widespread transition from ammonium nitrate fuel oil (ANFO) to ammonium nitrate emulsions (ANE) commencing in the 1980s. The in-hole energy profile that conventional ANE systems deliver is limited by the physics of chemical gassing and simple control systems. This compromises the range, precision, and accuracy of energy delivery into the blasthole, and in some cases limits the length of charge that can be delivered. Orica's 4D™ bulk system removes many of these constraints, enabling better blasting performance across all underground mining methods. This paper reviews work completed at three mining operations across Australia, the United States, and Chile with a focus on how this system can improve mining operations and deliver sustained commercial and productivity benefits into the future. These benefits include improved ore recovery, reduced dilution from waste rock, and reduced overall explosives consumption. This paper also reviews how this new technology can be applied for planning future mining operations.

INTRODUCTION

Orica's 4D™ bulk system uses a new, smarter method to control the delivery of gasser and modified emulsion to blastholes. The technology expands the range of average blasthole energy available to designers and gives them new options to control the distribution of energy within blastholes that are impossible to achieve with conventional systems. 4D™ improves drill and blast outcomes in regular production designs and enables novel loading techniques like placing multiple energy products in the same column without multi-pass loading. The application of this technology for development mining and production mining environments and in multiple commodity types is ongoing with trials in Australia, the United States, and Chile.

The conventional bulk loading options for underground mining explosives include ANFO, gassed ANE, and solid sensitised ANE. ANFO is known for its cost-effectiveness and ease of use but is constrained by its limited energy range and lack of water resistance. ANFO can only be pour loaded to give a relative bulk strength (RBS) of 100 per cent, or blow loaded to give an RBS of around 130 per cent. Gassed ammonium nitrate emulsion (ANE) sensitised with a chemical gassing agent offers a greater range of energy and better water resistance, but hydrostatic pressure changes the density and energy of the column with depth, especially in downholes. Conventional delivery systems are limited to only two or three options of final gassed density, in a narrow energy range of RBS 80 (%) to RBS 140 (%). Solid sensitised ANE can deliver fixed energy and density profile, but it does not offer flexibility to change density or energy, and usually only one or two energy options are available to the blast design and execution team.

Orica's 4D™ technology transforms the energy delivery system by offering an extended energy range from RBS 50 to 170 at increments of 10 per cent – a significant increase at both ends of the energy spectrum. Another critical difference from conventional gassed ANE systems is 4D™'s ability to counteract the effect of hydrostatic pressure on the energy profile. The 4D™ explosive delivery and gassing technology ensures consistent collars, and the ability to specify and vary the energy level throughout the column without the requirement for decking or multiple pass loading. 4D™ enables precise energy delivery at various points within the blasthole, or the blast in general, as required.

REAL WORLD APPLICATION

As of September 2024, the 4D™ bulk system has had extensive lab and field testing and has been deployed at three underground mining operations. The following examples show either final project data from these three mining operations or initial inferences, and the potential applications.

Stope blasting optimisation at Dugald River Mine, Australia

Background

Globally, the first implementation of the 4D™ bulk system occurred at MMG's Dugald River zinc mine in north-west Queensland, Australia. This implementation was as a multi-phase project between August 2022 and December 2023. While the initial stages of the project were purely focused on validating and verifying the system, the later Phases 3 and 4 set targets for blasting performance against an established baseline to measure the effectiveness of the extended energy range in improving blasting performance. This paper will review the performance of 4D™ in Phase 4 of the project between August and December 2023, using the largest data set, and the availability and deployment of the full range of 4D™ energy from RBS 50–170.

The Dugald River management team selected several Key Performance Indicators (KPIs) where adjusting energy was likely to lead to improvements. All the metrics were recorded as part of the mine's business as usual activities, meaning the project did not require any additional work to collect data, and for which a statistically relevant baseline could be easily produced. The KPIs were as follows:

- Stope recovery. The percentage of ore recovered from within the stope shape, up to a maximum recovery of 100 per cent. Typically, this is hardest to achieve along the footwall contact.

- Hanging wall dilution. Overbreak outside the planned stope shape leading to mining unwanted waste rock with ore.

- Brow failure rate. Stopes in which there was unwanted overbreak at the brow of the extraction level, past the first row of ground support, were deemed to have a failed brow.

- Clean-outs and redrills: The number of metres of redrills and hole clean-outs required in stopes where the same downholes were loaded and fired in more than one event, or lift.

Baseline practice and performance

Conventionally, blast designers at Dugald River are restricted to choosing low, standard, or high-density ammonium nitrate emulsion (ANE). Changing the explosive energy within the ring is not simple. Hydrostatic pressure increases the density and energy at the toe of long columns and limits the maximum column length. Overall, the lack of energy control options for designers often leads to unfavourable blasting results. Typical stope recovery in comparable stopes was 90 per cent and typical dilution was 11 per cent.

The hanging wall contact between ore and waste is the main source of dilution at Dugald River. The stope geometry usually places a single, parallel longhole along this contact. To achieve the desired stope shape, it is necessary to consistently distribute just enough energy along the hanging wall to break the ore up to the point of contact and no further. Recovering all the ore on the footwall presents a different challenge. The ore is more confined and multiple holes intersect the contact at different angles, sometimes penetrating the barren host rock.

Brow failure is a common issue at the intersection with the extraction level. An existing improvement project was in progress, using a charge of approximately two metres of low density ANFO at the contact with the brow. Even with this method of reducing energy, the brow failure rate was 44 per cent. Hole clean-outs and redrills were common in any situation where the toe of the blastholes was loaded and fired in the first firing event, and the remaining portion of the hole was to be fired in a subsequent event. Excess energy from the explosive column caused damage to the upper portion of the hole and access drive, and this often resulted in rework to enable the subsequent loading of the holes.

4D™ solution

Using 4D™ to deploy a wide range of energy throughout the blast provided several different solutions for the problems described, without requiring any changes to drilling or other practices on-site. A joint workshop between MMG and Orica produced a 4D™ loading guideline document. A summary of the proposed loading for the different hole types, as well as conventional loading practices is shown in Table 1.

TABLE 1

Conventional and proposed 4D™ loading options by hole type.

Hole type	Conventional loading	Proposed 4D™ loading
Hanging wall holes	Low density ANFO *or* 0.8 g/cc ANE	50–70 RBS
Brow holes	Low density ANFO (toe charge)	50–60 RBS (toe charge)
Winze holes	0.8 g/cc ANE	70–80 RBS
End wall holes	0.8 g/cc ANE	90 RBS
Easer holes	1.0 g/cc ANE	110–130 RBS
Footwall stabbing holes	1.0 g/cc ANE	150–170 RBS

An illustration of some of the proposed loading in practice within a stope ring is shown in Figure 1.

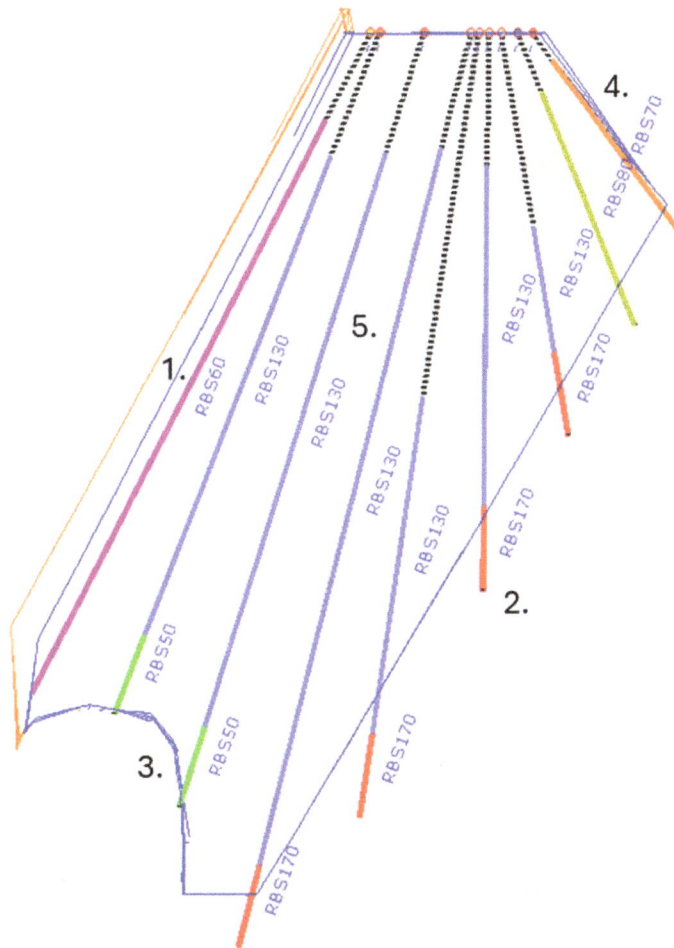

FIG 1 – Typical 4D™ loading in a Dugald River stope, where colour represents Relative Bulk Strength (as annotated).

Within the ring, the reasoning for each selection by application is as follows:

- Hanging wall holes used a consistent low energy charge of between RBS 50–70. Variance within this range was typically determined based on factors such as the drilled stand-off from the hanging wall, the existence of known faults or geological structures in the proximity of the hanging wall, and the performance of existing stopes within the same part of the mine. In narrow stopes, or stopes where the hanging wall was particularly susceptible to overbreak, a lower energy charge was occasionally used in the next hole to function as an additional buffer.

- High energy toe charges were loaded in the holes stabbing the footwall, where recovery was typically challenging to achieve. This ranged from RBS 150–170 depending on the orientation of the footwall, the length of the hole, the final toe spacing and burden on any given ring, and the amount of subdrill used. The energy reduces in that part of the hole in the middle of the stope. The ability to change energy quickly and easily at any point within a hole and load in a single pass gives the blast designer flexibility in making decisions about energy distribution.

- In the inverse of the high energy toe charge, holes drilled adjacent to the brow in the final ring used a low energy toe charge of RBS 50–60. This charge was typically 2–3 m in length and was intended to reduce the likelihood of unwanted damage and failure to the brow. The remainder of the hole was loaded with a higher energy charge, in this case RBS 130.

- A low energy charge, RBS 70, was also deployed in the stope shoulders, where maintaining the stope shape can sometimes be challenging. In this example, the adjacent hole was also loaded with less energy, in this case RBS 80. While this location within the stope was not a major source of overbreak or dilution, the ability to control it helped to further improve overall conformance to design.

- The bulk of the stope easers or production holes were loaded with RBS 130. This represents an increase in energy from the conventional 1.0 g/cc emulsion used on-site. This higher energy tended to improve fragmentation. The energy deployed within any given stope can be selected based on the requirements for the blast, but in most cases a charge of between RBS 90–130 was suitable for most easer holes within a stope.

Trial results

During the final phase of the trial, from August to December 2023, a total of 28 blasts across 23 stopes were loaded with the 4D™ system (most stopes consist of two or more blast events). The data regarding stope performance, brow control, and hole clean-outs was collected for these blasts as part of the mine's business as usual practices. A baseline data set of 24 similar stopes in the same mining areas that were loaded and fired within the same period as the 4D™ blasts was used to compare performance. The baseline and 4D™ data from the trial is presented in Table 2. By every measure, the 4D™ blast outcomes were better.

TABLE 2

Baseline and 4D™ performance at Dugald River Mine.

KPI	Baseline performance	4D™ performance	Relative change
Hanging wall dilution	11.0%	3.7%	- 66%
Stope recovery	90%	93%	+ 3%
Brow failure rate	44%	12%	- 73%
Clean-out metres per stope	44.5 m	23.3 m	- 48%
Redrill metres per stope	5.4 m	5.1 m	- 6%

As each stope firing was a learning process, best practices were refined throughout the duration of the trial. The most appropriate energy levels for the hanging wall hole and footwall stabbing toe charges were identified by poor blast performances occurring when the energy levels for hanging wall holes exceeded RBS 70, and when footwall stabbing holes were loaded with RBS 130, instead of the recommended RBS 150–170 range. After these blasts had been analysed and the loading parameters refined, performance was further improved. In stopes where loading followed the revised recommendations, hanging wall dilution averaged 2.6 per cent and stope recovery 94.9 per cent. The best performing stope during this phase recorded 1 per cent dilution and 99.3 per cent recovery.

There was also a significant improvement in brow failure rates, from a 44 per cent failure rate using conventional explosives to 12 per cent using the low energy 4D™ option. 4D™ eliminated the requirement to stock a speciality low density ANFO packaged product for this specific application, with a simplified magazine inventory, reduced freight requirements, and reduced manual handling. Hole re-drills were unaffected, with a very slight improvement observed, but hole clean-out requirements were approximately halved. This was attributed to the improved control of the final uncharged collar when using the 4D™ system. This represents a considerable time and cost saving through reduced rework.

Conclusions

While the improvements in hanging wall dilution and stope recovery are positive in isolation, it is also important to consider them in relation to one another. In the baseline data set it was evident that while many stopes performed well in one or the other of these metrics, they tended to perform poorly in the other. For example, the best performing 50 per cent of baseline stopes achieved a stope recovery of 95 per cent, but a hanging wall dilution performance of 8.5 per cent, while the top 50 per cent of 4D™ trial stopes recorded an average of 96.4 per cent recovery with only 2 per cent dilution. The top 50 per cent of baseline blasts in relation to hanging wall dilution achieved 2.7 per cent dilution, but only 93 per cent recovery, while the 4D™ performance was 1.3 per cent dilution at 94.6 per cent recovery. The ability to reliably achieve improved results in both metrics in the same stope represents a move towards optimising the drill and blast processes. This is visualised in Figure 2, which shows a ranking of all baseline and 4D™ trial stopes by their stope recovery performance. The best performing stopes are on the right-hand side, and the worst performing stopes are on the left side. The respective hanging wall dilution results for the same stopes are also plotted on the same axis. The absence of any large variations in hanging wall dilution, and consistently high stope recovery performance, demonstrates the benefit of controlled energy distribution in critical parts of the blast.

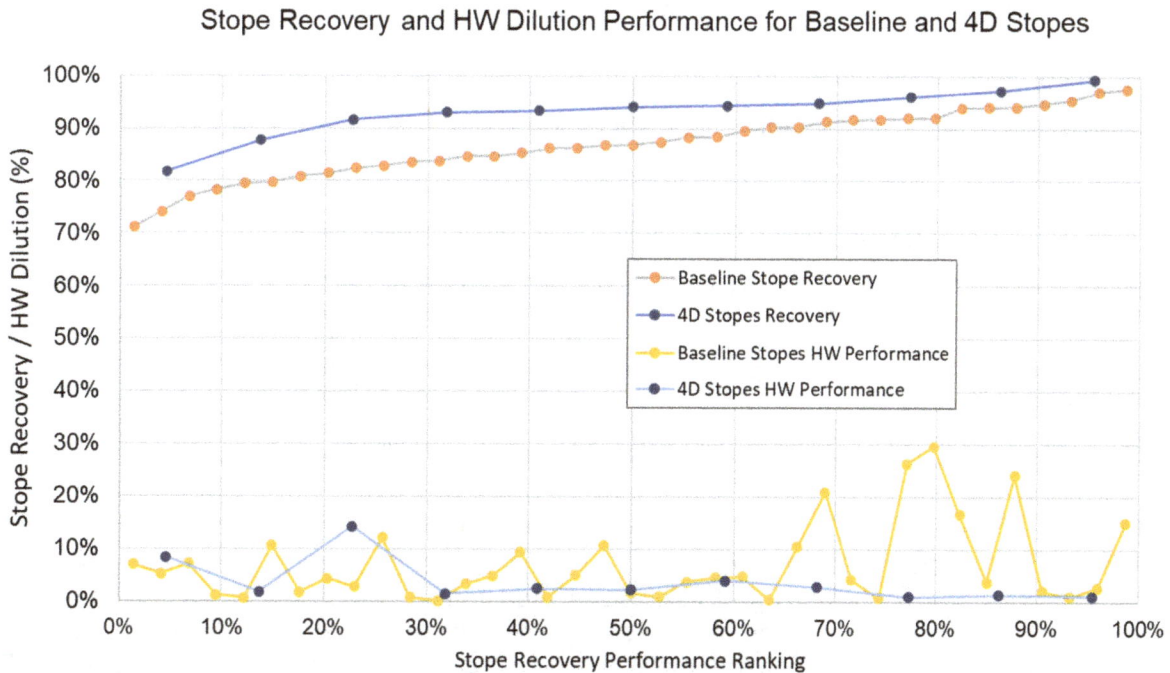

FIG 2 – Stope recovery and hanging wall dilution performance for baseline and 4D™ trial stopes.

The improvements in various performance metrics during the 4D™ system trial were significant. While a monetary value was not assigned to each of the metrics, for the two primary KPIs of hanging wall dilution and stope recovery, a joint value estimation was made by MMG and Orica. Scaling the improved stope recovery to a full year's mine production is equivalent to an uplift in Net Smelter Returns (NSR) of A\$23.5 million, and the reduction in dilution was equivalent to a saving in load and haul costs of A\$1.4 million. There was no monetary value assigned to cycle time efficiency improvements through improved conformance to stope shape, the reduction of the brow failure rates, and the reduction in hole clean-outs – but since all these metrics showed large improvements, it is likely that there would be significant economic benefits in these areas.

It is acknowledged that bulk explosives are just one of many key factors that will affect drill and blast performance in underground mining. Technological improvements in other fields such as the ability to derive geological information from the drilling process, as well as accurately mapping the deviation of the as-drilled hole are now becoming more frequent and commercially available to underground mining operations. Combining this technology with the ability to select the optimum energy levels from an extended range will increase the ability of the drill and blast engineer to design a blast for a desired outcome, and for the operations personnel to achieve this outcome.

Production blasting at Pucobre Mine, Chile

Background

Early-stage trials of the 4D™ bulk system commenced at the Pucobre copper mine in Chile, located in the Atacama region approximately 20 km south of Copiapó, commenced in mid-2024. The applications at this operation target many of the same metrics as Dugald River, and the experiences from that trial have been used to guide energy selection in the stope design process.

Leveraging this prior experience, during this trial it is intended to explore the potential to increase the number of in-hole energies within the same hole to three or more. In particular, there is a requirement for low energy at both the contact with the hanging wall, and close to the brow, while maintaining high energy for the bulk of the stope to ensure that fragmentation is kept below a critical level for downstream ore handling. Unlike the Dugald River mine example, the contact with the hanging wall tends to be in the form of holes perpendicular to the contact, rather than parallel with it. This creates a challenging situation when loading with conventional explosives. It is intended to use the 4D™ system to deliver multiple energy options within the same hole. While this is similar in application to some of the work previously completed at Dugald River, the scale and complexity of the stope energy requirements will require multiple in-hole energies in a much wider variety of holes. An example of this is shown in Figure 3.

The key design decisions implemented here are as follows:

- Use of RBS 50 close to the brow, to maintain stability at the extraction level.

- Use of a low energy toe charge of RBS 50 at the hanging wall contact.

- Moderate to high energy levels, RBS 130 used within the bulk of the stopes, to maintain fragmentation and ensure recovery.

- High energy toe charge of RBS 160 at toe of longer holes intersecting with upper level holes, to ensure recovery.

- A moderate to low energy RBS 100 buffer hole adjacent to the hanging wall hole.

Technology integration

The Pucobre operation is technology-focused and has the capability to gather data relating to blasting both pre- and post- loading of explosives. Additionally, the ability to gather data on the orebody prior to the completion of the blast design process combined with the flexibility of the 4D™ system enables the operation to increase the likelihood of achieving the desired blast outcomes. The ability to monitor aspects of the blast performance upon the completion of blasting and the excavation of blasting material enables improvements to be made for future blasts. The proposed workflow for this data use is shown in Figure 4.

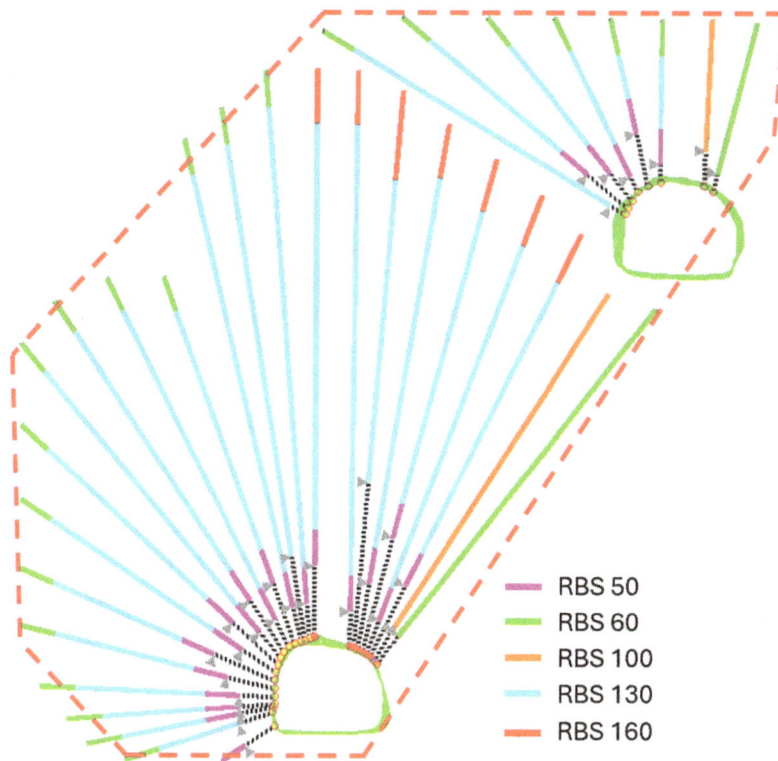

FIG 3 – Proposed loading of multiple energies in complex production environment, with the red dashed line showing the intended stope shape.

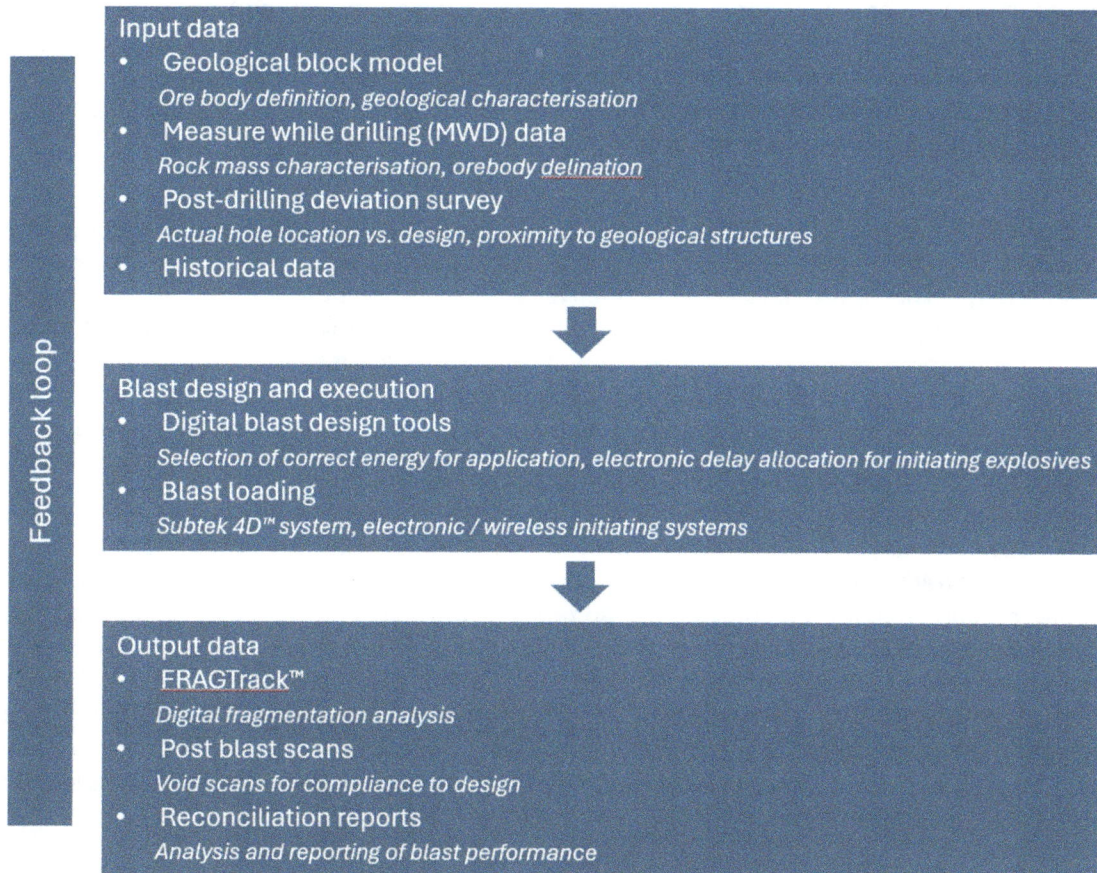

FIG 4 – Integrated data workflow and feedback loop.

The integration of multiple data inputs and outputs combined with a flexible blasting system is envisaged to enable significant improvements in the efficiency of the mining operation. While the project is still in preliminary stages, initial performance at the site appears to be positive.

Development blasting applications

Background

Early-stage trials at a mine in North America during early 2024 focused on the capability of the 4D™ system in development mining. These trials provided valuable insights into the system's performance under real-world conditions, particularly in the challenging environments typical of development mining. The trials demonstrated the system's ability to deliver precise energy distribution, which is crucial for achieving optimal blast outcomes. The results showed that the 4D™ system can significantly enhance the efficiency and effectiveness of development mining operations, paving the way for broader adoption of this innovative technology.

Geology, drilling, and explosives have the greatest influence on blast performance. Geology is most influential because the rock type, structure, and condition directly affect how the blast will propagate and break the material. Understanding the geological conditions allows for better planning and execution of the blast. Drilling is the next most influential element, as precise drilling patterns and hole placements are crucial for achieving the desired blast outcome. Accurate drilling ensures that the energy from the explosives is distributed effectively throughout the rock mass. Finally, the choice and application of explosives are essential for optimising the energy release and achieving the desired fragmentation. While all three factors are interdependent, geology defines the underlying conditions, drilling defines the framework, and explosives deliver the energy within these two constraining factors.

Development blasting is especially challenging because there is no free face for relief. The correct application of energy in development mining is critical to ensure that the cut and blast pull to full depth across the entire face, particularly in the burn cut and shoulder holes. Proper energy distribution ensures that the perimeter does not experience overbreak or underbreak, which can lead to structural instability and increased operational costs. Excess energy can be just as problematic as insufficient energy. By precisely controlling the energy delivered to each blasthole, the 4D™ system helps to achieve a uniform blast profile by reducing damage to the surrounding rock mass. This precision not only enhances safety and efficiency but also contributes to the overall success of the mining operation by reducing waste and optimising resource recovery.

Conventional development practices in underground mining are often constrained by the limited options available for energy distribution. Typically, these practices rely on a few different energy bulk products, specialty packaged products, and strategic hole placement as the primary means of controlling energy. This approach can be restrictive, as it does not allow for the fine-tuning of energy levels needed to optimise blast performance. The inability to precisely control energy distribution can lead to suboptimal blast outcomes, such as uneven breakage, overbreak, or underbreak, which can compromise the efficiency and safety of the mining operation. Relying on a limited range of energy products can increase operational costs and reduce flexibility in responding to varying geological conditions.

An example of a blast design using conventional explosives, and the blasting outcomes from the blast is shown in Figure 6 (the example shown is not from the trial site but is indicative of typical issues facing development mining). This highlights typical issues that may be encountered in development mining: failure to achieve a full advance in the burn (1), hole butts indicating failure to achieve a full advance throughout the face (2), dishing at the extremities of the blast indicating the buffer row has failed to break to the perimeter (3), and a poor perimeter profile (4), with both excessive overbreak (5) and underbreak (6). A revised loading plan using 4D™ is shown in Figure 5, with the relative energy compared to the standard design shown. While detailed surveys have not yet been completed in development blasts using 4D™, initial performance appears to be significantly improving blasting performance.

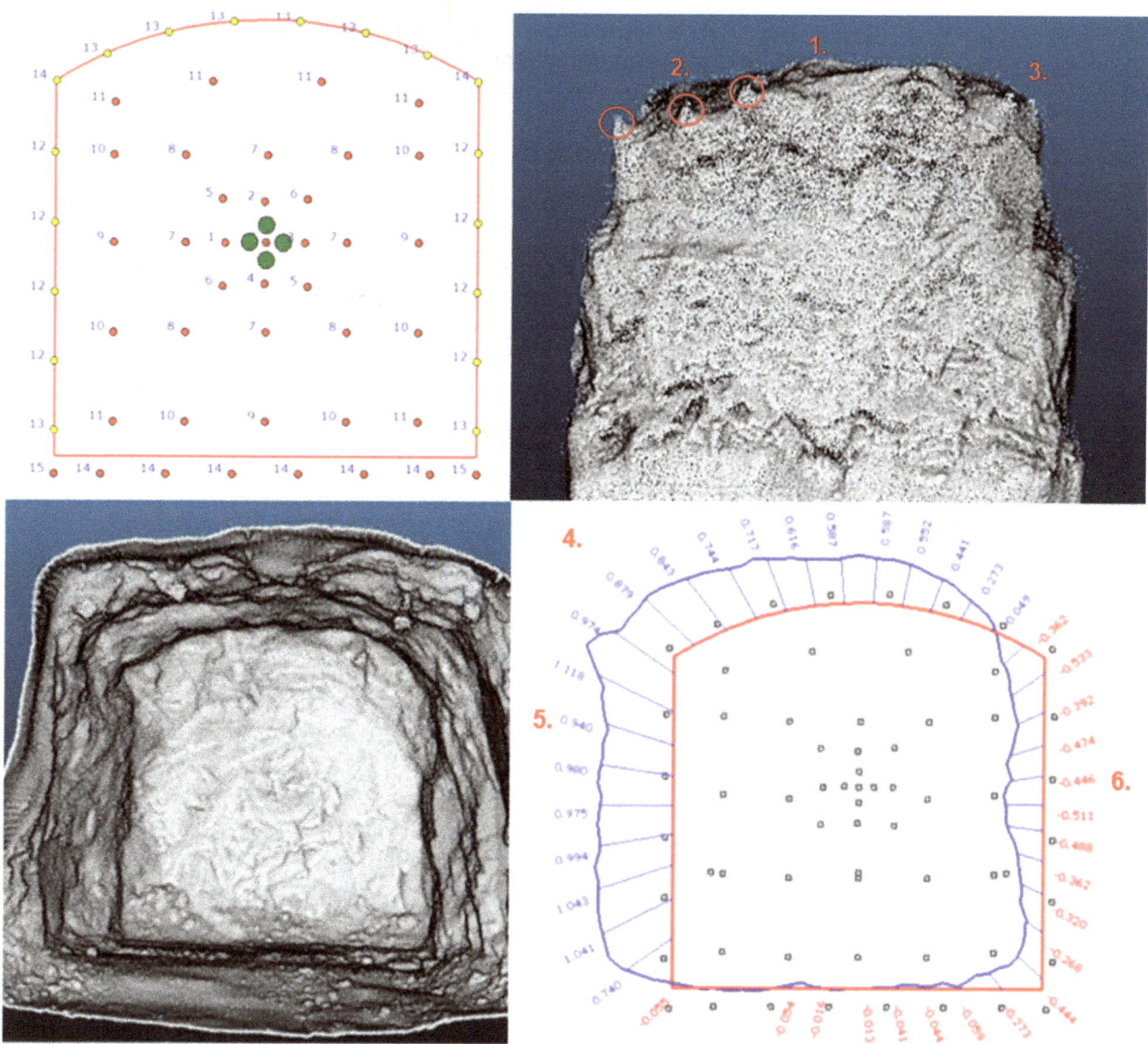

FIG 5 – From top left clockwise: a typical blast design for development heading, a plan view of the post-blast 3D scan, a 2D slice from the heading scan showing profile overbreak and underbreak, and a front-on view of the 3D scan.

4D™ applications

4D™ enables improvements in development mining performance by applying many different energy levels throughout the blast. High energy can be allocated to highly confined parts of the blast, such as the lifters and corner holes, where it can be challenging to achieve a full advance with conventional explosive energy levels. In the burn cut where the holes are close together, reliable detonation under extreme conditions is essential to create the void for the box to break into, setting up success for the rest of the blast. Conversely, lower energy can be used in the perimeter holes to maintain the profile and prevent overbreak. This precise control over energy distribution enables more effective and efficient blasting, reducing waste and improving overall blast outcomes.

In addition to potential improvements in general blasting performance, the ability to adjust energy as required will be useful in many other situations. In the event of blocked or missing holes, adjacent holes can be loaded with increased energy, or in the event of very weak rock mass conditions lower energy can be used in multiple rows of holes, besides just the perimeter holes. Additionally, it is possible to consider the potential to expand patterns – typically challenging given the small hole diameters used in development mining.

An example of the potential alternative distribution of energy is shown in Table 3 and visually represented in Figure 6. This highlights the increased flexibility of a system for selective distribution of energy throughout a blast to achieve specific goals, without the requirement to source and carry speciality products. It is also possible for operators to adjust energy distribution at the face to consider hole conditions. The ability to tailor energy distribution is likely to lead to improved advance efficiency (blasted metres achieved relative to drilled metres) and perimeter performance.

TABLE 3

A comparison between typical conventional explosive selection and proposed 4D™ options, with relative energy difference highlighted.

Hole type	Typical conventional explosive selection	Estimated RBS	4D™ explosive selection	Change in RBS
Burn holes	1.0 g/cc ANE	100%	140 RBS	+40%
Easer holes	1.0 g/cc ANE	100%	110 RBS	+10%
Lifter holes	1.0 g/cc ANE	100%	140 RBS	+40%
Buffer holes	1.0 g/cc ANE	100%	80 RBS	-20%
Perimeter holes	0.8 g/cc ANE	67%	60 RBS	-7%

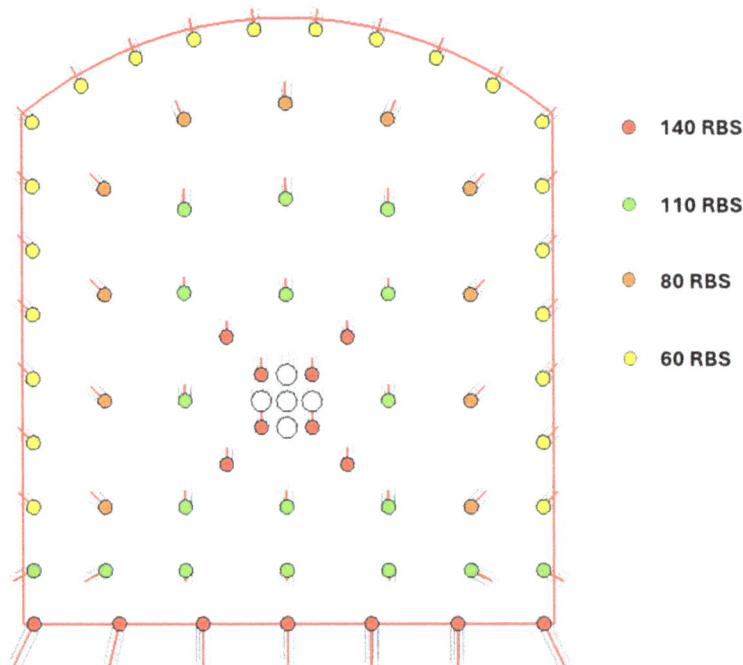

FIG 6 – Visual representation of proposed 4D™ energy distribution.

CONCLUSIONS

While to date there has only been one completed large scale trial with the 4D™ system, there have been many useful learnings that are now being applied at two other trial sites. Both trials are currently at initial stages and progressing with the work outlined above. However, it is evident from all three operations that this technology represents a major update to the capability of underground bulk explosive loading. It will function as an enabler for blast designers to target specific outcomes, and for operations personnel implementing blast designs to achieve the design and outcomes.

Future technology integrations

The future of blasting envisioned by the authors includes an integrated design-for-outcome system. The system will combine data collected during drilling and mucking earlier blasts, combined with design and as-charged digital records, interpreted by design algorithms. The data will include rock strength and condition, water levels and the as-drilled hole track and collar position. The algorithm will adjust the density and mass of each charge within and between each hole, based on a desired blast outcome. The system will manage constraints including vibration, geometry, and water.

Blastholes will be primed and charged autonomously with minimal human supervision. Wireless initiation and delivery system automation are enabling technologies essential to achieving this vision. Continuously variable energy within and between blastholes and decks will contribute to the adaptability of this system in production and development mining in upholes and downholes in all types of ground, and for all blasting scenarios. 4D™ brings this vision closer by providing an expanded range of charging options from a single delivery system, using one emulsion and gasser formulation.

ACKNOWLEDGEMENTS

Orica would like to acknowledge all of those who have worked on the development of the 4D™ system, and its deployment to sites across the globe. Orica would also like to acknowledge and thank the foundation customers at the sites on which these products have been trialled, in particular MMG and Pucobre who have given permission to publish data from the trials.

Hatch Mine VisionAI for underground productivity tracking

V Fitzmaurice[1] and J Selby[2]

1. Regional Mining Director, AUA Mining, Hatch, Brisbane Qld 4000.
   Email: vic.fitzmaurice@hatch.com
2. Systems and Process Control Specialist, Mining, Hatch, Sudbury CA P0H, Canada.
   Email: jennifer.selby@hatch.com

ABSTRACT

Construction projects in mining are often delayed resulting in cost overruns. Workforce productivity challenges are a key contributor alongside operational complexities such as deeper deposits and higher working temperatures.

Accurately measuring workforce productivity for comparison against plan is challenging, especially in underground mining projects. Manual and paper-based tracking systems are outdated, prone to errors and data gaps which limit real-time decision-making in project management.

Computer vision is a field of artificial intelligence (AI) that uses machine learning and neural networks to derive meaningful information from digital images and videos. Use of computer vision is widespread, and its applications are constantly evolving. Many industries such as manufacturing, healthcare and transportation have seen significant transformation due to its implementation.

Productivity tracking on mining projects may also be served by computer vision to provide real-time data and insights with minimal infrastructure and sensors. An extensive distributed system is not required for installation throughout the mine as is the case with other systems such as location tracking.

Using standard surveillance cameras, a computer vision model can be trained to detect any item within an image. These items are then tracked frame by frame and backend logic is applied to contextualise the scene and output the actual activity taking place. This automates the process of productivity tracking in an underground setting.

Hatch has developed a web-based application called Hatch Mine VisionAI, which includes event-based video playback and dashboards for decision-making. It provides stakeholders with real-time productivity insights to maintain timeline and budget. The solution is designed to:

- Identify tasks being performed.
- Track task duration and count.
- Support identification of safety risks.

Hatch Mine VisionAI facilitates sustained increases in productivity, reducing project deviations and cost overruns as enhanced monitoring leads to direct improvements in daily performance. It allows for increased project estimation accuracy and planning, reducing risk of cost overrun. Safety risk is also mitigated through reduced reliance on manual surveillance to perform time studies, allowing project personnel to focus on higher value activities.

INTRODUCTION

Mining construction projects are complex and frequently experience significant delays, often resulting in cost overruns. A key contributor to these delays is workforce productivity challenges, particularly as mining operations extend to deeper deposits where workers face elevated temperatures and harsher conditions. Tracking productivity in such environments becomes increasingly difficult. According to a Deloitte report, workforce inefficiencies are a primary factor in project delays within the mining sector (Deloitte, 2014). Traditional paper-based tracking systems, long used in the industry, are no longer adequate. They are prone to errors, data gaps, and reporting delays, hampering effective decision-making. As a result, there is an urgent need to adopt advanced technologies like artificial intelligence (AI) and computer vision (CV) to enhance the accuracy and timeliness of productivity tracking.

PROBLEM STATEMENT

Accurately measuring workforce productivity in underground mining is especially challenging due to harsh working conditions and reliance on manual, paper-based tracking methods. These traditional approaches are inefficient and prone to errors, often leading to discrepancies between reported and actual productivity. This limits real-time decision-making, increasing the risk of delays and cost overruns. The unique challenges of underground mining, such as limited visibility and confined spaces, further complicate the use of conventional monitoring systems. This has driven the demand for more automated, efficient solutions that can track productivity and ensure projects remain on schedule and within budget.

COMPUTER VISION IN MINING

Computer vision, a branch of AI, offers a powerful solution for addressing the challenges of productivity tracking in underground mining. It processes and interprets visual data from images and videos, providing insights that would otherwise require manual observation. While computer vision has been successfully implemented in industries like manufacturing (Gao *et al*, 2024), healthcare (Esteva *et al*, 2019), and transportation (Dilek and Dener, 2023), its application in mining is still emerging. However, it shows significant potential by addressing the limitations of traditional monitoring systems. Unlike sensor-based systems that require extensive infrastructure, computer vision can operate with existing surveillance cameras, capturing video footage that is analysed in real time to track tasks and monitor progress.

Currently, computer vision in mining is broadly limited to:

- Intelligent lithology.

- Autonomous loading perception (loader bucket and truck).

- Anti-collision systems.

HATCH MINE VISIONAI SOLUTION

Hatch Mine VisionAI is a web-based application designed to automate productivity tracking in underground mining using existing camera infrastructure. By processing video footage through machine learning algorithms, it detects tasks, monitors durations, and identifies safety risks. This real-time data is presented via dashboards, helping project managers make informed decisions. Additionally, the system offers event-based video playback to support task analysis, further improving efficiency and safety.

BENEFITS OF HATCH MINE VISIONAI

Mine VisionAI provides several critical benefits:

- Operational Efficiency: Real-time monitoring highlights delays and bottlenecks, allowing for prompt interventions.

- Resource Management: Tracks personnel, machinery, and materials, ensuring optimal resource utilisation.

- Safety and Compliance: Enhances safety by automatically tracking adherence to safety protocols.

- Cost Management: Early detection of inefficiencies helps reduce costs and avoid delays.

By automating productivity tracking and reducing reliance on manual processes, Mine VisionAI streamlines operations and improves project management. Real-time data enables better control over timelines and budgets, reducing the risk of cost overruns. Moreover, Mine VisionAI mitigates safety risks by detecting hazardous activities in real time, ensuring a safer work environment. This not only enhances worker safety but also reduces the likelihood of costly project delays.

GENERAL COMPUTER VISION APPLICATION WORKFLOW

Broadly speaking, progress tracking refers to the continuous monitoring and evaluation of various operational activities to assess productivity and efficiency. The goal is to ensure that tasks such as drilling, mucking (removing blasted rock), hauling, and other key operations are carried out according to planned sequences and durations.

The operational activities to be tracked for progress and comparison against schedule are initially defined. Video cameras are installed at suitable locations underground to ensure generation of high quality activity images for subsequent processing in the application workflow.

Cybersecurity and privacy risk review

Before deploying computer vision systems in an underground mining environment, conducting an initial cybersecurity and privacy risk review is critical to ensure the safety, privacy, and integrity of the data collected. This review addresses potential vulnerabilities in both the hardware and software components of the system and ensures compliance with regulatory standards. Key considerations include:

- Data Security: Since the system captures and processes sensitive operational data (eg video feeds of mine activity, personnel movement, equipment usage), secure encryption protocols must be in place to secure the data both at rest (stored data) and in transit (data being transmitted between devices and servers).

- Access Control and Authentication: Strong authentication mechanisms, such as multi-factor authentication (MFA), must be enforced for accessing the system, especially for administrators and other high-privilege users. Role-based access control (RBAC) should ensure that users have access only to the data necessary for their roles, preventing unauthorized access to sensitive data.

- Privacy Concerns: The system must respect the privacy of individuals captured in video feeds. This includes ensuring compliance with privacy regulations by anonymising or obfuscating personal data when necessary (eg blurring personnel faces in footage). Additionally, policies should dictate how long video data is stored and when it should be deleted.

- Physical Security: Underground cameras and edge computing devices must be protected from tampering. Physical security measures, such as tamper-resistant housings or access control for underground hardware, should be implemented to prevent sabotage or unauthorized removal of equipment.

- Network Security: The system must operate within a secure network architecture to prevent unauthorized access, data breaches, or cyberattacks. Measures such as firewalls, intrusion detection systems (IDS), and virtual private networks (VPNs) should be considered to protect the system from external threats.

Hardware selection

Selecting the appropriate video cameras and lighting is essential for ensuring accurate and reliable progress tracking in the underground mining environment. Due to the harsh conditions in mines, including poor lighting, dust, and varying temperatures, the right hardware choices directly impact the quality of the data collected.

Video camera considerations include:

- Low-Light Performance: Cameras must have strong low-light capabilities. Infrared (IR) or thermal cameras may be required for environments where visibility is extremely limited.

- Durability and Environmental Protection: Cameras must be rugged and able to withstand the challenging conditions in a mine, such as dust, moisture, and vibrations. Cameras rated with IP (Ingress Protection) certifications (eg IP66 or higher) are recommended to ensure they are waterproof and dust-resistant.

- Resolution: High-resolution cameras (at least 1080p or higher) are necessary to capture clear and detailed footage for accurate object detection and tracking, especially when personnel and equipment are distant or partially obscured.

- Field of View (FoV): Wide-angle lenses may be needed in areas where the camera must cover large zones, such as drifts or shaft stations. However, for areas requiring precision, narrower fields of view are preferred to reduce distortion.

- Frame Rate: Cameras should support a high enough frame rate (eg 15 fps or higher) to capture the fast movements of machinery and personnel during mining activities, ensuring smooth and continuous tracking.

- Power: PoE (Power over Ethernet) cameras are an option for providing both power and data over a single cable.

Since many areas of the mine may not have sufficient natural or artificial lighting, supplemental lighting systems should be installed. Lighting should be adjustable to prevent overexposure or glare in video feeds. This ensures that the cameras capture a balanced image, enabling more accurate detection of objects like vehicles, personnel, and equipment.

Data collection and preprocessing

Object classes such as personnel, mobile equipment and shotcrete, and specific mining tasks (eg drilling, mucking, hauling) are defined for detection, based on the operational activities in question. Accurately identifying these classes enables the tracking of operational progress and the automation of time studies.

Data is collected from video feeds installed in key areas of the underground mine, including drifts, remucks, and shaft stations. These feeds capture the movement of equipment and personnel during various phases of mining operations. Given the challenging conditions underground, such as low lighting and dust, data diversity is critical. Video frames must capture different environmental conditions, such as varying light sources, obstructions, and moving objects.

Preprocessing involves improving the quality of these raw video frames. Techniques like noise reduction, contrast enhancement, and resizing ensure that the images meet the model's input specifications. Data augmentation techniques (eg brightness adjustments, rotations, scaling) may be used to artificially expand the data set. Furthermore, class imbalance is common in mining (eg more frequent detection of trucks than personnel), and strategies like oversampling, undersampling, or generating synthetic data can be employed to address this.

Data annotation

Images and video frames are annotated with bounding boxes, polygons, or masks to define the objects of interest. For example, loaders or personnel working in drifts are highlighted, along with any visible equipment or safety markers. Annotation is performed either manually or with the assistance of pre-trained models that help detect key objects within underground environments.

To ensure high-quality annotations, periodic reviews or automated checks are implemented. This prevents errors that could result from the challenging conditions underground, such as misidentifying machinery due to poor lighting. Each annotation focuses on visible parts of objects, avoiding interpretive bounding that could introduce bias or errors in model training. The annotated data is stored in version-controlled data sets, which track changes and updates over time, allowing for continuous monitoring model's performance.

Model training

The model used for the underground mining application is a neural network architecture designed for real-time object detection, such as YOLOv5. YOLOv5 has the advantages of low computational complexity, fast inference time, memory efficiency and ease of deployment across CPUs, GPUs and the cloud.

Model training is conducted using annotated data from the mining environment. Hyperparameters such as learning rate, batch size, and the number of epochs are tuned to ensure optimal performance. Mean Average Precision (mAP) and loss functions are used to measure how well the model is detecting objects under the unique underground conditions.

Techniques such as cross-validation, regularisation, and early stopping are employed to avoid overfitting and ensure that the model generalises well to new data, even when the operational environment changes (eg after blasting or during shifts). Retraining occurs as new operational data is collected or when there are significant changes in mining equipment or processes.

Application development

The object detection model is integrated into a software system that tracks mining operations. The application manages the full data life cycle, from capturing video feeds in the mine to processing them for object detection and progress tracking. The frontend provides a user-friendly interface for operators or managers to monitor activities in real-time.

The application also includes object tracking, allowing detected objects (eg personnel or equipment) to be tracked across multiple video frames. This ensures continuous tracking of mining operations, enabling the system to automatically record how long specific activities take, such as drilling, mucking, or hauling muck. Each detected object is assigned a unique ID to maintain a consistent record of its location and movement throughout the mine.

Scalability is a key consideration, as the application must be able to handle multiple video streams from various sections of the mine. The system integrates with existing mine systems and databases through APIs, providing seamless communication. Additionally, edge computing may be utilised for video inference at the mine site, reducing the reliance on cloud servers and improving response times for real-time analysis.

Application deployment

Development environment

The development environment consists of both local and cloud resources, enabling model development and testing. Local machines are used to develop and test the initial models. Virtual environments are employed to standardise dependencies across different development stages, ensuring consistency between local and cloud environments. The entire model and application are frequently tested, verifying that the system can handle real-time video processing and object detection in the underground environment.

Production environment

Once the application is ready, it is deployed in a production environment. Cloud servers are utilised for high-availability tasks, such as centralised monitoring, while on-site edge devices may process video streams from the underground cameras in real time.

Orchestration tools manage the containerised services, providing scalability and load balancing as needed. Continuous integration/continuous deployment (CI/CD) pipelines ensure that any updates to the system, whether model retraining or application feature enhancements, are deployed without disrupting ongoing operations. Security is paramount, with firewalls and authentication protocols protecting access to the system, ensuring only authorized personnel can access video feeds and reports.

Application validation

Application validation is critical in an underground mining context, where accuracy and reliability directly impact operational efficiency and safety. The model outputs are validated by comparing them with manually annotated video logs. This ensures that the detection system is correctly identifying key events, such as the movement of machinery, personnel, or equipment usage.

Performance metrics, such as accuracy, precision, recall, and F1 score (used when one class occurs much more frequently than others), are used to quantify the system's performance. Cross-validation techniques are employed to ensure that the model generalises across different mining conditions. If necessary, the model is retrained with additional data or optimised through hyperparameter tuning to improve its accuracy.

Result data analytics and visualisation

A key component of the system is its ability to visualise mining progress and operational efficiency in real time. Dashboards are created to provide access to relevant stakeholders, displaying key performance indicators (KPIs) related to equipment usage and personnel movement for example. These dashboards are designed to overlay detection results on live video feeds from underground, with bounding boxes highlighting detected objects such as vehicles or personnel.

The system provides real-time tables and graphs that display mining progress, comparing actual operational times with planned schedules. This data is crucial for automating time studies, helping mining managers track the duration of various tasks and identifying bottlenecks in the operation. The frontend uses secure login and authentication protocols, ensuring only authorized personnel can access operational data. Cloud-based servers host the backend models, ensuring continuous communication between the video feeds, detection models, and dashboards. These details are shown in Figure 1.

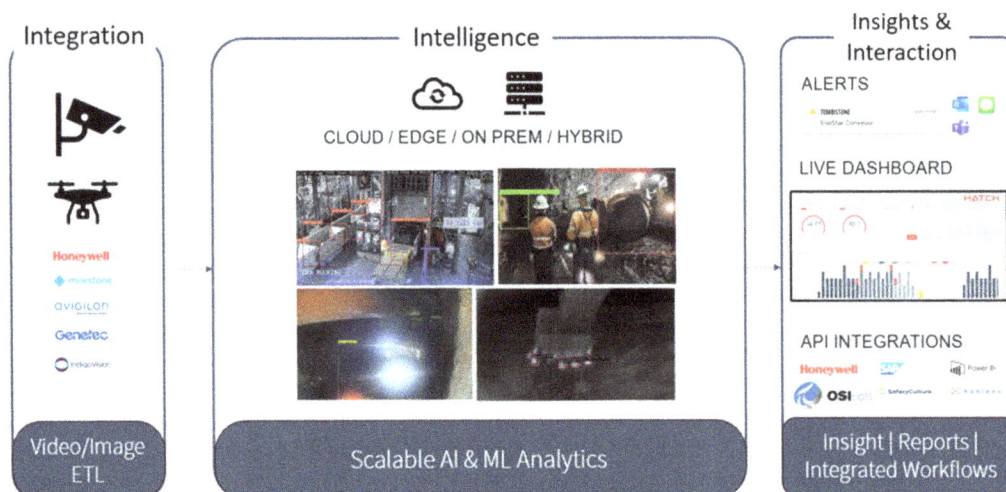

FIG 1 – Computer vision application overview images.

CASE STUDIES AND RESULTS

Shaft sinking cycle tracking

Overview

Shaft sinking in an underground hard rock mining setting is a complex process used to create vertical access to mineral deposits. The shaft is the main entry point for personnel, materials and equipment and the key steps in sinking are:

- Drilling: drilling of explosives holes.

- Blasting: fracturing of rock at the shaft bottom via explosives detonation.

- Mucking: removal of blasted rock using hoisting systems.

- Ground Support: installation of support systems to stabilise the vertical shaft walls during sinking.

- Lining: installation of a permanent shaft lining (typically concrete) for prevention of rock bursts.

The main steps outlined above make up the overall shaft sinking cycle. The planned cycle time is estimated by the sinking contractor, based on project parameters such as shaft diameter, ground support requirements and liner design.

To track shaft sinking project progress it is advantageous to compare planned and actual cycle times. This allows for identification of cycle steps for which the actual duration exceeds planned. Once identified, efforts may then be made to reduce duration. The comparison exercise may also highlight idle time within the sinking cycle. Reductions in sinking cycle step duration and idle time improve productivity and can decrease the potential for project cost overruns.

Determination of actual shaft sinking cycle time is typically a paper based exercise whereby sinking personnel manually record step durations. As previously discussed, such methods are prone to inefficiency and inaccuracy.

This case study involves use of computer vision for automated shaft sinking cycle time determination using the workflow outlined above.

System configuration

Video cameras were placed at the top shaft level (ie collar), enabling image capture of items into and out of the shaft. Video cameras were also placed at the hoist dumps to obtain images of mucking and at the tag board to determine personnel loading. Face blurring was employed to obscure personnel identity as required by a risk review conducted with the project team. The start and end indicators for each cycle step are shown in Table 1 along with an example of the overall cycle time determination in Figure 2. Live video was streamed to the cloud where the computer vision application workflow was executed and a web-based dashboard, accessible by project personnel was used for visualisation.

TABLE 1
Cycle steps with start and end indicators.

Cycle step	Start indicator	End indicator
Load	Emulsion loader enters shaft	Emulsion loader exits shaft (start/end of overall cycle)
Blast	Emulsion loader exits shaft (start/end of overall cycle)	First muck bucket dumps
Muck	First muck bucket dumps	Bolt basket enters shaft
Bolt	Bolt basket enters shaft	Second last muck bucket dumps
Drill	Second last muck bucket dumps	First transmixer enters hopper area
Concrete	First transmixer enters hopper area	Last transmixer enters hopper area

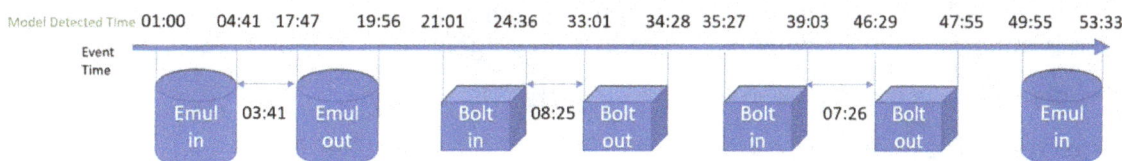

FIG 2 – Shaft sinking – overall cycle time and image examples.

Following commissioning of the hardware, data collection and annotation, model training and application development and deployment, the project team at site validated the outputs. This was done by manually reviewing video footage on high speed and recording cycle step duration and counts using the same logic outlined above. As part of the validation process, discrepancies were noted between the application output and a paper based logbook tracking system that was implemented by the shaft sinking contractor.

System use

The application was released for use by relevant project stakeholders including Hatch personnel, the client and the shaft sinking contractor. A view of the real-time dashboard is shown in Figure 3.

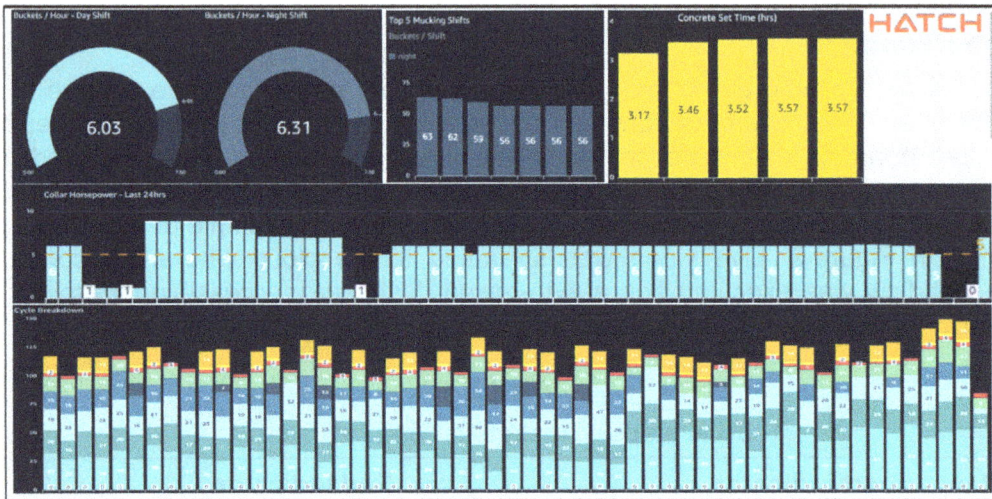

FIG 3 – Shaft sinking performance tracking dashboard.

Through use of the application the following was determined:

- Optimised personnel loading for concrete pours.

  On average eight-hour reduction in concrete time with optimised personnel loading and immediate delay response

- Impact of round length on cycle time.

- Variation in cycle time by crew.

- Validation of contractor cycle time estimates.

- Improved forecasting through use of accurate cycle time data.

- Identification of continuous improvement initiatives.

The system operated continuously with zero application specific downtime, while network and power connectivity issues resulted in minor instances of camera outage.

Service cage utilisation monitoring

Overview

In an underground mine, the service cage is a vertical transport system that is used to move personnel, equipment and materials between the various mine levels. Maintaining high cage utilisation is imperative to achieving mine productivity targets. This may be accomplished through:

- Proper scheduling of the cage for various operational needs.

- Maximising the number of cage trips while minimising downtime.

- Ensuring the cage is operating at or near capacity on each trip.

Depending on the extent of mining excavation and scope of supporting infrastructure construction, the required cage utilisation may be higher compared to the maximum allowable utilisation. Maximum allowable utilisation is impacted by downtime required for legislated maintenance and electrical inspections and the shift schedule employed.

This case study involves a mining project for which high cage utilisation was required to maintain the overall project schedule critical path. As such the client placed high priority on comparing planned cage utilisation with actual. While the planned cage utilisation was easily quantified via the owner issued weekly schedule, there existed no method to accurately measure actual utilisation.

System configuration

The workflow outlined above was followed to develop an computer vision application with the intention of quantifying cate utilisation. A series of video cameras were placed at operational mine level shaft stations, enabling image capture for personnel, equipment and materials moving onto and off of the cage. This was done following a risk review with the client's operations and Information and Technology (IT) teams. Similar to the case for shaft sinking, face blurring techniques were implemented. Functionality was also included whereby the planned daily cage use is input to the application for comparison against the actual use as determined by the computer vision application. The actual use was determined through correct identification of the cage contents (personnel, materials, shotcrete) and of the state of the cage doors (opened or closed). The loading and unloading of personnel and materials onto and off of the cage was used to establish cage utilisation.

Following implementation of the application workflow, the project team at site validated the outputs. This was done by querying the site data historian to retrieve data such as cage position and payload along with manual review of video footage on high speed to determine cage use. Similar to the case for shaft sinking discrepancies were noted between the application output and a paper based tracking system implemented by the cage operations contractor. Cage use for personnel movement was found to be underreported by the contractor.

System use

The application was released for use by relevant project stakeholders including Hatch personnel, the client and the shaft sinking contractor. A view of the real-time dashboard is shown in Figure 4, with the following features:

- Calendar for date selection.

- Live view of four shaft station cameras: Selection of individual level cameras with scrubbing bar for quick navigation to events.

- Cage use timeline by level with comparison of overall actual and plan values.

- Cage use plan input.

- Actual cage use editing: Ability to overwrite application assigned cage use for instances of misassignment.

- Stacked bar chart of total planned and actual cage use times.

- Seven day trend of daily cage utilisation along with period average for planned and actual.

- Table comparing planned and actual cage use values by type.

- Planned and actual shotcrete and concrete bag count by level (2).

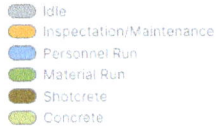

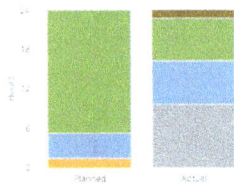

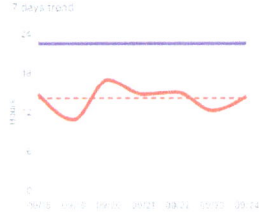

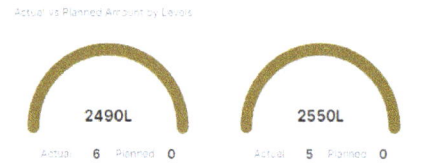

Operation Items ⊙ 24 Sep 2024

Production Utilization ⊙
7 days trend

Cage Activities Hours ⊙

Operation Items	Hours	
	Planned	Actual
Idle	0	9.64
Tripped	0	0.00
Gas Check	0	0.00
Inspectation/Maintenance	1.5	0.00
Personnel Run	3.75	6.67
Material Run	18.75	6.45
Shotcrete	0	1.24
Concrete	0	0.00
Mobile Equipment	0	0.00
Production Utilization	22.5	14.36

Shotcrete Bags ⊙
Actual vs Planned Amount by Levels

2490L — Actual 6 Planned 0

2550L — Actual 5 Planned 0

Concrete Bags ⊙
Actual vs Planned Amount by Levels

2490L — Actual 0 Planned 0

2550L — Actual 0 Planned 0

FIG 4 – Service cage utilisation dashboard.

The scrubbing bar functionality is shown in Figure 5. In this instance, the 2550L camera was selected by clicking on the box highlighted in red. Upon selection, a larger view of the 2550L shaft station is displayed on the screen along with the scrubbing bar. The bar allows the user to quickly navigate the video playback to view events of interest, as indicated by the Door Open (DO) and Door Closed (DC) text on the scrubbing bar. Cage use for shotcrete and personnel movement can be observed accordingly.

Upon implementation of the computer vision application, a significant deviation in how the cage was actually used compared to planned was observed. This was particularly pronounced for the movement of personnel as outlined in Figure 6. Personnel movement was initially over 100 per cent higher than planned, with a reduction observed in the months following commissioning of the application. Actual idle time was also significantly higher compared to planned with ongoing efforts to reduce.

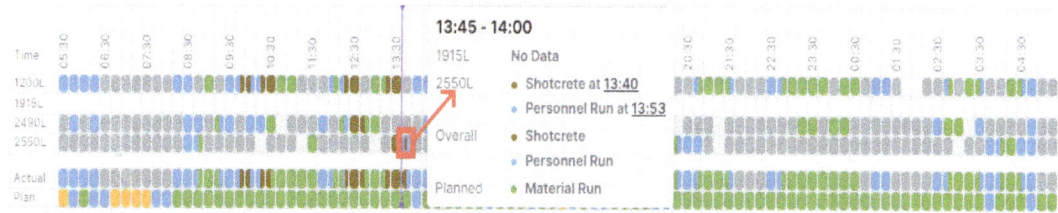

FIG 5 – Shaft station level camera with scrubbing bar.

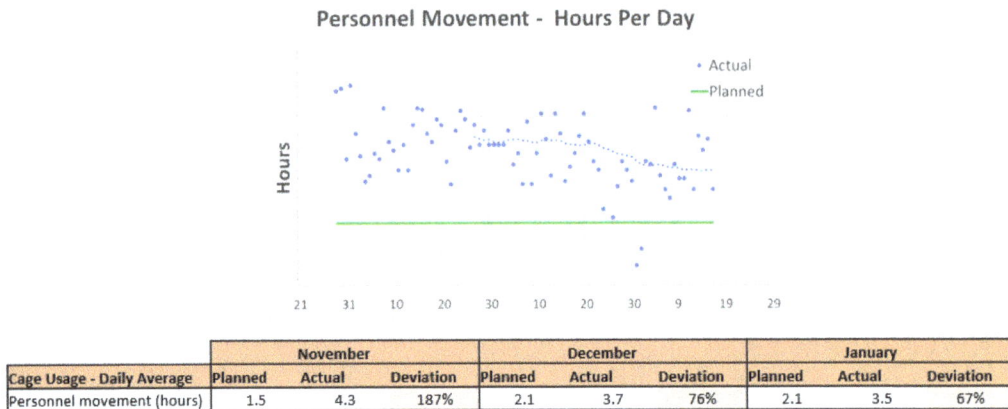

Personnel Movement - Hours Per Day

Cage Usage - Daily Average	November			December			January		
	Planned	Actual	Deviation	Planned	Actual	Deviation	Planned	Actual	Deviation
Personnel movement (hours)	1.5	4.3	187%	2.1	3.7	76%	2.1	3.5	67%

FIG 6 – Personnel movement – planned and actual.

The identification of unplanned cage use and idle time allowed the project team to focus efforts on reducing both, to increase actual use of the cage for material movement, as required to maintain project critical path.

CONCLUSION

Hatch Mine VisionAI represents a breakthrough in underground mining productivity tracking. By harnessing the power of computer vision, the system overcomes many limitations of traditional manual tracking methods, offering an efficient, cost-effective solution. The real-time data provided by Hatch Mine VisionAI allows project managers to make informed decisions that drive improvements in productivity, reduce cost overruns, and enhance safety. As mining operations grow more complex, adopting technologies like Mine VisionAI will be critical for the industry's future success.

RECOMMENDATIONS

To further capitalise on the benefits of productivity tracking and safety management in underground mining, the following recommendations are proposed:

- Expand AI Integration: Mining companies should look beyond productivity tracking and explore AI-driven solutions in areas such as geological analysis, equipment health monitoring, and hazard identification.

- Integrate IoT Technologies: Combining computer vision with Internet of Things (IoT) sensors would create a comprehensive system for tracking environmental conditions, equipment status, and operational efficiency.

- Promote User Adoption and Training: Success depends on personnel understanding and adopting the system. Comprehensive training should be provided to ensure users fully leverage Mine VisionAI's features to enhance decision-making and operational efficiency.

By adopting these recommendations, mining companies can enhance the return on their investment in digital transformation technologies, maintain high safety standards, and improve project timelines and budgets.

REFERENCES

Deloitte, 2014. Between a rock and a hard place – Addressing distress in the mining industry, report. Available from: <https://www2.deloitte.com/content/dam/Deloitte/ca/Documents/energy-resources/ca-en-energy-and-resources-between-a-rock-and-a-hard-place.pdf>

Dilek, E and Dener, M, 2023. Computer vision applications in intelligent transportation systems: A survey, *Sensors*, 23(6):2938. https://doi.org/10.3390/s23062938

Esteva, A, Robicquet, A, Ramsundar, B, Kuleshov, V, DePristo, M, Chou, K, Cui, C, Corrado, G, Thrun, S and Dean, J, 2019. A guide to deep learning in healthcare, *Nature Medicine*, 25:24–29. https://doi.org/10.1038/s41591-018-0316-z

Gao, R X, Kruger, J, Merklein, M, Mohring, H-C and Vancza, J, 2024. Artificial Intelligence in manufacturing: State of the art, perspectives and future directions, *CIRP Annals - Manufacturing Technology*, 73(2):723–749. https://doi.org/10.1016/j.cirp.2024.04.101

Advancing In-Place Recovery (IPR) – innovations in underground leaching operations for sustainable mineral extraction

A Hassanvand[1], P Dare-Bryan[2], Z Wang[3], D Hunter[4] and A Scott[5]

1. Senior Research Engineer, Orica Australia, Kurri Kurri NSW 2327.
   Email: armineh.hassanvand@orica.com
2. Senior Research Fellow, Orica Australia, Perth WA 6007. Email: peter.dare-bryan@orica.com
3. Research Fellow, School of Chemical Engineering, The University of Adelaide, Adelaide SA 5000. Email: zhihe.wang@adelaide.edu.au
4. Senior Metallurgist, Core Resources, Albion Qld 4010. Email: dhunter@coreresources.com.au
5. Lead Principle – Think and Act Differently, BHP, Brisbane Qld 4000.
   Email: andrew.j.scott@bhp.com

ABSTRACT

Underground mining operations are adopting innovative technologies to address the growing need for critical minerals while upholding sustainability goals. As a result, underground leaching operations are gaining more attention in recent years due to their capacity for low cost and low environmental footprint mineral extraction. Among these, In-Place Recovery (IPR) targets hard rock formations with limited natural permeability. IPR involves blasting a stope, using a lixiviant to dissolve metal-bearing minerals, and pumping the solution to the surface for processing. Blasting plays a critical role in generating small rock fragments to increase contact between the ore and the lixiviant.

Supported by a Cooperative Research Centres Projects (CRC-P) grant, this project assesses the viability of IPR at the BHP Prominent Hill mine site. A test area has been defined, and representative rock samples have been collected, mineralogical analysis completed ahead of leaching kinetics testing. The rock mass has also been characterised for blast modelling and a suite of blast designs has been simulated to determine suitable designs that maximise fragmentation. In parallel, computational fluid dynamic models coupled with leaching reaction has been developed to simulate the performance of a leaching column while varying influential parameters such as ground temperature, lixiviant diffusion through the rock fragment, and fluid flow rate.

This cross-functional collaborative project involves foundational research on blast optimisation for IPR, effective mineral leaching methods, and hydro-thermo-chemical fluid flow modelling, prior to in proof-of-concept field trials. The outcome of this work provides a framework for operations to assess the feasibility and viability of IPR for low-grade deposits which are often left un-mined, achieving increased copper production while meeting sustainability goals.

INTRODUCTION

Population growth, rise in living standards, and the transition to net zero through electrification are driving an unprecedented demand for copper and other minerals. Meeting this demand is increasingly challenging as miners confront more complex and deeper orebodies at lower grades. Surface mining methods with large environmental footprints are becoming economically unviable and socially unacceptable, leading to a shift towards underground (UG) operations. However, to meet the rising demand while adhering to the ESG (Environmental, Social, and Governance) requirements, the UG mining industry must adopt more sustainable approaches with enhanced safety, minimised waste, and reduced energy consumption.

In contrast to conventional mining methods, which involve transporting blasted ore to the surface for mineral processing, underground leaching operations such as *In situ* Recovery (ISR) and In-Place Recovery (IPR) have the potential to offer a more sustainable alternative. ISR involves injecting the lixiviant through inject wells, allowing it to circulate through a permeable ore deposit, creating a pregnant leach solution (PLS) and collecting the PLS via recovery wells. Extraction of the target element from the PLS is via standard hydrometallurgical processes similar to that of heap leaching operations. ISR has been proven successful in the uranium industry, accounting for nearly 60 per cent of the global production, as the mineral is hosted in naturally permeable sandstone. IPR, on the other hand, is applicable to hard rock deposits with minimal natural permeability. An IPR operation consists of fracturing a stope of the deposit via blasting, applying a lixiviant at the top and allowing it to percolate through the broken rock mass, and collecting the PLS at the bottom of the stope. The PLS is then pumped to the surface for processing. To blast a full stope, the initial removal of a small portion of the stope is necessary to create a void to blast the stope into. IPR offers increased safety by avoiding the creation of open stopes and by reducing personnel exposure to hazardous areas. Minimising rock haulage and eliminating the need for beneficiation circuits significantly reduce energy consumption, decrease water usage and eliminate tailings.

The authors have conducted a comparison study of the capital and operating cost of IPR versus Sub-Level Open Stoping (SLOS) for greenfield operations at low copper grades (Dare-Bryan and Hassanvand, 2023). The economic assessment showed that, at 1 per cent Cu, IPR is more profitable than SLOS due to its lower opex and capex. However, early adoption of the method will likely be as a hybrid approach where high-grade ore is mined conventionally, and adjacent low-grade ore is recovered through IPR (Mousavi and Sellers, 2019). This makes IPR a valuable addition to existing UG mining operations to increase production and/or to extend the mine life.

While IPR appears as an attractive mining method to unlock low-grade deposits, currently there is a lack of comprehensive understanding of lixiviant percolation and ore extraction through IPR. This CRC-P project is designed to explore and gain insight into all aspects of the IPR operation.

PROJECT SUMMARY

Problem statement and objectives

IPR presents several significant challenges that must be addressed prior to full-scale production in an operation. While blasting enhances permeability and increases the potential for higher metal recovery, IPR still suffers from lower recovery rates compared to conventional metal extraction methods. This is due in part to the larger average particle size of fragmented ore, which cannot be processed through crushing and agglomeration as in heap leaching, and the presence of gangue and clay materials that are not separated as they are in beneficiation circuits. In addition to that, operational challenges such as short circuiting of the lixiviant (ie where the lixiviant takes unintended shortcuts), and solvent entrapment in dead-end zones of the stope can further limit the recovery. These issues are exacerbated in the unsaturated environment of an underground leaching operation, where maintaining uniform flow of fluid(s) is particularly challenging.

Hence, prior to proceeding with a field trial at a mine site, the proposed project aims to address these challenges through a comprehensive study that includes laboratory experiments and fundamental modelling. The knowledge gained from the study will pave the way for planning an IPR trial in a safe and methodical manner. The objectives of the CRC-P IPR project include:

- Optimising the hydrometallurgical extraction through laboratory work.

- Developing blast fragmentation and heave modelling for IPR.

- Creating a hydro-thermo-chemical IPR fluid flow model.

- Establishing a multi-physics IPR simulation tool.

- Seeking necessary regulatory approvals and conducting a proof-of-concept field trial.

The success of this project hinges on collaboration between subject matter experts from Orica, the University of Adelaide, Core Recourse, and BHP, brought together under a Cooperative Research Centres Projects (CRC-P) arrangement, with $2.4M funding support from the Australian Government. In this project, each partner will contribute specialised knowledge and resources, from blasting techniques to fluid flow modelling and metallurgical test work, ensuring that the project not only addresses the challenges of IPR but also positions it as a viable and sustainable mining method.

Basic methodology outline

Test area allocation at Prominent Hill mine site

Through collaboration with BHP, a test area for the field demonstration of IPR was allocated in the Prominent Hill (PH) mine. Prominent Hill is a copper, silver and gold mine located 650 km north-west of Adelaide in South Australia, with a mineral reserve of 180 Mt at 0.9 per cent Cu and 0.8 g/t Au. The allocated test area is in the non-economic zone of the Ankata region and will have minimal impact on the site's production (see Figure 1).

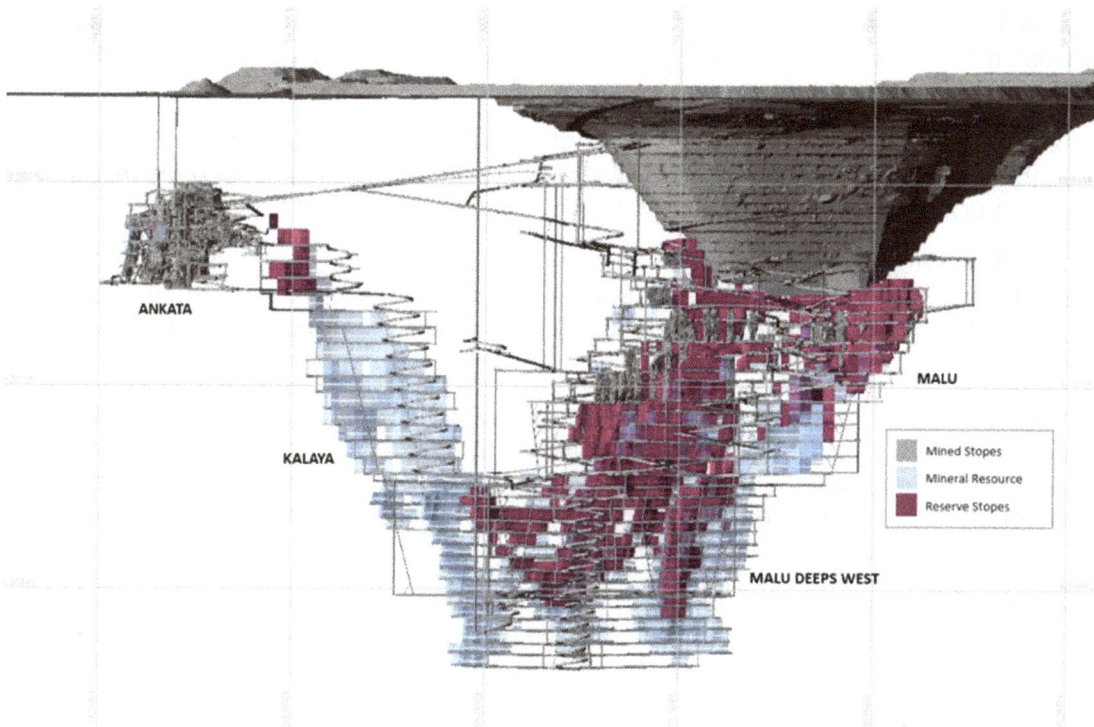

FIG 1 – Prominent Hill long section looking north, image source: 2022 Mineral Resource and Ore Reserve Statement from OZ Minerals.

The PH site team provided the geomechanical data for the designated test area and collected representative rock samples for characterisation and hydrometallurgical test work.

Leaching copper sulfides

Copper from copper oxide minerals can easily be dissolved as cupric ion (Cu^{2+}) under mildly acidic conditions. However, copper sulfide minerals are abundant at PH and extracting copper from such minerals is significantly more challenging than copper oxides. While a range of technologies are emerging to improve copper recovery from sulfide ores, the most common leaching chemistry consists of using both acidic conditions and an oxidising agent (Schlesinger *et al*, 2011). For secondary sulfides like chalcocite (Cu_2S), this process is achieved using ferric ion (Fe^{3+}) as the oxidising agent, as shown below:

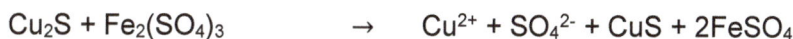

$$Cu_2S + Fe_2(SO_4)_3 \rightarrow Cu^{2+} + SO_4^{2-} + CuS + 2FeSO_4$$

Covellite, CuS, can be similarly oxidised by ferric ions:

$$CuS + Fe_2(SO_4)_3 \rightarrow Cu^{2+} + SO_4^{2-} + 2FeSO_4 + S^0$$

The ferrous ion (Fe^{2+}) produced is then re-oxidised to ferric ion (Fe^{3+}) in the presence of oxygen and acidic conditions in a separate reaction:

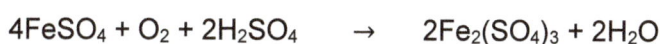

$$4FeSO_4 + O_2 + 2H_2SO_4 \rightarrow 2Fe_2(SO_4)_3 + 2H_2O$$

This leaching mechanism is a mature process and widely applied in heap leaching operations in Chile (Ghorbani, Franzidis and Petersen, 2016). Hence, in this project, we are utilising this technique for extracting copper.

Hydrometallurgical test work plan

Hydrometallurgical test work for this program is being undertaken at Core Resources' laboratories in Albion, Queensland. The program itself will be split into four distinct phases as follows:

- Phase 1: Complete mineral characterisation, including inductively coupled plasma optical emission spectroscopy (ICP-OES), X-ray fluorescence spectroscopy (XRF), quantitative X-ray diffraction (QXRD), and sequential copper analysis.

- Phase 2: Benchtop leach testing under different conditions on finely ground ore samples. Variables investigated in this phase will include the effects of acid concentration, ferric dosing, temperature and residence time on the extraction of copper from the ore.

- Phase 3: Intermittent bottle roll testing, including the investigation of different particle sizes on the leaching performance. These tests will be conducted on relatively course particles of different sizes over an extended period with test conditions based upon the outcomes of Phase 2.

- Phase 4: Based on outcomes from the previous test phases, column leach testing will be conducted on ore samples with a particle size distribution derived from the fragmentation modelling. These tests will be carried out over an extended period (eg 4–6 months) with monitoring and control of ferric concentration and free acid. Regular assays will be carried out to monitor the leach performance over time.

The outcomes of the test work program will provide a means of calibrating and then validating the hydro-thermo-chemical fluid flow model while providing crucial input into the design and engineering of the on-site program.

RESULTS AND DISCUSSION

In this section, the progress of the project is summarised in three key areas: hydrometallurgical laboratory work, blast fragmentation modelling, and hydro-thermo-chemical fluid flow modelling.

Mineralogical analysis and experimental data to date

Sample preparation

The ore samples from the Ankata region at PH were blended and crushed to a nominal top-size of 100 mm. The material was then representatively subdivided into two subsamples. One subsample was crushed to a nominal top-size of 50 mm and then further divided into two subsamples. One of these was subsequently crushed to a nominal top size of 25 mm. Table 1 summarises the mass split of subdivided samples.

TABLE 1

Subsample mass percentage in each nominal top size relative to total mass of sample.

Nominal top size (mm)	Mass percentage of total (%)
100	50
50	25
25	25

The 25 mm top size sample was rotary subdivided to produce 12 kg subsamples. One of them was further crushed to a nominal top-size of 3.35 mm before being ground to a top size of 600 μm. This sample was then subdivided further to produce samples for head grade characterisation and Phase 1 test work.

Head grade characterisation

Key results from elemental assays (ICP-OES and XRF) are given below in Table 2. Results show a head copper concentration of approximately 0.8 wt per cent. QXRD results, shown in Table 3, indicate that the primary copper bearing phases are chalcopyrite (0.7 wt per cent), chalcocite (0.5 wt per cent) and bornite (0.3 wt per cent). This generally corresponds well with sequential copper analysis which gave 21.9 per cent extraction using acid leaching, an additional 41.9 per cent extraction with cyanide leaching and 36.2 per cent remaining in the residue.

TABLE 2

Elemental results from ICP-OES and XRF on a mg/kg basis for key elements in ore sample.

Element	Units	ICP-OES	XRF
Al	mg/kg	67 500	65 363
Ca	mg/kg	6660	9005
Cu	**mg/kg**	**8060**	**7860**
Fe	mg/kg	192 000	191 000
K	mg/kg	29 200	35 528
Mg	mg/kg	6980	7659
Si	mg/kg	218 000	220 177
S	mg/kg	5070	5800

TABLE 3

QXRD Results from head characterisation.

Phase (-)	Concentration (wt%)	Phase (-)	Concentration (wt%)
Quartz	29.1	Microcline	1.3
Hematite	20.1	Calcite	1.3
Illite	17.8	Biotite	0.9
Muscovite	16.0	**Chalcopyrite**	**0.7**
Magnetite	4.9	**Chalcocite**	**0.5**
Dolomite	2.4	**Bornite**	**0.3**
Clinochlore	2.4	Pyrite	0.3
Albite-Ca	1.7	Siderite	0.3

Initial leach testing results

Phase 2 test work is currently underway and aims to explore the effects of temperature, acid concentration and ferric concentration on copper leaching from the ore samples provided. So far, three tests (LT1, LT2 and LT3) have been completed, and test conditions and results are given in Table 4.

Based on the interim results presented in Table 4, the copper recovery is maintained at a high level irrespective of the ferric concentration. This suggests that the acid concentration and temperature can be lowered in the following benchtop leach tests with minimal adverse effect on the copper recovery.

TABLE 4

Phase 2 interim test conditions and results.

Variable	Units	LT1	LT2	LT3
Test Conditions				
Temperature	°C	60	60	60
Pulp Density	wt% Solid	20	20	20
Residence Time	hr	8	8	8
Acid Concentration	g H_2SO_4/L	120	120	120
Fe^{3+} Concentration	g/L	33.5	16.8	8.4
Final Conditions				
Acid Concentration	g H_2SO_4/L	99	98	102
Fe^{3+} Concentration	g/L	28.2	12.4	8.0
Fe^{2+} Concentration	g/L	4.4	4.7	4.9
Extraction Extent (1-Tail/Head Basis)				
Al	%	15.8	16.8	23.9
Ca	%	74.8	76.9	78.8
Cu	%	65.4	65.3	66.6
Fe	%	1.7	5.9	4.4
K	%	5.6	5.8	12.9
Mg	%	36.6	36.9	43.7

Blast design fragmentation

In conventional underground metalliferous operations, blast designs are typically developed to produce a blasted muck pile with sufficient swell and fragmentation to ensure efficient loading and hauling to the surface. For IPR, significantly finer fragmentation is required compared to conventional operations to increase ore/lixiviant contact. Numerical modelling was conducted to determine the fragmentation achieved from typical production blast designs at PH, and then a series of design changes were modelled to determine what fragmentation is practically possible for IPR.

The Mechanistic Blasting Model (MBM) is based on Elfen, a large dynamic finite/discrete element code (Owen, Munjiza and Bicanic, 1992). MBM simulates the non-ideal explosive loading of the blasthole wall and the subsequent fracturing and bulk motion of the surrounding rock mass due to stress/strain effects and the influence of dynamic gas loading in the blasthole and throughout the fracture network (Minchinton and Lynch, 1996; Dare-Bryan, Mansfield and Schoeman, 2012). Fracturing is handled by a strain-rate-dependent softening Rankine plasticity model. Under sufficient fracturing, discrete elements are formed as separate, distinct polygonal elements made up of one or more finite elements. As such, these discrete elements are fully deformable and can support stress and strain. The detonation of the explosives in the blasthole is modelled using data from a non-ideal detonation model (Kirby, Chan and Minchinton, 2014) to derive the influence of the rock confinement and blasthole diameter on the resulting velocity of detonation (VoD) and blasthole wall pressure-time profile. The post Chapman-Jouguet pressure loading of the rock mass is handled dynamically by a gas flow model.

Within MBM the mesh is only allowed to fracture down to the element size. Therefore, in this study, as the fines portion of the fragmentation is of particular importance, models were run with small (20 mm) elements at the expense of long simulation run times. However, there is still a bias within the simulation particle size distribution (PSD) curve at the element size; therefore, a size distribution function that is known to reliably represent the full PSD is fitted to the larger passing size distribution data from MBM. The most widely used size distribution curve in blasting is the Rosin-Rammler curve, which follows a Weibull distribution and is used in the empirical Kuz-Ram fragmentation model (Cunningham, 2005). However, the more recent Swebrec function used in this work has been shown to fit blast size distribution data more accurately over the complete size range (Ouchterlony, 2010).

While MBM can be run in 3D, it is not tractable for this scale of modelling, therefore, the models will be run as 2D plane strain, taking a horizontal section through the charged portions of the blastholes within a ring. Due to the variable spacing between charges in ring design, multiple sections at different horizons will be modelled, and the individual fragmentation data combined to produce an overall PSD curve for the ring design.

High fidelity numerical models such as MBM require detailed information on the rock mass. The elastic/plastic properties for the ore bearing Haematite Breccia that was modelled in this study are listed in Table 5. The sonic velocities in Table 5 are used as inputs to the model for coupling the explosive loading to the mesh and are derived from the other elastic properties. The structural data provided in Table 6 are used to create discrete fracture networks within the model geometry.

TABLE 5
Ore elastic/plastic rock properties.

Rock type	Haematite breccia (HMBX)
Density (g/cc)	3.1
Young's modulus (GPa)	70
Poisson's ratio	0.25
P-wave Velocity (m/s)	5205
S-wave Velocity (m/s)	3005
UCS (MPa)	99
Tensile Strength (MPa)	9
K_{IC} (MPa.m$^{1/2}$)	1.95
Fracture Toughness (N/m)	55

TABLE 6
Ore structural data.

		Joint sets	
Set	Dip (°)	Dip direction (°)	Nominal spacing (m)
1	70	350	2
2	35	090	5
3	80	280	5
4	65	065	5
5	70	220	5
6	35	245	5

The typical ring design parameters utilised in production stope blasts are listed in Table 7 and ring layout with individual hole charging, denoted by the red portions of the holes, is shown in Figure 2. Also shown in Figure 2 is the baseline model geometry, taking a horizontal section through the toes of the holes, comprising of three rings and six holes from each ring. The geometry includes a representation of the existing structure within the rock mass (Table 6), and a void to provide sufficient relief for the rings to blast into so as not to impede fracture formation while limiting movement to make post-simulation fragmentation analysis easier.

TABLE 7

Typical production ring blast design parameters.

Blasthole Diameter (mm)	89
Burden (m)	2.6
Toe Spacing (m)	3.1
Nominal Powder Factor (kg/t)	0.3–0.4
Nominal Energy Factor (MJ/t)*	1.0–1.4

* Derived from ideal detonation data.

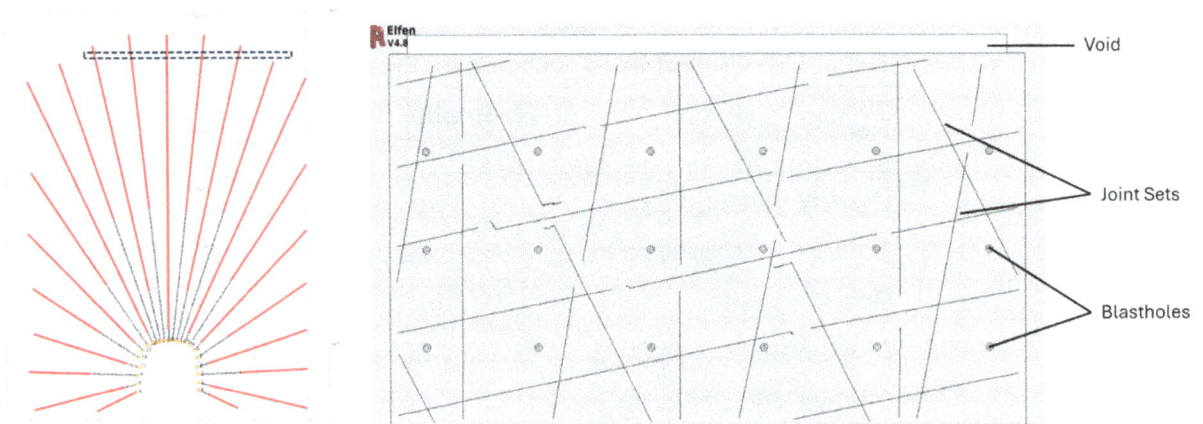

FIG 2 – Typical ring design (left) and the model geometry of the top portion of the ring design with a 2.6 m × 3.1 m blasthole pattern.

As the ring design produces an uneven distribution of the explosives through the stope due to the variable hole spacing, it results in a corresponding uneven explosive energy distribution. Three different hole spacing model geometries were created to generate representative fragmentation data across the range of charge spacings within the ring design (Figure 3). The ring layout was blocked out into regions based on three hole-spacing models (3.1 m, 2.5 m and 2.0 m) to create a weighted average fragmentation PSD for the whole ring. The simulation outputs in Figure 3 highlight the even damage pattern between holes in the 2.6 m × 3.1 m hole pattern, while for the tighter hole spacings of 2.5 m and 2.0 m there is significantly more damage between holes in each ring versus inter-ring within the burdens. This indicates that for this ring design, increased fragmentation, particularly in the coarser size fractions, can be achieved by reducing the ring burden over reducing the hole spacing. For each simulation the region denoted by the broken line was analysed to create a raw PSD curve, the sampled regions are inside the perimeter holes to eliminate edge effects from the analysis.

The raw fragmentation curves for the three simulations are shown in Figure 4, along with the weighted average of the three raw curves. Not surprisingly, the weighted average curve is very similar to the middle spacing (2.5 m) curve, therefore, it is considered reasonable to take the 2.5 m spacing model fragmentation as representative of the whole ring. This will significantly streamline the modelling and analysis of different design changes that maintain the original ring spacings. Figure 4 also includes a Swebrec function curve fitted to the raw 2.5 m fragmentation data. Note that, despite the model creating fragments down to the 20 mm mesh element size, there is a bias in the percent passing at this size such that the raw data overestimates the fragmentation below approximately 50 per cent passing. However, there is still sensitivity in the simulations at these finer passing sizes, as shown by the variation in percent passing at the 20 mm passing size across the three different hole spacing models.

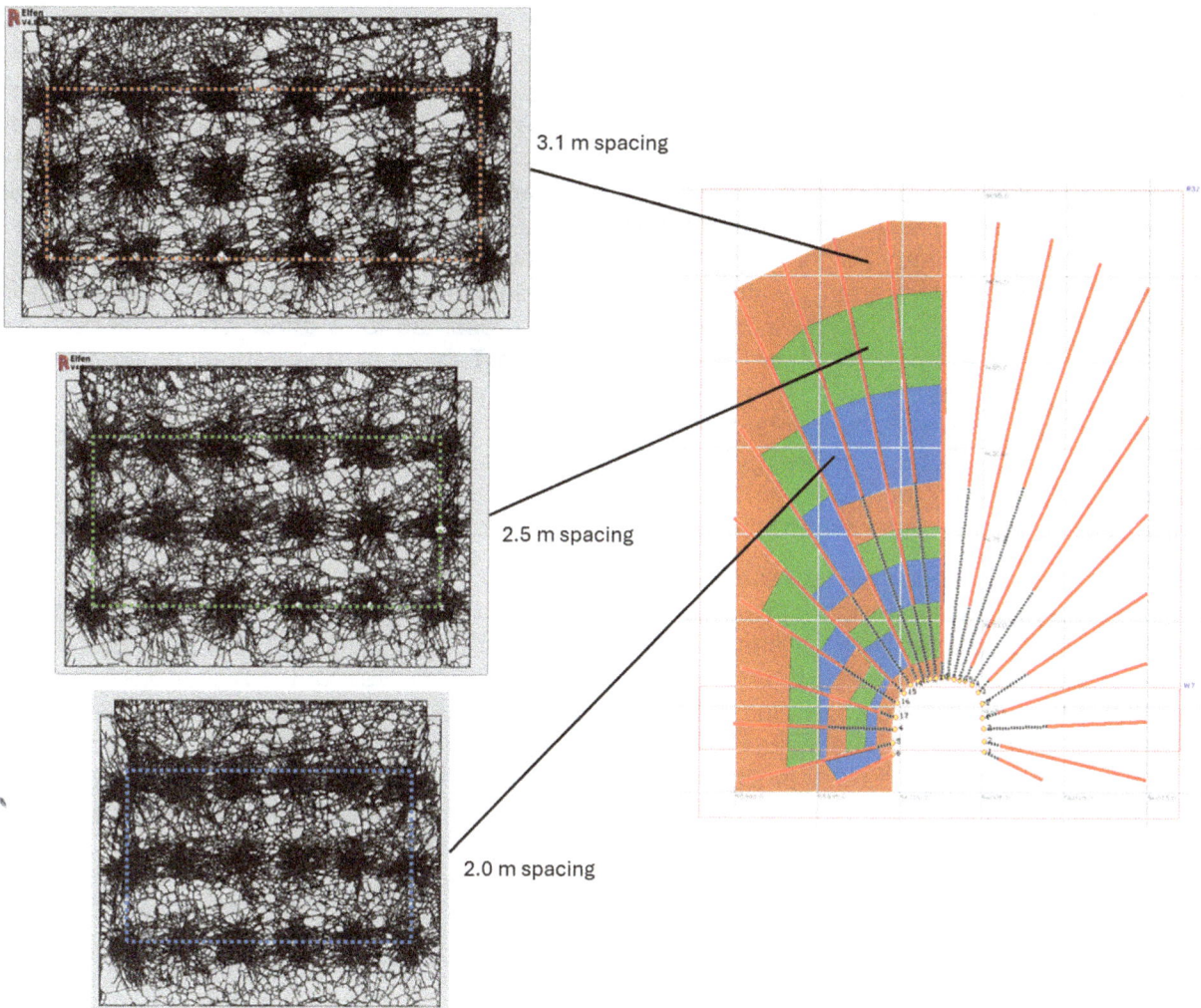

FIG 3 – The ring design (right) with half of the holes blocked out according to three different charge spacings, and simulation outputs for the three charge spacings (left).

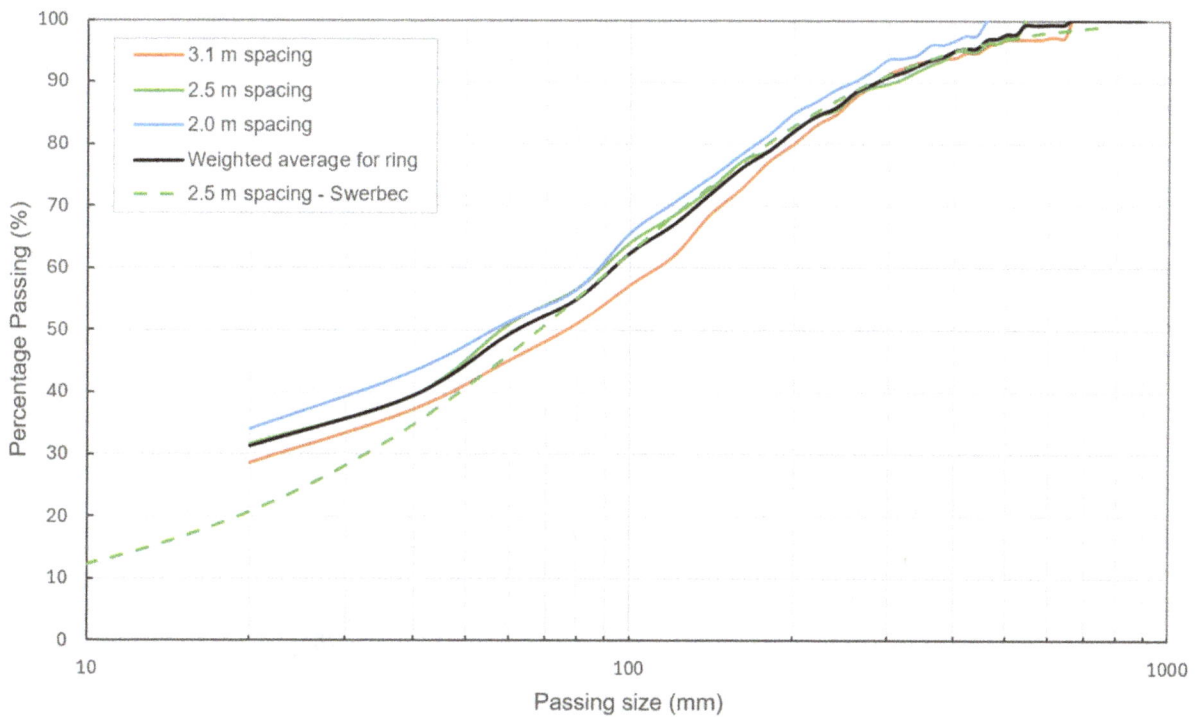

FIG 4 – Comparison of raw MBM PSD curves for different charge spacings, the average raw curve for the ring, and a Swebrec curve fitted to the 2.5 m spacing MBM data.

A series of simulations were run to determine the impact of reducing the design burden, increasing the blasthole diameter, and utilising a higher energy bulk explosive, on the resulting fragmentation (Table 8). Increasing the hole diameter not only increases the charge weight in the hole but also increases the efficiency of the detonation, increasing the VoD and detonation pressures. Orica's new Subtek™ 4D™ explosive allows for a higher in-hole density of the bulk explosive over the conventional Subtek™ product, and utilises a higher energy formulation, both contributing to higher VoD's and detonation pressures (Evans and Taylor, 2024). While the simulations used non-ideal detonation data, the Energy Factor derived from ideal detonation data is listed in Table 8 to provide guidance on the explosive energy within the modelled designs. Fragmentation data, derived from Swebrec functions fitted to the raw simulation data, are presented for a range of percent passings from 10 per cent (P_{10}) up to 99 per cent (P_{99}). The data in Table 8 is grouped by burden and then hole diameter. Within each burden, the data, including hole diameter and bulk explosive, is listed by ascending Energy Factor. It is interesting to note that while the general trend shows finer fragmentation with increased Energy Factor, there are instances where the 4D™ product outperforms the energy increase from a larger hole diameter. This indicates that a higher energy explosive can be more efficient than increasing the mass of explosive in creating finer fragmentation.

The fragmentation modelling indicates that with design changes, that are relatively easy to implement on-site, a blast design with a 2.2 m burden, 3.1 m toe spacing, 114 mm diameter blasthole and charged with Subtek™ 4D™ can reduce the blasted rock P_{80} from 178 mm in the current baseline design to 69 mm, a reduction of 60 per cent. The PSD curves for the higher energy design and the baseline design are shown in Figure 5 and compared to an example PSD for a heap leach (Maghsoudy, Bakhtiari and Maghsoudy, 2022), which has a much more uniform size distribution. For the higher energy design, the P_{50} is significantly finer than the heap leach curve used in the initial hydro-thermo-chemical fluid flow modelling. This bodes well for the leach performance for this achievable fragmentation in the field trial.

TABLE 8

Blast design changes and fragmentation data for the series of simulations.

Design				Passing Size (mm)					
Burden (m)	Hole diametre (mm)	Explosive	Energy factor (MJ/t)	P_{10}	P_{20}	P_{50}	P_{80}	P_{90}	P_{99}
2.6	89	Subtek™	1.29	8	19	68	178	279	739
		Subtek™ 4D™	1.51	7	16	54	144	232	689
	102	Subtek™	1.69	7	17	58	153	243	674
		Subtek™ 4D™	1.98	6	13	45	118	188	539
	114	Subtek™	2.11	6	14	48	129	207	600
		Subtek™ 4D™	2.48	5	12	38	95	146	377
2.4	89	Subtek™	1.39	7	16	58	152	239	634
		Subtek™ 4D™	1.64	6	14	47	121	191	522
	102	Subtek™	1.83	6	14	48	123	193	527
		Subtek™ 4D™	2.15	5	12	38	99	161	499
	114	Subtek™	2.29	5	12	38	98	156	468
		Subtek™ 4D™	2.68	5	10	31	78	124	367
2.2	89	Subtek™	1.52	6	15	49	128	204	581
		Subtek™ 4D™	1.78	5	12	39	105	174	578
	102	Subtek™	2.00	5	12	40	103	164	470
		Subtek™ 4D™	2.34	5	10	33	88	145	479
	114	Subtek™	2.50	5	10	32	81	128	370
		Subtek™ 4D™	2.93	4	9	27	69	110	333

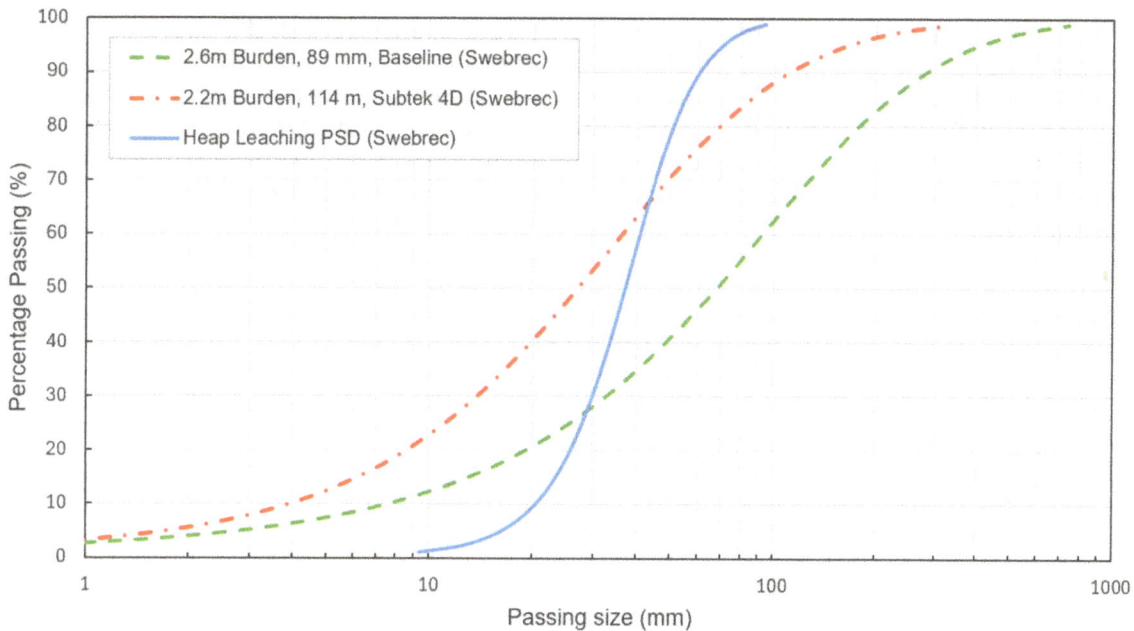

FIG 5 – Comparison between the baseline PSD curve and the highest energy modelled blast PSD and a heap leach PSD.

Development of a predictive model

Proper design and operation of IPR projects rely on a prior understanding of the leaching process and accurate prediction of leaching performance. This can be achieved by modelling comprehensively the leaching process of the full-scale operation and assessing the predicted recovery. In this project, a coupled hydro-thermal-chemical modelling framework was developed that covers relevant processes in mineral grain-scale, fragment-scale, packed fragment-scale, and full stop-scale (see Figure 6). The model incorporates material properties and leaching kinetics from laboratory testing (eg mineralogical analysis and column leaching) and packed fragment structures (eg fragment size distribution and void space distribution) from blast modelling.

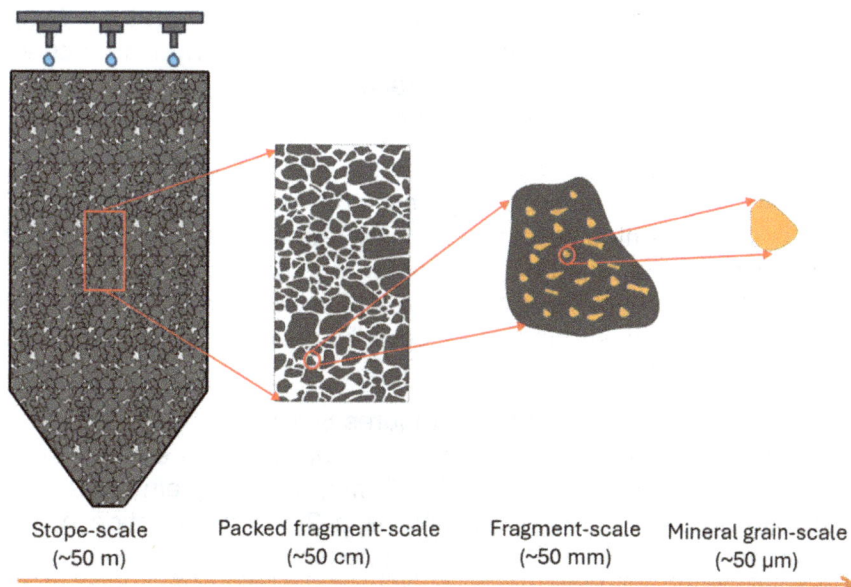

FIG 6 – Illustration of different scales relevant to IPR.

Leaching of low-grade copper ores in the blasted stope, consists of: fluid flow and mass transport that govern how the lixiviant and dissolved minerals travel through inter-fragment macro-pores and intra-fragment micro-pores; chemical reactions that describe the dissolution of copper-bearing minerals and gangue minerals; and heat transfer that considers the variation in temperature associated with reaction-generated heat and the effect of convection (see Figure 7). We consider the fluid flow to be governed by Darcy's law, in which permeability is related to liquid saturation through van Genuchten's retention model for unsaturated conditions. Mass transport is described by the convection-diffusion equation for dissolved substances (in aqueous solution) and oxygen (in air) to travel in both macro- and micro-pores. Dissolution of copper-bearing minerals is quantified by a double shrinking core model for reactions at both mineral grain-scale and fragment-scale. Heat transfer through conduction and convection in fluid and solid phases is considered by the energy balance equation.

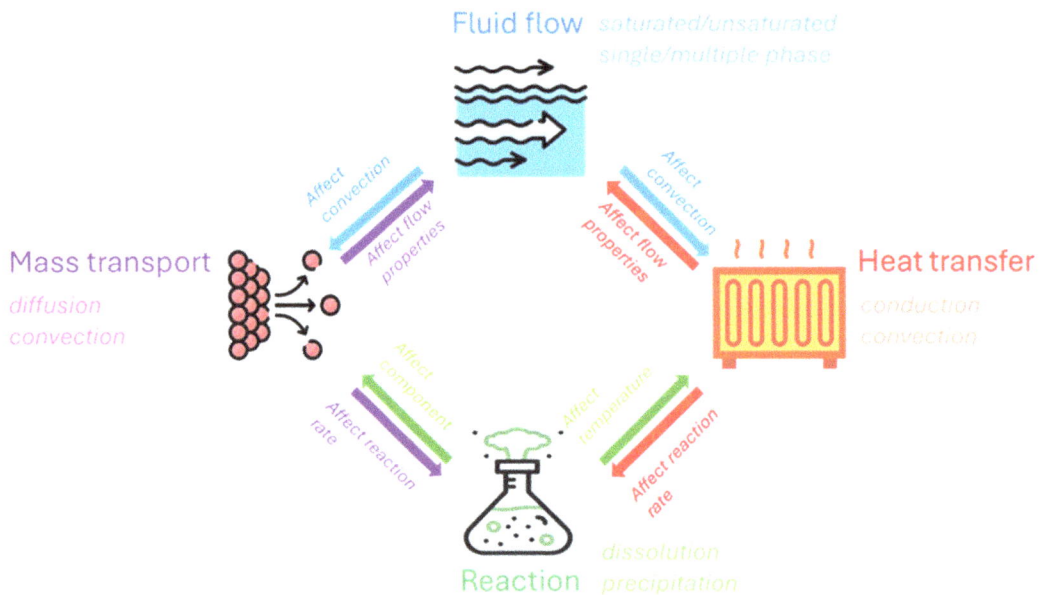

FIG 7 – Coupling of the physical and chemical processes.

By defining the relevant material properties and boundary conditions, these governing equations are coupled and solved with the COMSOL Multiphysics finite element code. Material properties, such as mineralogy, mineral content, gangue content, mineral grain size distribution, rock fragment size distribution, permeability (both fragment and stope), diffusion coefficient (both fragment and stope), porosity (both fragment and stope), lixiviant composition, and reaction rate, are inputs determined by blast modelling, lab measurements, and/or previous public data. Boundary conditions, including irrigation rate and duration, aeration rate and duration, and underground temperature, are set according to the site conditions and leaching schemes designed for the project.

Simulation results at the packed fragment-scale are presented in Figure 8 to demonstrate the capability of the model to capture different physical and chemical processes relevant to stope leaching. In the simulations presented here, the leaching rate expressions, which are defined based on a two-stage chalcocite dissolution model, and other material properties derived from published research (Bennett *et al*, 2012; Ortega-Tong *et al*, 2023). These rates and material properties will be further refined as experimental measurements, especially the leaching kinetics, currently underway in the laboratory, become available.

Overall, the simulations and results presented in Figures 8–10 are obtained for column leaching of chalcocite-bearing ores (0.6 per cent Cu) with a lixiviant application rate of 1×10^{-5} m/s, a diffusion coefficient of 5×10^{-14} m^2/s, fragment sizes of 1.2–12 mm, mineral grain sizes of 20–200 µm, and operation temperatures of 30–60°C. Figure 8a shows a 2D representation of a packed column (Maghsoudy, Bakhtiari and Maghsoudy, 2022) with dimensions of 190 mm in width and 350 mm in height, for which the coupled model has been developed. Lixiviant containing 0.1 M Fe^{3+} at pH 2 was applied at the top of the column for a duration of 360 days.

Figure 8b and Figure 8c depict the simulated steady-state flow velocity and liquid saturation in the inter-fragment macropores, demonstrating clear flow channelling and heterogeneous saturation. Figure 9a shows the recovery of copper in the ore fragments at 0.1 days and 360 days. Due to a low fragment diffusion coefficient, a typical diffusion-limiting process can be observed with the reaction front moving progressively from the fragment surface to the centre at 360 days. Therefore, smaller fragments are generally leached out much faster than larger fragments, as long as they have comparable access to the lixiviant. The corresponding distribution of lixiviant concentration (Figure 9b) is in agreement with the recovery, with lixiviant hardly penetrating larger fragments due to a low diffusion coefficient.

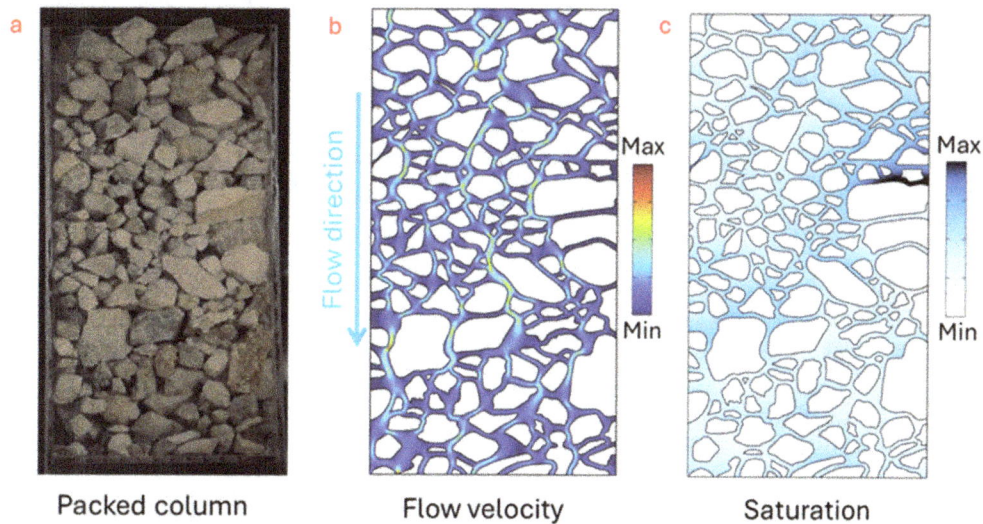

FIG 8 – Packed column (a), with simulated flow velocity (b), and saturation (c).

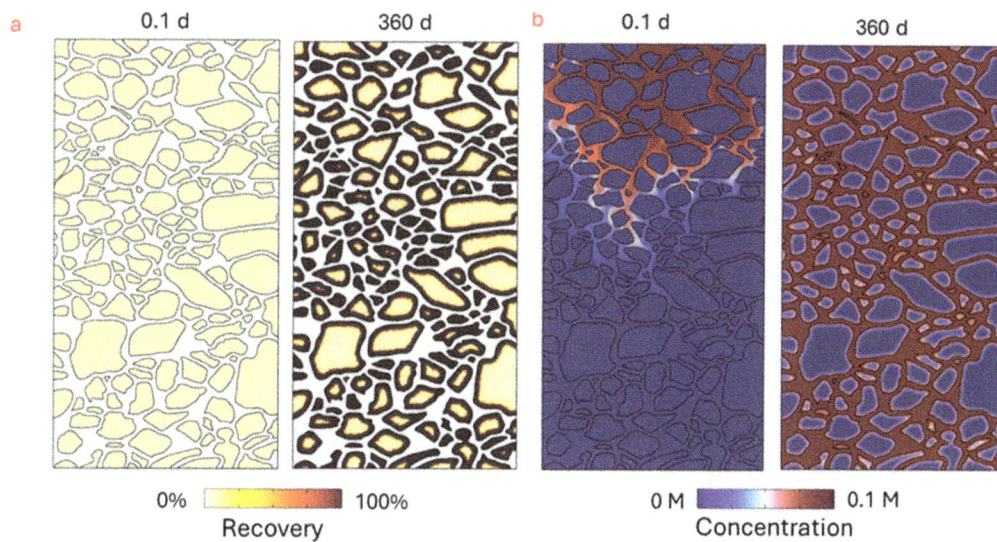

FIG 9 – Packed column leaching with simulated copper recovery (a), and lixiviant concentration (b) after 0.1 days and 360 days.

The model is also capable of demonstrating how different material properties and/or leaching conditions will affect recovery. To illustrate, copper recovery was examined at three different ground temperatures (eg 30°C, 45°C and 60°C). As shown in Figure 10, temperature shows an evident effect on recovery as the reaction kinetics are promoted by higher temperatures, which leads to increased copper extraction. This suggests that establishing IPR operations at greater depths in an underground mine will enhance copper recovery due to the higher ground temperatures.

The above results are obtained based on both lab measurements and published data, and a comprehensive validation program will be conducted to ensure the model is producing accurate results at different scales as the project progresses. At the mineral grain scale, the leaching kinetic parameters measured by batch testing with fine mineral particles are used to verify and calibrate the rate expression for the shrinking core model of copper-bearing minerals. At the fragment-scale, the leaching kinetics parameters measured by the static soaking test on single ore fragments are used to calibrate and verify the rate-limiting and diffusion-limiting reactions of the shrinking core model. At the packed fragment scale, the leaching kinetics parameters measured in the column leaching tests are used to calibrate and verify the permeability and diffusion coefficient of macro-pores. The final full stope-scale model will be tested against field trial operations as the final model calibration and validation.

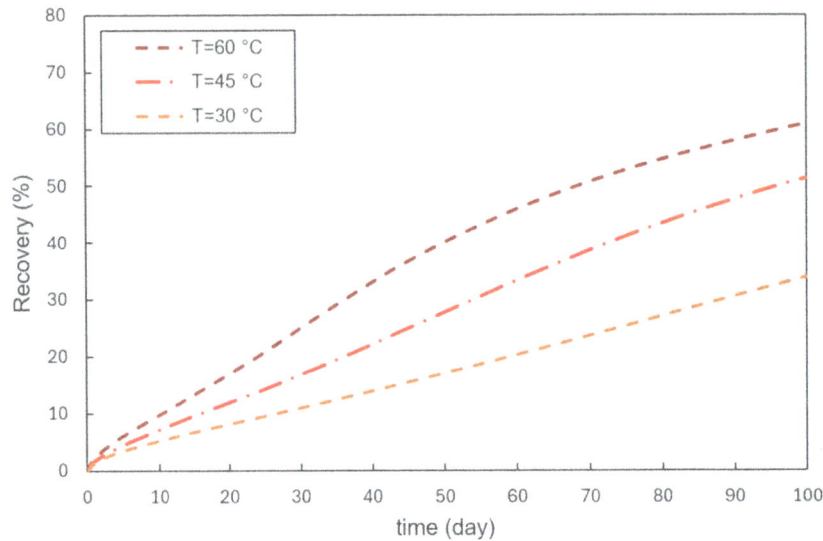

FIG 10 – Recovery from packed column leaching different ground temperatures.

CONCLUSIONS AND FUTURE WORK

Summary of the key findings

Through fostering collaboration between industries and research institutes, this project is exploring the feasibility of hard rock underground leaching. A test area has been allocated at PH mine site and representative rock samples have been collected. The test area is within a currently uneconomic region of the deposit to ensure minimal impact on day-to-day operations. Head grade characterisation has confirmed the low-grade at approximately 0.8 per cent Cu and identified the dominant copper-bearing minerals as chalcopyrite, chalcocite and bornite. To optimise the leaching kinetics of copper sulfide ores a benchtop test regime has been developed and the interim results presented here achieved approximately 65 per cent copper extraction with ground rock. The results demonstrate that despite being challenging, extraction of copper from the copper sulfide ores is achievable. In the next phase of the hydrometallurgical laboratory work, rolling bottles and leaching columns experiments will be conducted on larger particle sizes and limited agitation to understand their impacts on the rate of extraction.

The blast fragmentation modelling has also proved to be a powerful tool in assessing the impact of the drill and blast design on the resulting PSD. Through the modelling of a suite of blast design changes it was determined that with readily applicable changes to the design burden, hole diameter and bulk explosive it is possible to reduce the P_{80} from 178 mm to 69 mm, corresponding to greater than a sixfold increase in the contact area between the lixiviant and the rock.

The coupled model that has been created for this project is a reliable tool to combine all physics involved in an IPR operation. The hydro-thermo-chemical fluid flow model developed here operates at a range of scales, from grain to fragment to packed fragment and will be eventually extended to model at stope scale in an underground mine. The coupled model allows the user to monitor the operation at different scales, for instance to evaluate the flow within the broken rock mass, or track the mineral recovery of the entire stope and specific fragments, to investigate the effect of a wide range of parameters on the leach performance. It is worth adding that while the model is currently using a narrower particle size distribution, the next phase of the project involves building the coupled model based on the output of the blast fragmentation model. Additionally, while some of the data presented here are based on the publicly available data on leaching kinetics, the information will be updated as more real-life data becomes available from the hydrometallurgical laboratory work.

Field trial at PH

Conducting an in-field trial is a critical step toward scaling up to a full-scale IPR operation. The data gathered from these trials will be instrumental in validating our simulation tools and ensuring they are reliable and reflective of real-life operations.

The laboratory results and modelling tools created through this project will play a crucial role in planning the small-scale demonstration at Prominent Hill. Further advanced blast modelling will be used to assess the heave characteristics of the blasted rock and the formation of broken stock within the stope. Through this work blast designs can be developed to increase swell within the broken stock and ensure even porosity through the stope. Suitable blast designs can be developed to achieve a target fragmentation and swell, and the fluid flow rate can be determined using the hydro-thermo-chemical fluid flow model, and lixiviant condition and its composition can be decided based on the leaching test results.

Despite having accurate forecasting tools, the team will implement the proof-of-concept field demonstration in a staged approach to verify the desktop results and identify any potential operational risks. To illustrate, partial rock withdrawal between blasts allows for particle size distribution verification, or irrigation with water (containing no leaching chemicals) will allow validating the permeability of the stope post-blasting.

A key concern for an IPR operation is ensuring the containment of the lixiviant within the stope to prevent environmental contamination. However, since the target application for IPR is a hard rock stoping environment, where the host rock is predominantly impermeable, the lixiviant is expected to remain confined within the stope. This containment should facilitate the complete recovery of the irrigated fluid at the base of the stope, minimising environmental impact.

In the upcoming phase of the project, we will employ advanced blast modelling to simulate the damage accumulated in the rock mass in the stope walls. This will allow us to evaluate the effects of blasting on existing discontinuities (ie joints, bedding planes, faults etc) and determine the likelihood of generating new joints and/or faults. The data obtained from this analysis will be integrated into a sophisticated multi-physics model to simulate fluid flow at the stope wall. The fluid flow simulation results can be validated on-site during the field trial using tracers, where benign chemical markers, dissolved in water, are irrigated at the top of stope and its movement is traced by collecting samples at the base of the stope, in the surrounding areas, and lower levels. Fluid flow modelling and the use of tracers will provide evidence-based information to mitigate risks and build public trust, which is essential for securing the social license to operate.

Our vision

The transition to clean energy technologies is expected to significantly increase the demand for critical minerals including copper. IPR offers a solution to rapidly boost production from existing underground operations, in conjunction with ongoing operations. This hybrid approach allows for the simultaneous extraction of valuable minerals from high-grade ore through conventional mining methods and low-grade ore through IPR, thereby increasing production, extending mine life, and enhancing safety by reducing rock movement and personnel exposure in hazardous areas. The goal of this project is to gain in-depth understanding of the IPR operation, identify technical solutions to operational challenges, and ultimately pave the way for the commercial implementation of IPR.

ACKNOWLEDGEMENTS

The authors acknowledge the Australian Government for its contribution through the funding of the CRC-P grant. We also extend our appreciation to Associate Professor Chaoshui Xu and Professor Peter Dowd at The University of Adelaide, Rob Coleman and Jon Loraine at Core Resources, and BHP Think and Act Differently (TAD) Team for their ongoing support of the project. We wish to thank BHP team at Prominent Hill, especially Anne-Marie Ebbels, for providing data and rock sample collection, and for approval to publish this work.

REFERENCES

Bennett, C R, McBride, D, Cross, M and Gebhardt, J E, 2012. A comprehensive model for copper sulphide heap leaching: Part 1 Basic formulation and validation through column test simulation, *Hydrometallurgy*, 127–128:150–161.

Cunningham, C V B, 2005. The Kuz-Ram fragmentation model 20 years on, in *Proceedings 3rd EFEE World Conference on Explosives and Blasting* (ed: R Holmberg), pp 201–210, (UK European Federation of Explosives Engineers: Reading).

Dare-Bryan, P C and Hassanvand, A, 2023. Economic and environmental assessment of underground in-situ leaching processes utilising drill and blast to achieve high permeability, *ALTA In-Situ Conference*, Australia, 1–5 May.

Dare-Bryan, P C, Mansfield, S and Schoeman, S, 2012. Blast optimisation through computer modelling of fragmentation, heave and damage, in *Proceedings 10th International Symposium on Rock Fragmentation by Blasting*, pp 95–104 (AA Balkema: Rotterdam).

Evans, S and Taylor, B, 2024. Subtek™ 4D™ – Optimised blasting performance through the application of new underground bulk explosive technology, *International Future Mining Conference,* pp 143–154 (The Australasian Institute of Mining and Metallurgy: Melbourne).

Ghorbani, Y, Franzidis, J-P and Petersen, J, 2016. Heap leaching technology—current state, innovations and future directions: a review, *Mineral Processing and Extractive Metallurgy Review*, 37(2):73–119.

Kirby, I, Chan, J and Minchinton, A, 2014. Advances in predicting the effects of non-ideal detonation in blasting, in *Proceedings 40th Annual Conference on Explosives and Blasting Techniques*, pp 301–314 (International Society of Explosives Engineers: Cleveland).

Maghsoudy, S, Bakhtiari, O and Maghsoudy, S, 2022. Tortuosity prediction and investigation of fluid flow behavior using pore flow approach in heap leaching, *Hydrometallurgy,* 211.

Minchinton, A and Lynch, P M, 1996. Fragmentation and heave modelling using a coupled discrete element gas code, in *Proceedings of the 5th International Symposium on Rock Fragmentation by Blasting*, pp 71–80 (AA Balkema: Rotterdam).

Mousavi, A and Sellers, E, 2019. Optimisation of production planning for an innovative hybrid underground mining method, *Resources Policy*, 62:184–192.

Ortega-Tong, P, Jamieson, J, Kuhar, L, Faulkner, L and Prommer, H, 2023. In Situ Recovery of Copper: Identifying Mineralogical Controls via Model-Based Analysis of Multistage Column Leach Experiments, *ACS ES&T Engineering*, 3(6):773–786.

Ouchterlony, F, 2010. Fragmentation characterisation; the Swebrec function and its use in blast engineering, in *Proceedings of the 9th International Symposium on Rock Fragmentation by Blasting*, pp 3–22 (AA Balkema: Rotterdam).

Owen, D R, Munjiza, J A and Bicanic, N, 1992. A finite element-discrete element approach to the simulation of rock blasting problems, in *Proceedings of the 11th Symposium on finite elements in South Africa*, pp 39–58 (Centre for Research in Computational and Applied Mechanics: Cape Town).

Schlesinger, M E, King, M J, Sole, K C and Davenport, W G, 2011. *Extractive Metallurgy of Copper*, Elsevier.

Unlocking potential of the Gold Fields St Ives Gold Mine using wireless blasting technology

B Jesionek[1] and R Massabki[2]

1. Unit Manager: Technical Services, Gold Fields, Perth WA 6000.
   Email: bart.jesionek@goldfields.com
2. Senior Engineer – WebGen, Orica, Perth WA 6000. Email: ricardo.massabki@orica.com

ABSTRACT

The Gold Fields, St Ives Gold Mine (SIGM) uses Orica's WebGen™ wireless initiation system to realise safety, productivity, and cost benefits at its underground operations. This wireless initiation system uses magnetic induction to wirelessly transmit an initiation signal in lieu of a physical connection. The absence of any in-hole or surface wiring allows stopes to be broken down into firings of any size, shape or sequence, to be charged in one campaign, and then fired off independently on demand without the need to ever re-enter the stope. This 'pre-charging' of stopes enables new novel mining methods while increasing safety and productivity in more traditional extraction styles.

Wireless initiation was first introduced at St Ives for the safety and productivity benefits it brings by eliminating personnel exposure to the brow. The use of wireless initiation has since broadened to include designs that achieve:

- Reduced development: eliminating costly lateral development in transition and transverse stopes.

- Optimal slot placement: relocating the rise and slot to the optimum point for stope performance without concerns for ground support rehab, re-entry and personnel.

- Eliminating hole cleaning: stope pre-charging negates requirements for mechanical hole cleaning, ensuring continuity of production while reducing exposure to stope brows.

- Operational flexibility: pre-charging blasts calculated to optimum void ratios and increasing recovery with better blast geometry, particularly on small, narrow stopes.

- Improve ore recovery: Expanding stope shapes to include stranded ore, reverse firing and changing firing directions to throw blasted material around corners.

As of August 2024, these applications have eliminated 330 m of development, unlocked scheduling, generated stope turnover gains, realised production uplift and reduced production ore unit costs across 32 stopes and 69 individual blasts.

The mining methods enabled by the wireless initiation have been fully incorporated into the mine's production schedule and Orica continues to support Gold Fields in its journey to unlock safer, more reliable and cost-effective production from the St Ives asset.

INTRODUCTION

Historically, underground blasting operations require a physical connection between an initiation source, be it a flame, detonation pressure front or electrical current, and the explosive agent. Enabling this physical connection can be particularly hazardous near open voids, but also forms a limiting factor on blasting and excavation design and scheduling in mining operations (Wicks and Lovitt, 2017), adding to mine development costs (Hawtin et al, 2021).

By removing this physical connection requirement with a wireless through rock signal, stopes can be completely pre-charged, and areas that would otherwise be stranded, can be initiated remotely. This materially reduces exposure of personnel to risk associated with open voids and unlocks the potential of novel mining methods seen in literature (Piercey, Grace and Robinson, 2018; Melbourne et al, 2020; Hawtin et al, 2021; Small et al, 2023).

The WebGen™ 200 wireless initiation system works via a magnetic induction signal (Orica, 2024) that penetrates rock, air, and water, initiating primers in specific groups to create optimum sized blasts tailored to void and operational requirements.

ST IVES GOLD MINE

St Ives Gold Mine is located 20 km south-east of Kambalda, and 80 km south of Kalgoorlie in Western Australia. The mining complex has extracted more than 10.5 million ounces of gold over its lifetime, and currently operates two underground areas and two open pits, supporting a centralised mill.

The underground mining operations use a mix of mining methods including longitudinal narrow vein (1–5 m wide) sublevel stoping, longitudinal bulk (5–20 m wide) sublevel stoping, and transverse (20 m wide) sublevel stoping. Cemented paste fill is used to backfill stopes.

With an increased number of bulk and transverse stopes, St Ives undertook a critical evaluation of the efficiency of available resources at the Invincible Underground Complex to determine if activities were aligned to maximising value. Findings pointed to a higher-than-expected resource allocation to tasks associated with production, such as slot drive development, increased drilling density, and re-drilling/hole cleaning post blast, among others. This prompted investigation into alternatives which would allow greater value to be realised from the project.

WIRELESS BLASTING TECHNOLOGY

The wireless technology concept has been researched since the 1940s and was first commercialised with the introduction of the first generation of WebGen™ in 2017 (Wicks and Lovitt, 2017). Since then, it has expanded in commercial use across the globe. WebGen™ uses ultra-low frequency magnetic inductive waves to transmit signals from an antenna located underground to wireless primers in blastholes (Figure 1). Communication between the antenna and a blaster tablet at a designed firing point can be made via various remote communication protocols.

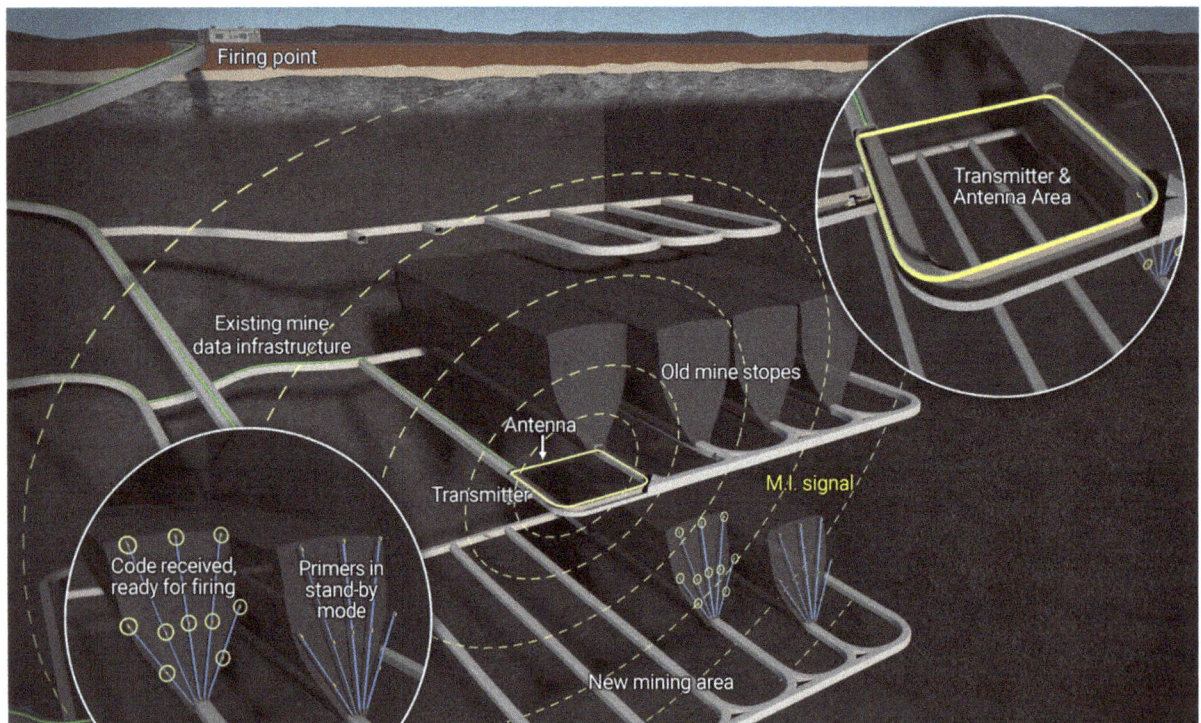

FIG 1 – WebGen™ uses a magnetic inductive signal to wirelessly initiate blastholes (Orica, 2024).

The fact that there is no physical connection between in-hole primers and the blaster means that there is no risk of cutting off initiation leads and no need to retain physical access to blastholes between blasts. The immediate benefits include reduced firing delays, but greater value is unlocked by using novel mining methods that take advantage of pre-charging.

USE OF WIRELESS INITIATION SYSTEM TO 'PRE-CHARGE' BLASTHOLES

Generally, conventional (wired) initiation systems are loaded, connected and fired in one blast, commonly termed a 'firing', where many firings are required to complete a stope. Blasthole loading does not extend beyond the area planned for each firing and the only period loaded blastholes 'sleep', is whilst waiting for loading to be completed, at which point all charged holes are fired together.

In contrast, as wireless initiation can be remotely initiated and does not require any connection at the blastholes, all the individual firings of a stope can be loaded in a single campaign. Blastholes are charged before their firing is required (pre-charged), then sleep waiting to be individually initiated in the desired sequence. The firings of a wireless stope can therefore be broken down further (smaller) to manage void ratios and directional firing requirements without impacting production rates.

SAFE PRODUCTION DRIVING INNOVATION

The ongoing commitment and pursuit of safe production by Gold Fields and the broader mining industry have seen evolutions of historically acceptable work methods to mitigate, as far as practicable, the risk to personnel to material unwanted events (fatality and major disabling injury). Over time this has resulted in more stringent stope brow controls, application of tele-remote technologies and expansion of exclusion zones. Conventional (wired) blasting methods restrict production processes (drill and blast) once these controls are implemented. This adds schedule complications, delay, and cost. Wireless blasting technology enables novel approaches to production processes while maintaining or improving on operational safety requirements, compressing, and simplifying operating schedules, and reducing operational costs.

CHALLENGES IN USING WIRELESS INITIATION SYSTEMS

The introduction of wireless initiation systems to St Ives Gold mine have not come without its challenges.

The change management of introducing a new system included relevant risk assessments, training of charging crews to handle the initiation system, as well as an update on drill and blast guidelines, to reflect the new mining method applications.

Additionally, Initial stopes were charged according to their 'blasting sequence' to mitigate the risks of incorrect charging. This procedure generated additional charging delays due to several charging rig set-ups to follow the desired sequence.

Once additional controls were put in place to manage this risk, a change in charging procedures allowed the improvement of the charging process, allowing for an optimised 'ring by ring' charging operation. A representation of an example of both sequences can be seen on Figure 2.

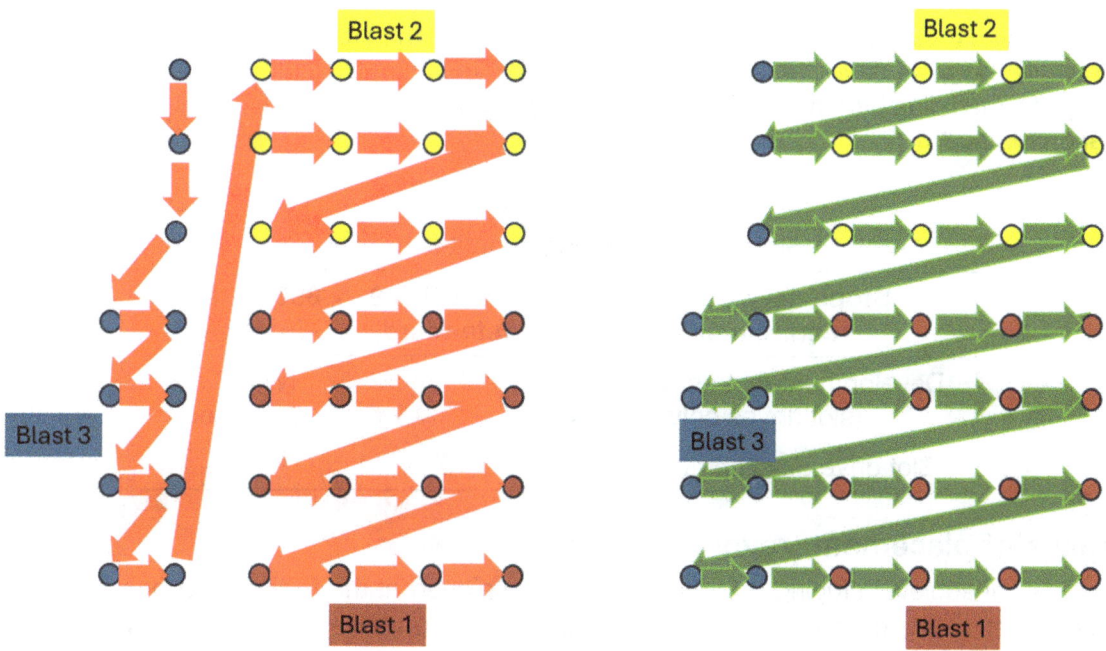

FIG 2 – Charging sequence on initial stopes by blast sequence (left), as opposed to optimised charging sequence by ring (right).

MINING METHODS ENABLED BY WIRELESS BLASTING

Since 2023, Goldfields and Orica have been exploring opportunities to improve safety and productivity by using the wireless initiation. This change in drilling and blasting paradigms has enabled novel mining methods that are now embedded in the mine production schedule.

Reducing lateral development (slot drive elimination)

Some stoping methods require a 'T' shaped intersection at the end of each cross-cut or extraction drive. These are often called *slot drives* or *hammerheads*. The slot drive gives access to establish the opening void (rise and slot), into which the first production rings are fired. Pre-charged blastholes initiated with wireless primers can eliminate the slot drive by progressively enlarging a rise, as seen in Figure 3. Once the material from the rise (F1) is excavated, subsequent transverse blasts (F2, F3 and part of F4) can be fired to create the slot, before firing the main production rings.

1) Rise F1 (Red) is extended, using drive void to increase blast span

2) Firing F2 (Yellow) and F3 (Purple) are fired transversally into the void in the centre

3) Firing F4 (Blue) is fired, blasting remaining volume on floors and far side of stope

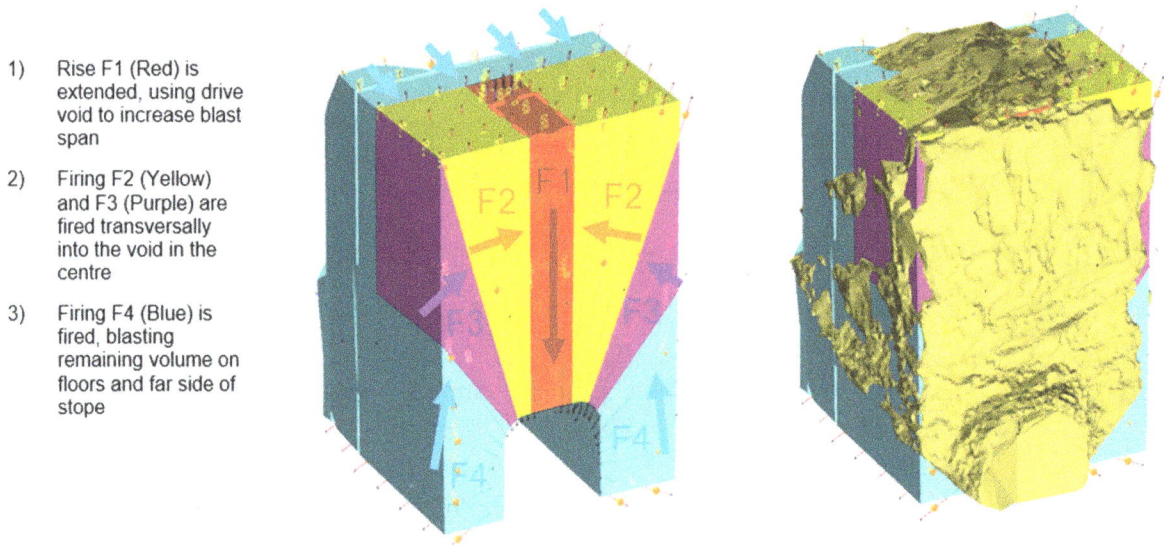

FIG 3 – Slot drive elimination sequence and CMS.

By removing the slot drive development, the associated development drilling, blasting and ground support costs are also removed, offsetting the cost of the extra production drilling and wireless initiation systems (Hawtin *et al*, 2021).

The method also brings considerable scheduling benefits, as the stope can be brought to production sooner, while the mine development crew can be deployed to other mining areas. Table 1 shows a summary of scheduling benefit of two similar stopes mined with the standard, and wireless, slot drive elimination method.

TABLE 1

Slot drive elimination benefit summary.

Stope development requirements	8932 GIB 9 (standard)	8962 GIB 12 (wireless)
Development metres required (slot drives only)	11 m	Nil (13 m estimated)
Slot drive development days	7	Nil

Optimal slot placement (reverse firing)

In conventional sublevel stoping, the rise and slot are placed at the furthermost point of the panel, with mining retreating towards the access point. In areas on the edge of the orebody, or where dilution control is critical, this may not be desirable. However, when blasting with traditional wired initiation systems, safe access must be maintained for loading and tying-in blastholes between blasts, so there is no alternative.

Using pre-charged wireless blasting, the slot and rise can be placed anywhere along the extraction drive. Placing it in the middle of the stope and closer to the final brow position may be advantageous for optimising initial void and reducing dilution from the hanging wall. A pre-charged wireless sequence is demonstrated in Figure 4.

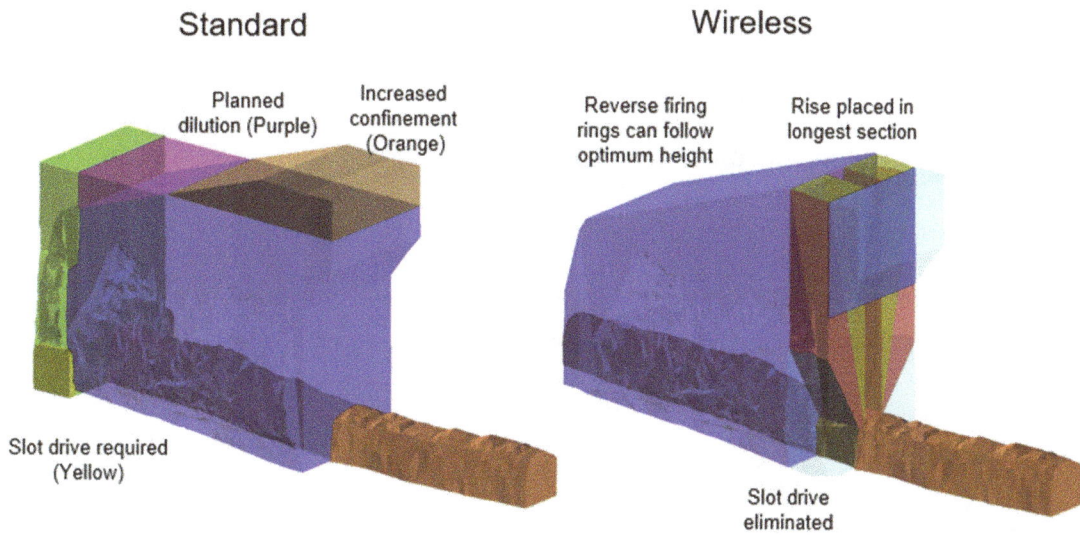

FIG 4 – Comparison of wired and wireless approach for slot/raise placement and blasting sequence.

When the stope is opened at some point midway along the extraction drive, rings on the far side will eventually be fired towards the drawpoint, opposite to conventional firing direction and therefore 'reverse fired'. Before firing, this volume can be used as a temporary pillar to control dilution from the contact or backfilled adjacent stopes (Piercey, Grace and Robinson, 2018; Small et al, 2023).

When eventually fired, the reverse fired volume casts towards the drawpoint, reducing the volume of less efficient tele-remote bogging. Table 2 shows a comparison of 3D modelled volumes from a standard 45° rill angle from the brow and a cast shape based on the profile seen in Figure 5.

TABLE 2

Increased manual bogging tonnes.

Manual bogging tonnes	Standard 45° rill angle	Wireless (modelled from Figure 5)
Estimated tonnes cast beyond brow	450	850

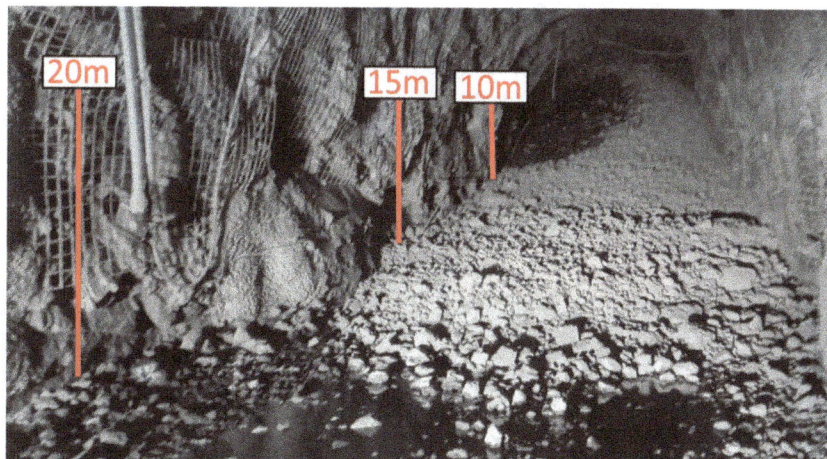

FIG 5 – Material cast towards the access after reverse firing.

Improving recovery (transverse firing)

Pre-charging with wireless initiation can improve selectivity in irregular orebodies where void ratios restrict the volume that can be blasted in a single firing event. Pre-charging can drastically improve recovery confidence by firing smaller, more selective parcels relative to available void and ground conditions.

In the example shown in Figure 6, calculations of the original stope shape shown in pink, indicated that there would be insufficient void, resulting in a redesign (red shape) which sterilised 1.5 kt of high-grade ore (containing 155 oz of gold) from the mine plan. This shortfall would have placed additional stress on development and production to make up for this sterilised volume.

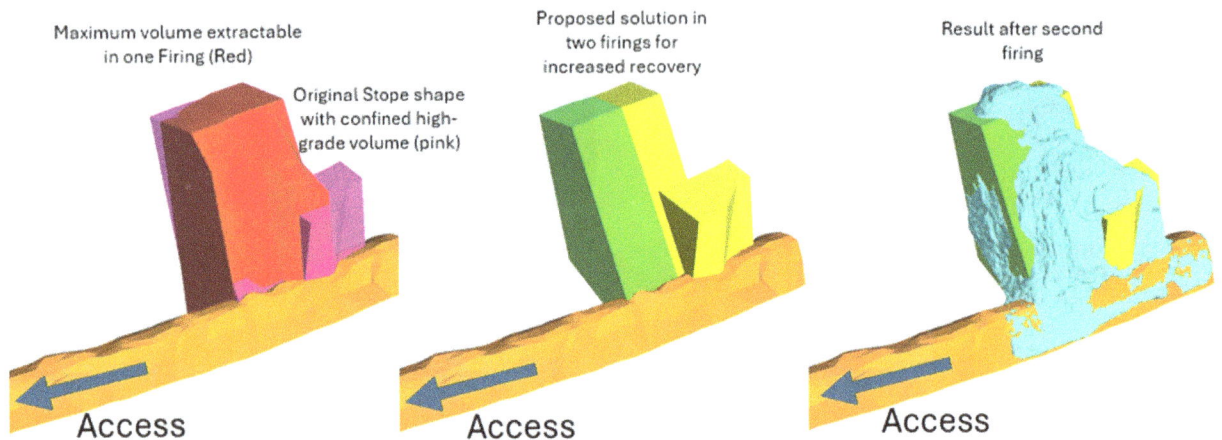

FIG 6 – Pre-charging to achieve required void.

Introducing wireless initiation systems permitted the extraction of the otherwise sterilised ore by dividing the stope into two blasts. Charged in the same campaign, the rear section of the stope was pre-charged, or sleeping, and fired later when sufficient void was available. Since the rise firing will take place when there is still access for physical connection, it can be fired using standard initiating systems, reducing initiation systems costs.

Once the blasted material is bogged, the sleeping, pre-charged wireless blast can be fired towards the stope access longitudinally and transversally. This increase in void enabled a greater recovery of the confined high-grade area, allowing the stope to achieve the targeted ounce production.

A summary of the recovery results is displayed in Table 3.

TABLE 3
Increased recovered ounces.

Recovery (stope 8830 FTN 36, Figure 6)	Mined tonnes	Ounces
Target (based on original shape in red)	3998	383
Actual (based on blue CMS shape)	5618	446 (116.5%)

ONGOING USE OF WIRELESS INITIATION SYSTEMS

Wireless initiation systems come at an increased cost compared to conventional initiation systems, however, this increased cost can be offset in its entirety by the reduction of development (Hawtin *et al*, 2021). Furthermore, the downstream benefits associated with increased production enabled by increased bogging productivity, decreased geotechnical requirements, overall reduced stope cycle times and increased recovery have led St Ives to embed wireless initiation mining methods into their continuous production cycle, as seen in Figure 7.

Wireless initiation adoption

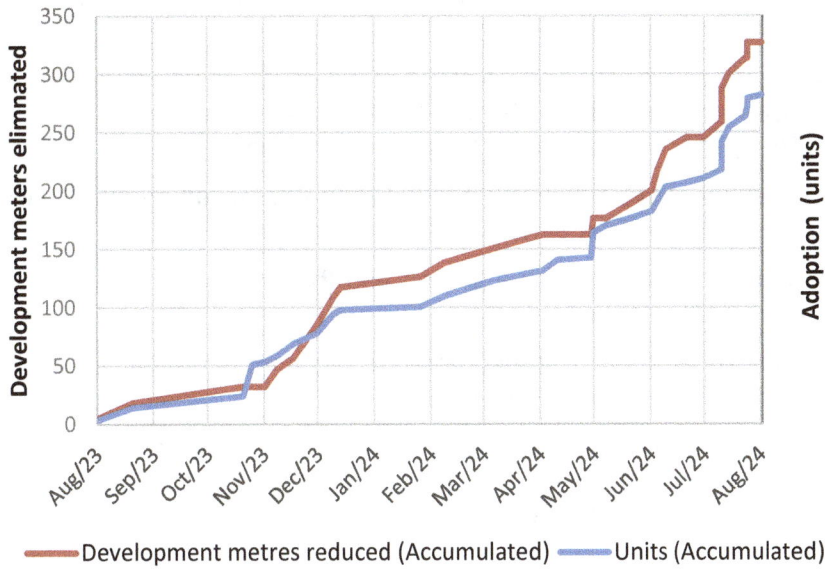

FIG 7 – St Ives directly correlated wireless primer consumption (right axis) with development metres eliminated (left axis).

The continuous development of novel mining practices at St Ives is allowing the site to combine several of the previous techniques mentioned in this paper. The example depicted in Figure 8 shows a stope currently in production that will combine the slot drive elimination, reverse firing, and transverse firing techniques, while using just one paste filling set-up to fill both sublevels.

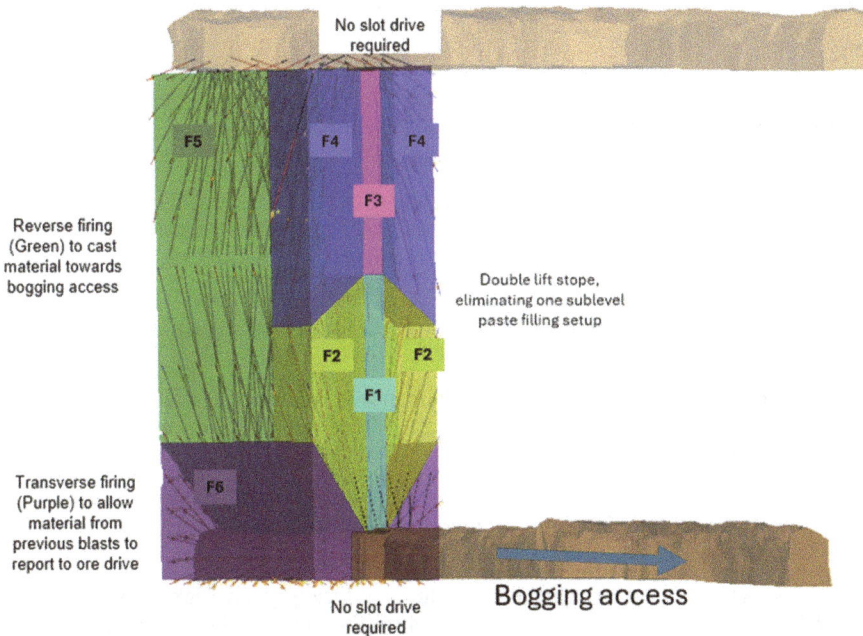

FIG 8 – Double lift sublevel stope blast in production.

This stope aims to capture the combined benefits of eliminating slot drives, transverse firing, and reverse firing. Further productivity gains will come from increasing manual bogging and using one paste fill set-up for two sublevels.

CONCLUSIONS

Removing the physical connections required by conventional initiation systems has allowed the development and pursuit of novel mining methods, to increase operational safety and productivity at St Ives Mine.

The application of wireless initiation technology has enabled St Ives underground operations to remove the slot drives on bulk stope development, allowing not only to reduce the costs of development, but also optimise the mine development schedule.

By allowing optimal rise placement, the wireless initiation decreased dilution, while also improving bogging productivity by reducing less efficient remote bogging.

It has also increased the profitability of the stopes and the mine's ore reserves, by improving the recovery in irregular orebody shapes.

These mining methods have been fully incorporated into mine's production schedule, and Orica continues to support Gold Fields in its journey.

As the mining activity progresses into new areas, the methods discussed continue to be combined and enhanced to increase the production benefits even further, unlocking safer, more reliable, and cost-effective production from the St Ives asset.

ACKNOWLEDGEMENTS

Permission to publish this paper and ongoing support by Gold Fields LTD and Orica LTD is respectfully acknowledged.

The authors also appreciate the help and support of Gold Fields St Ives Invincible Underground production teams and senior engineers Joshua Grinham, Jonathan Bernitz and Jobin Thomas.

The continued support of Gold Fields Invincible Underground technical teams and the Orica West Metals WebGen Team has made this project possible and their unvaluable contribution is also acknowledged.

REFERENCES

Hawtin, M, Say, J, Cann, P and Carlton, L, 2021. Reduced Development through improved Rise and Widening Firings using Orica WebGen™ at Ernest Henry Mine, in Proceedings of the Underground Operators Conference 2021 (The Australasian Institute of Mining and Metallurgy: Melbourne).

Melbourne, A, Andrade, W, Biulchi, F, Ribeiro, M, Lima, C and Santos, F, 2020. Pilar recovery with WebGen™, 100 wireless technology, Nexa Resources Vazante – Brazil, in *Proceedings Eighth International Conference and Exhibition on Mass Mining* (eds: R Castro, F Báez and K Suzuki), pp 86–90 (University of Chile: Santiago).

Orica, 2024. WebGen™, 200 – second generation Wireless Initiating System, Orica Limited. Available from: <https://www.orica.com/Products-Services/Blasting/Wireless/How-it-works/WebGen-200/webgen-200> [Accessed: 5 September 2024].

Piercey, S L, Grace, B and Robinson, H N, 2018. Reducing Dilution with Technology, A twist on the AVOCA mining method is made possible at Goldcorp's Musslewhite Mine utilizing Orica's wireless, through the earth initiation system, In CIM2018 Convention, CIM.

Small, B, Bermingham, A, Goodwin, R, Watson, S and Ireland, M, 2023. Reverse stoping using wireless initiation at CSA Mine, in *Proceedings Underground Operators Conference 2023*, pp 269–278 (The Australasian Institute of Mining and Metallurgy: Melbourne).

Wicks, B and Lovitt, M, 2017. A New Era of Blast Initiation Systems reducing Safety Risks, Costs and enabling Automation, in *Proceedings EFEE World Conference 2017*, pp 415–425 (European Federation of Explosive Engineers).

Author index

www.ingramcontent.com/pod-product-compliance
Lightning Source LLC
Chambersburg PA
CBHW081753220326
41597CB00056B/3963